CHEMICAL MICROSTRUCTURE OF POLYMER CHAINS

Chemical Microstructure of Polymer Chains

JACK L. KOENIG

Case Western Reserve University
Cleveland, Ohio 44106

A WILEY-INTERSCIENCE PUBLICATION
JOHN WILEY & SONS New York, Chichester, Brisbane, Toronto

6440 — 643X

CHEMISTRY

Library of Congress Cataloging in Publication Data:

Koenig, Jack L

  Chemical microstructure of polymer chains.

"A Wiley-Interscience publication."
Includes bibliographical references and index.
1. Polymers and polymerization. I. Title.

QD381.K595   547.8′4   80-15165
ISBN 0-471-07725-9

Printed in the United States of America

10 9 8 7 6 5 4 3 2 1

To my wife, Jeanus, and our children:
John, Robert, Stanley, and Lori

Preface

This book is addressed to the reader who wishes to become familiar with some of the basic ideas of determining the chemical microstructure of polymers and the relationship of this microstructure to the polymerization process. This text is based on lecture notes utilized in teaching a graduate course in the physical chemistry of polymers at Case Western Reserve University over a period of 15 years. The recognition of the need for a modern text on this important subject motivated the conversion and expansion of these lecture notes to a text that is suitable for students entering the field of high polymer research and for those engaged in industry who desire to know more about this rapidly changing subject. In addition, it is hoped that the book will also be of service to chemists who are already acquainted with aspects of this subject, but who would welcome a general perspective of this field and an updating of the new techniques of polymer characterization.

Let us consider the following situation. You or one of your colleagues has prepared a novel polymer or copolymer. Initial tests indicate that the polymer has unusual and useful mechanical, physical, and thermal properties. The novel polymer system presents all of the following questions: Is it a new polymer? What is its structure? What structure does the polymer have which gives rise to its unusual properties? Why does the polymer have the structure that it has? Can its useful properties be improved?

Even to the most experienced and sophisticated polymer scientist, these are challenging questions. One first turns to the standard methods of polymer characterization. Experience can provide some indication of when these methods should be used, but without this personal experience where does one turn? How can one determine which standard method is applicable? How can one determine which standard method is preferable?

Every polymer scientist needs some knowledge of what methods are available and what these methods can accomplish. Without personal experience he or she needs information illustrative of their use, as well as a knowledge of what has been attempted before and what suggestions can be made for dealing with his problem. But obviously if the system is entirely new, standard methods will not provide the necessary information, so he or she must have a theoretical and practical understanding of the chemical and physical methods of polymer characterization to determine which method or technique is applicable to his or her problem. Time and money do not allow the luxury of trying them all in a systematized or haphazard fashion. The polymer scientist must be able to deduce which technique or combination of methods might yield the necessary information about structure.

Even then, the person who is intelligent or lucky enough to select the proper combination of physical or chemical techniques must be able to relate the derived structure to the polymerization methods used in the preparation of the polymer. The structure must be related to the mechanical, physical, and thermal properties of the system.

This book attempts to serve as a guide to a number of these questions. No single monograph can attempt to answer all of these questions—but we focus particularly on a number of them. How does one determine the chemical microstructure of a polymer or copolymer? What methods, from the point of view of a current polymer knowledge, are most likely to be useful for a particular kind of system? How does one proceed to take full advantage of the power of each applicable technique? Assuming one has developed a concept of the structure of the system, how does one deduce how the structure was formed? What are the chemical reactions that lead to the structure? How does one optimize or control the structure of the system? What parameters are important or negligible in the polymerization process? Can one vary the polymerization conditions in such a way as to optimize the structure of the polymer system?

These questions cannot be simply or completely answered, but a guidance can be obtained from a road map or outline based on fundamental principles. This is our purpose: to present the fundamental precepts, illustrate their utility, and give a guide to their utilization. Such a road map or outline will not always be appropriate or successful, but the myriad of organic and inorganic polymer systems prevents the presentation of a systematized procedure for all possibilities. In fact, the recognition that one has encountered an entirely new problem is of great help in generating a proper solution. Therefore we develop the fundamental and enduring principles necessary for determining the structure of a polymer system.

We indicate the areas of potential usefulness and weakness of each method. Whenever possible, we refer to the pertinent review or monograph where the details can be sought by the interested.

Our intention is to direct the reader to the prevalent methods of determining the parameters of the polymerization. Tests are presented to serve as guideposts to one's progress or to evaluate one's hypothesis about the polymerization mechanism.

JACK L. KOENIG

Cleveland, Ohio
May 1980

Acknowledgments

I wish to thank many people for inspiring me to write this book. First, of course, are the students who have suffered through my efforts to formulate a course that would meet their ever-changing needs. Their sharp and inquiring minds are a continued stimulus to improvement in course content and presentation. Their test results are the ultimate evaluation of success or failure by a teacher. Some students taught me so much that they should be singled out by name, but nearly all of them taught me something about people or polymers.

My colleagues in the Department of Macromolecular Science have always been stimulating and helpful. The opportunity to exchange ideas with people like J. Reid Shelton, M. Litt, J. Lando, E. Baer, C. Rogers, and P. H. Geil is unique. To have known and taught with the best teacher I have ever known, the late Professor S. Maron, was a singular privilege. He inspired and taught not only his students but those around him as well.

Finally, I must acknowledge the support of my family, who felt "Daddy's book" was important and gave up their time accordingly. No words of appreciation are adequate for my secretary and friend of 15 years, Mrs. Barbara Leach. Her typing skills are exceptional, her organizational skills outstanding, and her patience like that of Job. Without her work, this book might have been written, but it would never have seen the typed page.

J.L.K.

Contents

1

Determination of Structure-Property Relationships in Polymers

1.1 STRUCTURE-PROPERTY RELATIONSHIPS OF POLYMER CHAINS

Staudinger, often called the father of polymer chemistry, said (1):

Macromolecular chemistry has opened a field of chemistry where molecules are 1000–100,000 times the size of those of the well known low molecular weight compounds. If new dimensions are opened in a field, new methods have to be found to work with the new phenomena.

Consequently, polymer scientists have continually been developing new methods to attempt to understand the relationship between structure and properties of macromolecules.

In 1953 Staudinger, in his Nobel lecture, stated the problem facing chemists in attempting to characterize macromolecules (2):

The task of the chemist is facilitated by the fact that the agreement of just a few properties of two substances is sufficient for regarding them as identical. This principle applies only for low molecular weight compounds in which, owing to the small size relatively large differences in the properties occur, i.e. they differ abruptly from one compound to another. This situation is different for macromolecular compounds: a plastic such as polystyrene can be consistent with another polystyrene in a number of essential properties and still differ in composition. It is almost impossible to charac-

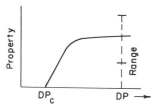

DEGREE OF POLYMERIZATION

Figure 1.1 The relationship between properties and the degree of polymerization for polymers.

terize and identify macromolecular substances by techniques which have contributed greatly to the rapid successes in low molecular weight chemistry.

The goal of most polymer research is to seek an understanding of polymer behavior. To achieve this objective it is necessary to gain some knowledge of structure-property relationships. Our purpose is to develop the theoretical and experimental basis for characterizing the structure of polymer chains at the molecular level.

For simple polymeric systems, it is only necessary to know the molecular weight or chain length to predict the properties of the polymer, since a general relationship exists between the degree of polymerization and the properties, as shown in Figure 1.1. However, in real systems a broad range of properties can be exhibited by polymers of the same degree of polymerization as shown in Figure 1.1. The questions are: What are the sources of these differences? Do they arise from differences in chemical composition? Differences in chemical structure? Differences in microstructure distribution of chemical structures? Are these differences a result of the polymerization reaction, the process variations, or fabrication variables? To answer these questions we need the ability to characterize polymer samples in detail, but the problem is complex, as indicated in the following section.

1.2 THE PROBLEM OF POLYMER CHARACTERIZATION

The structure of a single polymer chain can be written

$$\boxed{X}-\boxed{A}-\boxed{A}-\boxed{B}-\boxed{B}-\boxed{C}\cdots\boxed{A}\cdots\boxed{Y}$$

$$\begin{matrix} 1 & 2 & 3 & 4 & 5 & 6 & i & N \end{matrix}$$

where A is the dominant monomer structure and the alternate or defect structures are B, C, D, E, and so forth, are comonomer, geometric isomer, position isomer, stereoregular isomer, rearrangement isomer, branch, crosslink, X, Y = endgroups.

The total system contains many chains and nearly all possible combinations of units. Ideally, we would like to determine precisely the number and position of each structural element in each chain, but for large N this is impossible. For example (1),

A protein of molecular weight of 100,000 and composed of 20 different amino acids, the number of isomers is 10^{1270}. The size of this number becomes clear when compared to the number of molecules present in the seas of the earth—a mere 10^{46}.

We must resort to statistical considerations for polymers in order to seek answers. We have N units, each having several possible structures A, B, C, D. How many possibilities of arranging these units on the chain are there? Obviously,

The number of possibilities
for the first cell is N,
for the 2nd cell, $N-1$
and for the 3rd cell $N-2$, and so forth

$$\vdots \qquad \vdots$$

till for N, only 1 choice remains.

So the total number of structures is the product of these choices:

$$N(N-1)(N-2)\ldots 2,1 = N!.$$

But this assumes N possible different choices for each unit, whereas in fact, we have only a limited number of possible structures, so some of the arrangements are equivalent.

How many equivalent arrangements are there? Let us assume that the distribution is

n_1 cells have structure A,
n_2 cells have structure B,
$\vdots$
n_i cells have structure I.

Then these numbers $n_1, n_2, \ldots, n_i$ specify the composition of the system and, in general, are determinable. But the structure is not specified, since even when we know the numbers $n_1, n_2, \ldots, n_i, \ldots$, we still do not know just which of the N systems have the structure A, B, C, or J; we only know that n_1 of them are of structure A, n_2 of structure B, and so forth.

Thus we can consider that the n_1 items in the sequence form a set of n_1 structures. Now most of the $N!$ distributions differ from each other only in the sequence of items within a given set. For instance, there are $n_1!$ ways of forming the sequences which differ only in the order of the items. But we cannot distinguish between the identical items with any particular set, so there are $N!/n_1!$ possibilities when the first set is unordered and all other sets are ordered. Similarly for n $N!/n_1!n_i!$, and finally

$$\text{number of structures} = \frac{N!}{n_1!n_2!\ldots n_i!\ldots},$$

which can be written

$$\frac{N!}{\prod_i n_i!}$$

The total number of different structures of the system is given by the sum for all possible sets of numbers:

$$\text{total number} = \frac{\Sigma}{\left(\begin{array}{c}\text{all possible sets of}\\ \text{values of } n_1, n_2, \ldots\end{array}\right)} \frac{N!}{\prod_i n_i!}.$$

If the number of different structures is limited, the problem reduces but is still beyond our comprehension. For example, if only two different structures are allowed, the number of possibilities is 2^N, but for large N (10^4–10^6), the number of different structures available is impossibly large, so a statistical approach is required.

1.3 APPROACH

Our approach to this problem is diagrammed in Figure 1.2. We intend to apply probability theory to the polymerization process. Utilizing our chemistry knowledge, we can analyze different model polymerization processes (terminal, penultimate, etc.) in statistical terms. The chemical microstructure of the chain is related to the polymerization parameters for the different models. Then the real polymer is characterized in terms of its chemical microstructure, that is, its composition, sequence distribution, and so on.

The structural information acquired is compared with the theoretical predictions for the various polymerization mechanisms. From this comparison, a selection of the polymerization mechanism can be made and we can calculate any of the desired structural properties of the polymer. When these calculated structures agree with the measured values, one can assume that the polymer has been adequately characterized.

At this stage, correlations with engineering performance properties are feasible. The process for establishing structure property relationships is

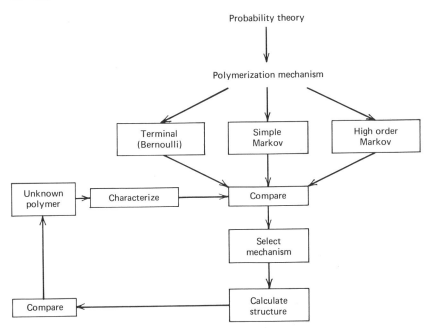

Figure 1.2 The logic process for the determination of the relationship between the micro-structure of a polymer and the polymerization mechanism.

diagrammed in Figure 1.3. The process begins with the preparation of a "new" polymer for which the initial tests indicate that the polymer has unusual and potentially useful mechanical, physical, or thermal properties. Sometimes it is possible to recognize the structural origin of the unusual properties of the "new" polymers. In this case, the procedure is to optimize the polymerization process, manufacture the product, sell it, and collect the profit. Most often, however, the structural origin is obscure, and one must decide whether to continue to study the system. If the decision is not to continue, then the "new" polymer is "junked" and a loss in dollars occurs as a result of the expense of the labor and chemicals used during this effort.

   If the decision is to continue, additional information is considered and various structural variables are postulated. Experiments are designed to vary the proposed structural variable, and suitable characterization procedures are instituted to determine the postulated structural variable. The results are correlated with the variation induced in the mechanical, physical, or thermal properties and, if a correlation is found with the postulated variable, it is optimized. If no adequate correlation is made, one must again decide whether additional work is justified. If the answer is affirmative, the cycle is repeated. If the answer is negative, the polymer is "junked." Hopefully, the number of cycles or structural variables to be followed is

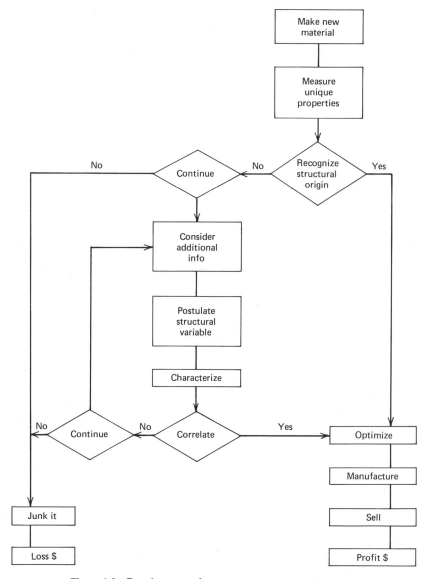

Figure 1.3 Development of structure-property relationships.

sufficiently low that the research effort has minimal cost and maximum success. Of course, as illustrated in Figure 1.4, the overall development program involves several additional important aspects, including patents, utility, and cost considerations.

In a given polymer field such as plastics, fibers, paints, elastomers, or packaging, a vast amount of information exists on the property requirements. Using this information, chemists can synthesize new polymers that

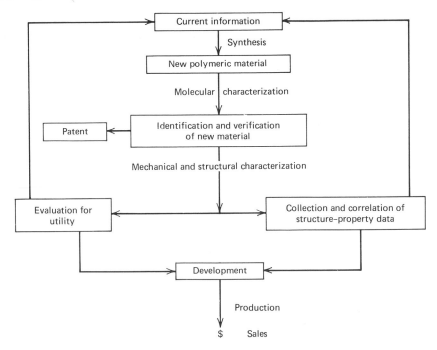

Figure 1.4   Nature of polymer development program.

are potential candidates for applications in a given area. First, molecular characterization is required to identify the new material, and in this case the initiation of patent procedures is desirable. Additionally, it is necessary to make detailed mechanical and structural characterization of the new polymer to determine the utility of the material and whether the establishment of correlations with structure can lead to optimization of the properties. Then there are additional tests for utility, including costs of synthesis, processing, and fabrication, and these results must be compared with those for competing polymers and other materials. Finally, if the above results are favorable, a development program can be initiated, with subsequent production and sale of the polymer.

Staudinger, in a lecture in Freiburg on September 13, 1954, summarized the above discussion in memorable fashion:

I have tried to point out the novelty of macromolecular chemistry by comparisons with buildings. Molecules as well as macromolecules can be compared with buildings which are built essentially from a few types of building stones: carbon, hydrogen, oxygen and nitrogen atoms. If only 12 or 100 building units are available, then only small molecules, or relatively primitive buildings can be constructed. With 10,000 or 100,000 building units, an infinite variety of buildings can be made: apartment houses, factories, skyscrapers, palaces and so on. Constructions, the possibilities of which cannot even be imagined, can be realized. The same holds for macromolecules. It is understandable that new properties will therefore be found which are not possible in low molecular materials. The number of possible macromolecular compounds is infinitely large.

The role of polymer characterization in the ultimate development of polymers is apparent, but even the most experienced and sophisticated polymer scientist is continually faced with new challenges in his efforts to characterize macromolecules. Obviously, a vast library of information exists. Standard methods are available for certain problems and they are utilized to the extent possible. However, every polymer scientist needs some concept of the new methods that are available, what these methods can accomplish, and, in absence of his personal experience, he needs information illustrative of their use. But obviously, if the system is entirely new, standard methods will not provide the necessary information, so he must have a theoretical and practical understanding of the chemical and physical methods of polymer characterization to determine which method or technique is applicable to his problem. Time or money does not allow the luxury of trying all the methods in a haphazard fashion. He must be able to deduce which technique or combination of methods might yield the required information.

But, even if he is intelligent or lucky enough to select the proper combination of physical or chemical techniques, he must be able to relate the derived structure to the polymerization methods used in the preparation of the polymer and he must be able to relate the structure to the mechanical, physical, and thermal properties of the system.

REFERENCES

1. H. Staudinger, *From Organic Chemistry to Macromolecules*, Wiley-Interscience, Publishers, New York, 1970, p. 91.
2. H. Staudinger, Nobel Lecture, December 11, 1953.

2
Probability Applied to Polymerization

Polymer chains are formed as a result of competing reactions, and their length and structure depend on the outcome of these processes. It is obvious that a broad range of lengths and structures is generated in the polymer system. The polymerization process is developing in time in a manner controlled by probabilistic laws, and these processes are called *stochastic processes*. Probability theory allows the study of these processes.

This chapter describes the probabilistic methods as they are applied to polymerization. For pedagogical reasons, the techniques are applied to chain length distributions, as the degree of polymerization is the simplest structural variable of the polymer chain. Consequently, the methods can be described for simple, physically understandable polymerizations. The application of these techniques to the chemical microstructure of polymers is described in Chapter 3.

2.1 ELEMENTS OF PROBABILITY THEORY (1)

A theory of probability is designed to deal with situations that have more than one possible outcome and seeks to answer questions of the likelihood of the various possible outcomes. The situation is called an *experiment*, and the outcome of the experiment is called an *event*. For our purposes, the experiment is a polymerization reaction and the events are the steps of propagation, termination, or chain transfer.

Probability theory answers questions such as: What is the possibility of a certain event occurring in a situation or experiment? Our state of

9

knowledge concerning the event for a given experiment allows us to evaluate the probability of that event. The two possible end points, that is, impossibility and certainty, are instances in which our knowledge is complete. If an event is either an impossibility or a certainty, we have all the knowledge needed. Probability is intended to measure our degree of ignorance about the possible outcomes. For some events we have more knowledge than for others, and the probabilities of these events are correspondingly greater. If for some events they differ in no known pertinent attribute, then they are equally likely, and our knowledge requires that these events be equally probable.

It is necessary to adopt some basis for assigning the probabilities. For our purposes we adopt the relative frequency basis, where the probability of an event E is given by the relative occurrence of event E compared to the total; that is,

$$P(E) = \frac{N_E}{N}, \tag{2.1}$$

where N_E is the number of E events occurring, and N is the total number of events.

This means that the probability of event E is equal to the number fraction of events (E) occurring. The terms *number fraction* and *probability* are often used interchangeably.

As a consequence of this choice, all probabilities are confined to the range of proper fractions, including the end points 0 and 1, which represent impossibility and certainty, respectively. The problem of assigning numerical likelihood or probabilities of events can be worked out. This convention applies, and is limited to, mutually exclusive events; that is, only one type of event can occur for each experiment. It further implies that the group of events is "complete," that is, that one or the other of the events *must* happen.

If the events can be considered to be either desirable or undesirable, they are said to be complementary. The probability of a desirable event may be designated $P(E) = p$, and the probability of the trial or experiment giving an undesired result is then

$$q = \frac{N - N_E}{N},$$

and necessarily

$$p + q = 1.$$

If several possible results $(i, j, k, l, \ldots)$ can occur, the probability of a particular result (i) becomes

$$P(i) = \frac{\text{Number of } i \text{ events}}{\text{Number of events}}. \tag{2.2}$$

Thus, if N_p is the number of propagations and N_t the number of terminations, the probability of propagation is

$$P = \frac{N_p}{N_p + N_t}. \qquad (2.3)$$

In termination we can have several processes producing termination, including coupling, disproportionation, and chain transfer. The probability of termination is the sum of the respective probabilities of coupling, disproportionation, and chain transfer.

In copolymerization we have chains ending in A and B. We need to consider the probabilities that compound events will occur; that is, what is the compound probability that the sequence AB will occur given that AA, BA, and BB can also occur? The compound event AB requires that the simple event A occurs first, followed by the B event. We can determine the probability of A forming. If N is the number of chains in the sample,

$$N = N_A + N_B,$$

where N_A and N_B are the number of chains ending in A and B, respectively. Then the probabilities of chains ending in A and B are

$$P_A = \frac{N_A}{N}, \qquad P_B = \frac{N_B}{N}.$$

In copolymerization we have addition of A and B to two different chain ends:

$$A* \begin{matrix} \xrightarrow{B} & AB* \\ \searrow_{A} & AA* \end{matrix},$$

$$B* \begin{matrix} \xrightarrow{A} & BA* \\ \searrow_{B} & BB* \end{matrix}$$

The question we desire to answer is: What is the probability of AB forming given that A* has occurred? One can use the relative frequency principle to calculate the compound probability that AB will occur by considering the total number of events occurring in an instant of time. The conditional probability, $P(B/A)$, that a B event occurs given that an A event has first occurred is

$$P(B/A) = \frac{\text{number of compound AB events}}{\text{number of simple A events}} = \frac{N_{AB}}{N_A}, \qquad (2.4)$$

where N_{AB} gives the number of AB events occurring. Similarly,

$$P(A/A) = \frac{N_{AA}}{N_A}, \qquad P(A/B) = \frac{N_{BA}}{N_B}, \qquad P(B/B) = \frac{N_{BB}}{N_B}$$

By dividing the numerator and denominator of these probabilities by N, we obtain

$$P(B/A) = \frac{N_{AB}/N}{N_A/N}.$$

But $N_{AB}/N = P(AB)$ and $N_A/N = P(A)$, so

$$P(B/A) = \frac{P(AB)}{P(A)}.$$

The compound probability that the event AB will occur is given by the product of the simple probability, $P(A)$, and the conditional probability, $P(B/A)$. Thus

$$P(AB) = P(A)P(B/A). \tag{2.5}$$

This equation is termed the *multiplication theorem*. Similarly,

$$P(AA) = P(A)P(A/A),$$

$$P(BA) = P(B)P(A/B),$$

$$P(BB) = P(B)P(B/B).$$

However, if A and B are independent events, the conditional probability $P(B/A)$ reduces to a simple probability, $P(B)$, so

$$P(AB) = P(A)P(B).$$

For independent events the conditional probabilities become

$$P(B/B) = P(B/A) = P(B),$$

$$P(A/B) = P(A/A) = P(A).$$

The probability that two independent events will occur sequentially is the product of their respective probabilities,

$$P(ij) = P_i P_j. \tag{2.6}$$

Since we are dealing with probabilities, we are not able to specify any of the variables exactly. We are only able to specify the ranges where the variable is preferred and other ranges where the variable is exceedingly rare. Such a variable is best thought of in connection with its distribution function, that is, a function representing the probabilities of all the possible outcomes. The set of numbers $(p_1, p_2, \ldots, p_n)$, which are the probabilities of the complete set of expected outcomes associated with the experiment, is called the probability distribution of the experiment, $P(x_j)$.

Using this distribution function, we want to be able to calculate an *expectation value* for a given set of experiments. It can be shown (1) that if x can take only the values $x_1, x_2, \ldots, x_a$ and zero, the probability of each being $p(x_1), p(x_2), \ldots, p(x_a)$ and $p(0)$, the mathematical expectation of x is

$$\epsilon_1(x) = \sum_{j=0}^{\infty} p(x_j)x_j. \tag{2.7}$$

From eq. (2.1) we can write

$$p(x_j) = \frac{n_j}{N},$$

and then

$$\epsilon_1(x) = \sum_{j=0}^{\infty} \frac{n_j}{N} \cdot x_j = \bar{x}$$

so the mathematical expectation value of the variable x is simply its average value,

$$\bar{x} = \sum_{j=0}^{\infty} p(x_j) x_j. \tag{2.8}$$

Distributions of this type can be treated by the method of moments. A point in the distribution curve has two coordinates associated with it, x_j and $p(x_j)$, usually as abscissa and ordinate, respectively, since x_j is the independent variable. In the statistical sense, the first moment of this point about the x_j origin is $p(x_j)x_j$, the second moment $p(x_j)x_j^2$, and the third moment $p(x_j)x_j^3$. The first moment about the origin of the distribution as a whole, v_1, is defined by the average summation of individual moments:

$$v_1 = \frac{\sum_j p(x_j) \cdot x_j}{\sum_j p(x_j)} \quad (= \bar{x}_n, \text{ incidentally}).$$

The second and higher moments about zero are

$$v_2 = \frac{\sum_j p(x_j)x_j^2}{\sum_j p(x_j)} = (\bar{x}_{v_2})^2,$$

and

$$v_3 = \frac{\sum_j p(x_j)x_j^3}{\sum_j p(x_j)} = (\bar{x}_{v_3})^3.$$

But the averages we are interested in are $\bar{x}_n$, $\bar{x}_w$, and $\bar{x}_z$, the number, weight, and z-average, respectively, which are preferred to represent the "centers of gravity" or balance points of their respective distributions. That is, the sum of the *first* moments of these distributions, $\bar{x}_n$, $\bar{x}_w$, or $\bar{x}_z$, must be zero. Accordingly, by definition and algebra,

$$\sum_j p(x_j)[x_j - \bar{x}_n] = 0, \qquad \bar{x}_n = \frac{\sum_j p(x_j)x_j}{\sum_j p(x_j)}, \text{ as before,}$$

$$\sum_j p(x_j)x_j[x_j - \bar{x}_w] = 0, \qquad \bar{x}_w = \frac{\sum p(x_j)x_j \cdot x_j}{\sum p(x_j)x_j} = \frac{\sum p(x_j)x_j^2}{\sum p(x_j)x_j},$$

$$\sum_j p(x_j)x_j^2[x_j - \bar{x}_z] = 0, \qquad \bar{x}_z = \frac{\sum p(x_j)x_j^2 \cdot x_j}{\sum p(x_j)x_j^2} = \frac{\sum p(x_j)x_j^3}{\sum p(x_j)x_j^2}.$$

Notice that the final algebraic formal statements of $\bar{x}_w$ and $\bar{x}_z$ look as though they were based on the second and third moments of $p(x_j)$ or x_j about zero. But $\bar{x}_w$ and $\bar{x}_z$ must be interpreted as the centers of gravity of the $p(x_j)x_j$ and $p(x_j)x_j^2$ distributions, respectively, on x_j; that is, the first moments of these distributions are $\bar{x}_w$ and $\bar{x}_z$ quite distinct from v_2 and v_3 above. Rather, $\bar{x}_w$ is comparable to a v_1 for a weight distribution, $p(x_j)x_j$, and $\bar{x}_z$ is a v_1 for a z-distribution $p(x_j)x_j^3$.

On the other hand, the first, second, and third moments about the *mean*, rather than about zero as for the v's, are convenient statistical measures of some characteristic properties of the $p(x_j)$ distribution on x_j. The first, second, and third statistical moments about the mean, v_1 or $\bar{x}_n$, are

$$\sigma_1 = \frac{\sum_j p(x_j) \cdot [x_j - v_1]}{\sum_j p(x_j)} [= 0],$$

$$\sigma_2 = \frac{\sum_j p(x_j) \cdot [x_j - v_1]^2}{\sum_j p(x_j)},$$

and

$$\sigma_3 = \frac{\sum_j p(x_j) \cdot [x_j - v_1]^3}{\sum_j p(x_j)}.$$

The second moment, σ_2, is the same as the statistical variance, and its square root is the statistical standard deviation. Thus σ_2 is always positive (or, rarely, zero), and the larger it is, the broader the distribution. The third moment is a measure of asymmetry or skewness. The direction of asymmetry is shown by the sign of the third moment, and the magnitude of the asymmetry by its value.

A finite distribution function of x_j up to n is completely determined by its first n moments because the x values are given by the solution to n equations in n unknowns:

$$\sum_{j=1}^{n} x_j^k = nv_k, \qquad k = 1, 2, \ldots, n.$$

An excellent numerical example of the magnitudes of these moments for polymerization has been given (2).

2.2 BERNOULLI OR RANDOM CHAIN PROCESSES

2.2.1 Probability Applied to the Addition Polymerization Process

What is the probability an active species propagates rather than terminates?

Consider an active species in a polymerizing system. The only reactions occurring are assumed to be the following:

$$[M_n^*] + M \rightarrow [M_{n+1}^*] \qquad \text{propagation,}$$

$$[M_n^*] + [M_m^*] \rightarrow [M_n] + [M_m] \qquad \text{termination (disproportionation).}$$

The initiation reaction is neglected, which is equivalent to assuming a high degree of polymerization. The probability of propagation can be written in terms of the number of reactions occurring in an interval of time:

$$p = \frac{\text{number of propagations occurring in time } dt}{\text{number of reactions occurring in time } dt} = \frac{N_p}{N}. \qquad (2.9)$$

Similarly, the probability of termination is

$$q = \frac{\text{number of terminations occurring in time } dt}{\text{number of reactions occurring in time } dt} = \frac{N_t}{N}. \qquad (2.10)$$

We calculate the number of occurring reactions from the kinetics of the reaction. For propagation of the species M_n^*, we have

$$[M_n^*] + [M] \xrightarrow{k_{pn}} [M_{n+1}^*].$$

Therefore the rate of disappearance of the reactive species M_n^* due to polymerization is

$$-d[M_n^*] = k_{pn}[M_n^*][M]dt = N_{np}$$

where N_{np} is the number of propagations of the nth species in the time dt.

Likewise for termination, the M_n^* species reacts:

$$M_n^* + M_i^* \xrightarrow{k_{tni}} M_n + M_i.$$

where the disappearance of the reactive species M_n^* due to termination then becomes

$$-d[M_n^*] = [M_n^*] \sum_i k_{tni}[M_i^*]dt = N_{tn}.$$

The sum over all the i species is necessary, since all possible reactions of the species i must be considered. The number of terminations of the M_n^* species is noted as N_{tn}. Thus the probability of propagation for the nth species is

$$P_n = \frac{N_{pn}}{N_{pn} + N_{tn}}.$$

In the polymerization mixture at any given time, there will be many active species of various chain lengths. We want to know the course of the *overall* processes occurring in the polymerization mixture, so we sum over all species in the mixture:

$$-\sum_n d[M_n^*] = \sum_n k_{pn}[M_n^*][M]dt,$$

$$-\sum_n d[M_n^*] = \sum_n \sum_i k_{tni}[M_n^*][M_i^*]dt.$$

To proceed it is necessary to make the assumption that the reactivity in the propagation and termination steps is independent of the chain length (3). This assumption is equivalent to the following:

$$k_{p_1} = k_{p_2} = k_{p_3} = k_{p_n} = k_p,$$

$$k_{t_{nm}} = k_{t_{n1}} \cdots k_{t_{n1}} = k_t,$$

and implies that all events are equal, independent of chain length, so the polymerization process is a random process. This allows the sums to be simplified:

$$-\sum_n d[M_n^*] = k_p \sum_n [M_n^*][M]dt,$$

$$-\sum_n d[M_n^*] = k_t \sum_n \sum_i [M_n^*][M_i^*]dt.$$

Then it follows that

$$-d[M^*] = k_p[M^*][M]dt,$$

$$-d[M^*] = k_t[M^*]^2 dt, \qquad (2.11)$$

where $M^* = \Sigma_n M_n^* =$ concentration of active species.

In the interval of time dt we can now obtain the following:

$$N_p = \text{number of propagations} = k_p[M^*][M], \qquad (2.12)$$

$$N_t = \text{number of terminations} = k_t[M^*]^2, \qquad (2.13)$$

$N =$ number of reactions = number of propagations and terminations

$$= k_p[M^*][M] + k_t[M^*]^2. \qquad (2.14)$$

The probability of propagation can be calculated in terms of the concentration of the reactants and the rate constants for propagation and termination. Substituting the results from eqs. (2.12), (2.13), and (2.14) in eq. (2.9) yields the probability of propagation as

$$p = \frac{k_p[M^*][M]}{k_p[M^*][M] + k_t[M^*]^2} = \frac{k_p[M]}{k_p[M] + k_t[M^*]}. \qquad (2.15)$$

Likewise, for the probability of termination we obtain

$$q = \frac{k_t[M^*]^2}{k_p[M^*][M] + k_t[M^*]^2} = \frac{k_t[M^*]}{k_p[M] + k_t[M^*]}, \quad (2.16)$$

and it is obvious that

$$p + q = 1.$$

In examining eq (2.15), one notes that in most polymerization systems $[M^*]$ is about 10^{-6} and $[M]$ approaches unity, so $k_p[M] > k_t[M^*]$.

The values of k_p have a wide range for the various monomers depending as it does on the reactivity of both the radical and the monomer. Likewise for the termination rate constants, but the termination rate constant, k_t, is usually 10^6 or 10^7 times faster, as termination involves reaction between two radicals. In the probability for propagation the term $k_t[M^*]$ involves a high rate constant but extremely low concentrations of radicals, whereas the k_p term is smaller although the monomer concentration more than compensates for difference.

The probability of polymerization approaches one, which is required for the formation of high polymer. The probability of propagation increases with increase in monomer concentration and decreases with increase in reactive species, which agrees with our chemical knowledge.

This probabilistic model is built on several assumptions, and these must be recognized. First, only two reactions were considered; all others are ignored. Also, p is assumed to be independent of the degree of polymerization and of polymerization time. Isothermal conditions are assumed. These are highly restrictive assumptions.

Now that we have obtained the probability of propagation and termination of our addition polymerization system, we can use this knowledge to calculate the nature of the chains in the mixture. However, it is to be mentioned that we are calculating average values for the various parameters of the system by utilizing the distribution function for the polymerization reaction.

2.2.2 Distribution Function of the Polymer Molecules

The first task is to derive the probability distribution function for the various possible chain lengths in the polymerization mixture. This is accomplished by calculating the probability for obtaining chains of length $n = 1, 2, 3$, and so forth until a general formula can be generated. The probability of having a molecule with chain length $n = 1$, which is symbolized $P(1)$, is

$$M_1^* \xrightarrow{q} M_1, \quad \text{so } P(1) = q = (1 - p).$$

Remember that the initiation step has been neglected. The reactions required to obtain a dimer are

$$M_1^* \xrightarrow{p} M_2^* \xrightarrow{pq} M_2,$$

and the probability is

$$P(2) = pq = p(1-p),$$

where we have made use of eq. (2.6) in calculating the probability of a propagation reaction being followed by a termination reaction sequentially as the product pq of the respective probabilities. The reactions leading to chain length $n = 3$ are

$$M_1^* \xrightarrow{p} M_2^* \xrightarrow{p^2} M_3^* \xrightarrow{p^2q} M_3, \qquad P(3) = p^2q = p^2(1-p).$$

Observe that, because we can now generalize the results by noting that for chain length n we require $n-1$ propagations and one termination, we can write

$$M_1^* \xrightarrow{p} M_2^* \xrightarrow{p^2} M_3^* \longrightarrow \overset{p^{n-1}}{\cdots} M_n^* \xrightarrow{p^{n-1}q} M_n, \qquad P(n) = p^{n-1}q = p^{n-1}(1-p).$$

This is the probability distribution function for the chain lengths in the polymerization mixture. It is related to a familiar quantity. The fraction of molecules in the mixture having chain length n may be written in terms of the respective probabilities:

$$X_n = \frac{P(n)}{\sum P(n)}.$$

It is required of the probabilities that

$$\sum P(n) = 1,$$

so

$$X_n = P(n).$$

Resorting to our familiar chemical terminology, we see from this equation that the mole fraction of polymer having chain length n is given by its probability of formation in the mixture. We thus call

$$P(n) = p^{n-1}(1-p) \tag{2.17}$$

the mole fraction distribution function representing the "most probable" or random polymerization process. It is to be observed that eq. (2.17) is a discrete and not a continuous function, and that it is limited to positive integer values of n. The discrete nature must be recognized, as some of the moment calculations may be in error. The redeeming feature of the function is that for large n very little error is introduced, whereas for small n considerable error is caused by integration rather than summing. Fortunately, practical cases involve large n.

2.2.3 Average Chain Length in the Polymerization Mixture

Using the mole fraction distribution function, we can calculate the average chain length or degree of polymerization of polymer in the mixture. The number average degree of polymerization can be written

$$\overline{DP}_n = \frac{\sum\limits_{n=0}^{\infty} nN_n}{\sum\limits_{n=0}^{\infty} N_n},$$

where N_n is the number of molecules of length n. Note that $\Sigma N_n = N =$ number of molecules in system. The number of molecules of $DP = n$ is given by $N_n = NP(n)$, since the number of molecules having a specific length corresponds to the probability of finding molecules of this length multiplied by the total number of molecules. Substituting, we get

$$\overline{DP}_n = \frac{\sum\limits_{n=0}^{\infty} nNP(n)}{N} = \sum\limits_{n=0}^{\infty} nP(n)$$

$$= \sum\limits_{n=0}^{\infty} np^{n-1}(1-p). \tag{2.18}$$

This sum is calculated by noting that

$$\sum\limits_{n=0}^{\infty} np^{n-1}(1-p) = (1-p)\frac{d}{dp}\left(\sum\limits_{n=0}^{\infty} p^n\right)$$

and

$$p\sum\limits_{n=1}^{\infty} p^{n-1} = p(1+p+p^2+p^3+\cdots p^n)$$

Note as well that

$$\frac{1}{1-p} = 1+p+p^2+\cdots \qquad \text{when } p<1,$$

so

$$p\sum\limits_{n=1}^{\infty} p^{n-1} = \frac{p}{1-p}$$

and

$$\frac{d}{dp}\left(\frac{p}{1-p}\right) = \frac{1}{(1-p)^2}.$$

Therefore

$$\overline{DP}_n = \sum\limits_{n=0}^{\infty} np^{n-1} = \frac{1}{1-p}. \tag{2.19}$$

This equation demonstrates that a measurement of the average degree of polymerization allows the calculation of the probability of propagation for this model polymerization.

2.2.4   Number of n-mers in the Polymer Mixture

Chemically, we know that the average chain length in the mixture is the ratio of the number of molecules at the start of the polymerization, N_0, to the number of molecules in the mixture at the time dt, which we have designated as N. Thus $\overline{DP}_n$ can also be written

$$\overline{DP}_n = \frac{N_0}{N} = \frac{1}{1-p},$$

so

$$N = N_0(1-p).$$

The number of molecules of length n, N_n, is simply the number of total molecules times the fraction of molecules of length n:

$$N_n = NP(n)$$

Substituting for N in terms of the initial number of monomer units yields

$$N_n = N_0(1-p)p^{n-1}(1-p),$$

$$= N_0(1-p)^2p^{n-1} \quad \text{"most probable" number distribution} \qquad (2.20)$$

This equation gives the number distribution of molecules of length n in the mixture, which is plotted in Figure 2.1. On a number basis, monomers are more plentiful than any other molecular species at all stages of the polymerization, and their relative number decreases as the probability of propagation increases.

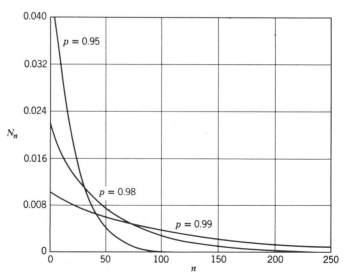

Figure 2.1 Number of molecules of n length in an ideal addition polymer for various probabilities of propagation.

2.2.5 Weight Fraction of n-mers in the Polymer Mixture

We are interested in what weight of the polymerization mixture has a definite chain length. The weight of molecules of length n is the product of the number of length n, or N_n, and the weight of a molecule of length n, which is nM_0, where M_0 is the molecular weight of one unit. The total weight of the polymerization mixture is N_0M_0. The weight fraction, W_n, of polymer molecules that are n units long is

$$W_n = \frac{nN_nM_0}{M_0N_0} = \frac{nN_n}{N_0} = np^{n-1}(1-p)^2. \tag{2.21}$$

On a weight basis the proportion of low molecular weight material is very small, decreasing as the average molecular weight increases, as shown in Figure 2.2.

The peak in distribution curve moves to longer chain length as the probability of propagation increases. Many of the flow and mechanical properties are a function of the weight fraction distribution.

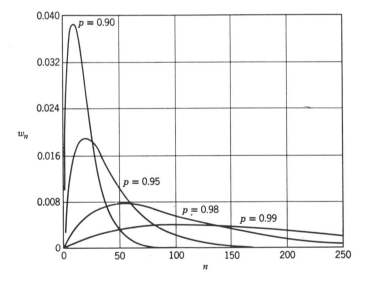

Figure 2.2 Weight fraction of molecules of n length in an ideal addition polymer for various probabilities of propagation.

2.2.6 Molecular Weight Averages and Polydispersity

When a procedure suitable for the determination of molecular weights is applied to a sample containing molecules of different sizes, it yields a value that we may call the *average molecular weight*. For a monodispersed system the molecular weight is

$$M_n = W/N$$

where W is the weight of the sample and N is the number of molecules. For a polydispersed system one must write $W = \Sigma \, N_n M_n$, $N = \Sigma_n \, N_n$, so

$$\bar{M}_n = \frac{\Sigma \, N_n \bar{M}_n}{\Sigma \, N_n} = \text{Number average molecular weight.}$$

Now $M_n = n M_0$, where M_0 is the molecular weight of a repeat unit. The number distribution function can be substituted for N_n in eq. 2.20, giving

$$\bar{M}_n = \frac{M_0 (1-p)^2 N_0 \sum n p^{n-1}}{(1-p)^2 N_0 \sum p^{n-1}} = \frac{M_0 \sum n p^{n-1}}{\sum p^{n-1}}$$

$$= \frac{M_0}{1-p}. \tag{2.22}$$

In the determination of molecular weights by the colligative properties of solutions, each molecule, large or small, makes the same contribution to the observed effect. This effect would remain unchanged if the total weights were shared equally among the molecules of the system, so one is obtaining a *number average* molecular weight.

For light scattering of polymer solutions, the effect is proportional to the molecular weight of the solute. The effect obtained with a polydispersed solute depends on its weight average molecular weight, $\bar{M}_w$. The average molecular weight is given by

$$\bar{M}_w = \frac{\displaystyle\sum_n W_n M_n}{W}$$

$$= \frac{\sum (n M_0 N_n)(n M_0)}{W}$$

$$= \frac{M_0^2 N_0 (1-p)^2 \sum n^2 p^{n-1}}{M_0 N_0} \tag{2.23}$$

$$= \frac{M_0 (1+p)}{1-p},$$

where $W_n = n M_0 N_n$ and $W = M_0 N_0$. A common measure of the breadth of a polymer distribution is the ratio of the weight average to number average molecular weights. Thus

$$\frac{\bar{M}_w}{\bar{M}_n} = 1 + p$$

and, since p is nearly unity, this ratio approaches 2 in the limit.

The second moment, σ_2, or the variance for this system, is

$$\sigma_2 = \frac{p}{(1-p)^2}, \tag{2.24}$$

and the square root of this quantity is the standard deviation, which is

$$\sigma = \frac{p^{1/2}}{1-p}. \tag{2.25}$$

The fraction of molecules lying within k standard deviations of the mean is at least $1 - 1/k^2$. Thus only 25% of the molecules can lie outside the range of two standard deviations, and no more than 11% outside the range of three. In this case, it is easily observed that the breadth of the distribution increases as a function of the degree of polymerization or the average size of the molecule. This result is generally applicable.

Another interesting measure of the breadth of a distribution is the coefficient of variation, defined as

$$\Delta = \left[\frac{\overline{DP}_w}{\overline{DP}_n} - 1 \right]^{1/2} = \frac{\sigma}{\overline{DP}_n}. \tag{2.26}$$

This index is particularly useful in dealing with either very broad or narrow distributions because of the square root relationship. For the distribution that we have been considering, the coefficient of variation is

$$\Delta = p^{1/2}, \tag{2.27}$$

which is less sensitive to the probability of propagation.

2.2.7 Calculation of Parameters from Probability Distributions

The mathematical expectation of any function denoted $f(x)$ is defined by

$$f(x) = \sum_{n=-\infty}^{+\infty} f(n)P_x(n)$$

where $P_x(n)$ is the appropriately weighed functional (x) distribution function. The fth moment of a distribution is the rth power of the variable

$$f(x) = X^r,$$

hence the first moment is the number average degree of polymerization:

$$\overline{DP}_n = \frac{\sum nP(n)}{\sum P(n)} = \sum nP(n),$$

and for the second moment

$$\overline{DP}_w = \frac{\sum nW(n)}{\sum W(n)} = \sum nW(n).$$

Table 2.1 gives a tabulation of some of the useful sums necessary to evaluate equations of this type, as well as some indication of their method of calculation. These are relations between the single moments and the higher ones. By evaluating a normalizing factor we obtain

$$W(n) = KnP(n)$$

subject to the condition that

$$\sum_{i=1}^{\infty} W(n) = 1.$$

Since K is not a function of n,

$$K = \frac{1}{\sum nP(n)} = \frac{1}{\overline{DP}}.$$

Then

$$\overline{DP}_w = \sum nW(n) = \sum n\frac{1}{\overline{DP}}nP(n)$$

$$= \frac{1}{\overline{DP}} \sum n^2 P(n)$$

$$= \frac{\overline{DP^2}}{\overline{DP}}.$$

Table 2.1 Tabulations of Summations Over an Index

A. $\displaystyle\sum_{m=1}^{\infty} p^{m-1} = 1 + p + p^2 + p^3 + \cdots = \frac{1}{1-p}$, $p \leq 1.0$

B. $\displaystyle\sum_{m=2}^{\infty} p^m = p \sum_{m=1}^{\infty} p^{m-1} = \frac{p}{1-p}$

C. $\displaystyle\sum_{m=1}^{\infty} mp^{m-1} = \frac{d}{dp} \sum_{m=1}^{\infty} p^m = \frac{d}{dp}\left(\frac{p}{1-p}\right) = \frac{1}{(1-p)^2}$

D. $\displaystyle\sum_{m=1}^{\infty} m^2 p^{m-1} = \frac{(1+p)}{(1-p)^3}$

    Let $X = \displaystyle\sum_{m=1}^{\infty} p^m = \frac{p}{1-p}$

    $\dfrac{dX}{dp} = \displaystyle\sum_{m=1}^{\infty} mp^{m-1} = \frac{1}{(1-p)^2}$

    $\dfrac{d^2X}{dp^2} = \displaystyle\sum_{m=1}^{\infty} (m^2-m)p^{m-2} = \frac{2}{(1-p)^3}$

    So $\displaystyle\sum_{m=1}^{\infty} m^2 p^{m-1} = \frac{2p}{(1-p)^3} + \sum_{m=1}^{\infty} mp^{m-1} = \frac{1+p}{(1-p)^3}$

E. $\displaystyle\sum_{m=2}^{\infty} p^{m-2} = \sum_{k=1}^{\infty} p^{k-1} = \frac{1}{1-p}$

F. $\displaystyle\sum_{m=2}^{\infty} mp^{m-2} = \frac{1}{p} \sum_{m=2}^{\infty} mp^{m-1} = \frac{1}{p}\left(\sum_{m=1}^{\infty} mp^{m-1} - 1\right) = \frac{[1+(1-p)]}{(1-p)^2}$

Similar arguments can be used for higher moments. These equations can be used for computational purposes.

However, there may be a simpler method of calculating these moments by using the moment-generating function (4), which can be obtained for a given probability distribution by evaluating a single infinite sum.

The moment-generating function $M_n(t)$ of an integral-valued variable n is defined as the mathematical expectation e^{tn}; that is,

$$M_n(t) = \overline{e^{tn}} = \sum_{n=1}^{\infty} e^{tn} P(n).$$

Note that

$$e^{tn} = 1 + tn + \frac{t^2 n}{2!} + \frac{t^3 n^3}{3!} + \cdots$$

so

$$M_n(t) = \sum e^{tn} P(n) = 1 + t + \frac{t^2}{(2)1!} + \frac{t^3}{(3)3!} + \cdots.$$

The dummy variable t has no physical significance. If the value of $M_n(t)$ is known, moments of the distribution can be generated.

The rth moment is the rth derivative (with respect to t) of the moment-generating function evaluated at the point $t = 0$; that is,

$$\overline{X^r} = \frac{d^r M_n(t)}{dt^r}\bigg|_{t=0}.$$

Use of cumulant-generating functions $K_n(t)$ may even be simpler than the use of $M_n(t)$. The function is defined as

$$K_n(t) = \ln[M_n(t)].$$

From this function cumulants of the distribution are obtained in a similar manner to moments:

$$K_r = \frac{d^r K_n(t)}{dt^r}\bigg|_{t=0}.$$

The first few derivatives of $K_n(t)$ frequently are more easily evaluated than are the corresponding derivatives of $M_n(t)$.

The moments are

$$\overline{X} = K_1,$$
$$\overline{X^2} = K_1^2 + K_2,$$
$$\overline{X^3} = K_1^3 + 3K_1 K_2 + K_3,$$
$$\overline{X^4} = K_1^4 + 4K_1 K_3 + 3K_2^2 + 6K_1^2 K_2 + K_4,$$

and

$$K_1 = X,$$
$$K_2 = \overline{X^2} - \overline{X}^2,$$
$$K_3 = \overline{X^3} - 3\overline{X}\,\overline{X}^2 + 2\overline{X}^3,$$
$$K_4 = \overline{X^4} - 4\overline{X^3}\overline{X} - 3(\overline{X^2})^2 + 12\overline{X^2}\overline{X}^2 - 6\overline{X}^4.$$

An example of evaluation for our simple addition system gives

$$M_n(t) = \frac{(1-p)e^t}{(1-pe^t)}$$

and

$$K_n(t) = \ln(1-p) + t - \ln(1-pe^t).$$

Successive differentiation of $K_n(t)$ with respect to t and evaluation of each result at $t = 0$ give for the first four cumulants

$$K_1 = \frac{1}{1-p},$$

$$K_2 = \frac{p}{(1-p)^2},$$

$$K_3 = \frac{p(1+p)}{(1-p)^3},$$

$$K_4 = \frac{p(1+4p+p^2)}{(1-p)^4}.$$

Thus for our idealized addition system the results are

$$\overline{DP}_n = \frac{1}{1-p},$$

$$\overline{DP}_w = \frac{1+p}{1-p},$$

$$\overline{DP}_z = \frac{1+4p+p^2}{(1-p)(1+p)},$$

$$\frac{\overline{DP}_w}{\overline{DP}_n} = 1+p,$$

$$\Delta = p^{1/2}.$$

2.2.8 Real Distributions and Generalized Molecular Weight Distributions

We have considered chain length distributions predicted on the basis of simple and idealized kinetic considerations. In practice these distributions are modified by a number of complicating factors. For example, reaction conditions may vary during the polymerization, the temperature may change, or the monomer concentration may drift with conversion. Impurities may react and change the nature of the polymerization. As a result, the polymer can be characterized by a superposition of the normal distributions obtained at any given time. The result is a broadening of the molecular weight distribution. In the ideal case

$$\frac{\bar{M}_w}{\bar{M}_n} \approx 2,$$

but commercial polymers often have a weight to number average ratio between 10 and 20, and this ratio may reach as high as 40, since competing reactions such as branching occur.

Thus we cannot rely on the applicability of the form of the distribution function based on kinetic considerations. It is therefore advantageous to utilize empirical functions derived from generalized molecular weight distributions (4). We are able to specify the averages and a breadth, but a limited number of parameters cannot uniquely define a molecular weight distribution. Fortunately, methods such as gel permeation chromatography allow a detailed analysis of the molecular weight distribution.

2.3 STATISTICS OF MARKOV CHAINS

2.3.1 Definition

The Bernoulli or random chain processes are defined as a series of independent events. The Markov process is a method of characterization of a series of *dependent* events. Random processes are no-memory processes whereas Markov processes do have memory, but the dependence of future events on past events is of a particularly simple nature. As illustrated later, these Markov chain methods have a wide range of applications in polymer analysis.

Consider a chain that may be described at any time as being in one of a set of states $S_1, S_2, \ldots, S_m$ which are mutually exclusive and complete. For a chain polymerization process, for example, the only two states are propagation (S_p) or termination (S_t). The system may undergo changes of state (or state transitions). We number the particular changes as a result of each transition as the first transition, second transition, and so on. So let $S_i(n)$ be the event that the system is in the state S_i immediately after the nth transition. The probability of this event may be written $P[S_i(n)]$. Each trial may be described by transition probabilities of the form

$$P[S_j(n)/S_a(n-1)S_b(n-2)S_c(n-3)\cdots]. \tag{2.28}$$

These transition probabilities specify the probabilities associated with each trial, and they are conditional on the entire history of the process.

If the transition probabilities for a series of dependent trials satisfy the Markov condition

$$P[S_j(n)/S_a(n-1)S_b(n-2)\cdots] = P[S_j(n)/S_i(n-1)] = P_{ij}, \qquad (2.29)$$

then a first order Markov condition pertains and the system depends on only the previous state.

The quantity P_{ij} is the conditional probability that the system will be in state S_j immediately after the next trial given that the state of the system is S_i. The quantity P_{ij} is a one-step transition probability function for a first order Markov chain. We also require $0 \le P_{ij} \le 1$ and

$$\sum_j P_{ij} = 1 \qquad \text{for } i = 1, 2, 3, \ldots, m$$

Higher order Markov chains are possible and follow the same process except that additional steps must be included.

It is convenient to display these transition probabilities as a transition matrix $\mathbf{P}$, for which P_{ij} is the entry in the ith row and the jth column corresponding to the initial and final states:

$$\mathbf{P} = \begin{bmatrix} P_{11} & P_{12} & \cdots & P_{1m} \\ P_{21} & P_{22} & \cdots & P_{2m} \\ \cdots & \cdots & \cdots & \cdots \\ P_{m1} & P_{m2} & \cdots & P_{mm} \end{bmatrix} \qquad (2.30)$$

The transition matrix can be used to calculate the averages and moments of stochastic processes. It is first necessary to determine the elements of the matrix for the process under investigation.

2.3.2   Initial Probability Vectors

From the preceding discussion it would appear that one can easily calculate the current state of the system as a function of its previous states, but we need to know the initial state of the system. The initial probabilities, $P_i^{(0)}$, give the probability of the system originating in the ith state. The initial function can be arrayed in vector form:

$$\mathbf{P}^{(0)} = \mathbf{P}^{(0)} = \{P_1^0, P_2^0 \cdots\},$$

and

$$\sum_i P_i^0 = 1.$$

Thus the probability that the system is in state k after n steps is

$$P_{ik}^{(n)} = \sum_i P_i^{(0)} P_{ik}^{(n)},$$

or

$$\mathbf{P}^{(n)} = \mathbf{P}^{(0)}\mathbf{P}^n. \tag{2.35}$$

2.3.3 Stationary Probability Vectors

For many systems, after many state transitions a "steady state" can be achieved such that the probability function reaches a stage where the population of any state is constant as $n \to \infty$. This implies that

$$\mathbf{P}^{(s)}\mathbf{P} = \mathbf{P}^s.$$

In other words a stationary probability vector $\mathbf{P}^s$ is obtained. Of course, $\Sigma_i P_i^s = 1$.

We might wonder what condition will guarantee the existance of a unique stationary probability vector. Physically, this corresponds to the steady state. If for all powers of the transition matrix there are only positive entries, the transition matrix is defined as a regular transition matrix. If $\mathbf{P}^{(0)}$ is the vector of initial probabilities, then as n increases the elements of $\mathbf{P}^{(n)} = \mathbf{P}^{(0)}\mathbf{P}^n$ approach the corresponding elements of $\mathbf{P}^s$. That is, $P_i^{(n)}$ approaches P_i^s, where $i = 1, 2, \ldots, k$. This result implies that if the matrix of transition probabilities is regular, then, regardless of the initial probabilities, the probability that the chain is in state i is very close to P_i^s after the series of experiments is performed a sufficiently large number of times. Hence we define $\mathbf{P}$ for our polymerization and determine $\mathbf{P}^s$ for the stationary states using

$$\mathbf{P}^s\mathbf{P} = \mathbf{P}^s.$$

These matrices are stochastic matrices and have special mathematical properties (5).

For example, if the Markov chain has only two states, the transition matrix is

$$\mathbf{P} = \begin{bmatrix} P_{11} & P_{12} \\ P_{21} & P_{22} \end{bmatrix},$$

and one can show that the stationary probability vector is

$$\mathbf{P}^s = \left[\frac{P_{12}}{P_{12} + P_{21}}, \frac{P_{21}}{P_{12} + P_{21}} \right].$$

This method is useful for systems with many accessible states and has the implications associated with the steady-state approximation in the kinetic approach previously outlined. For complicated systems the $\mathbf{P}^s$ vector represents the state of the system, and the determination of the $\mathbf{P}^s$ vector constitutes a complete description of the system.

We can use the transition probability matrix to predict the states of the system regardless of its initial state, because $\mathbf{P}^n$ approaches a limiting form. The elements of each row of $\mathbf{P}^n$ become identical, and those in each column are identical to the corresponding elements in $\mathbf{P}^s$. If the transition matrix $\mathbf{P}$ is multiplied by itself repeatedly, the elements of each row attain constant values that are the probabilities of the system being in the corresponding state. Normalization is not necessary. A good example of this result is demonstrated by Harwood and co-workers (7), who indicate the desirability of this approach for computer calculations.

Consider the matrix

$$\mathbf{P} = \begin{bmatrix} P(A/A) & P(A/B) & P(A/C) \\ P(B/A) & P(B/B) & P(B/C) \\ P(C/A) & P(C/B) & P(C/C) \end{bmatrix}.$$

For the convenience of numerical analysis let us arbitrarily give the conditional probabilities the following values:

$$\mathbf{P} = \begin{bmatrix} 0.5 & 0.1 & 0.2 \\ 0.3 & 0.6 & 0.4 \\ 0.2 & 0.3 & 0.4 \end{bmatrix}.$$

then

$$\mathbf{P}^2 = \mathbf{P} \cdot \mathbf{P} = \begin{bmatrix} 0.31999 & 0.16999 & 0.21999 \\ 0.40999 & 0.50999 & 0.45999 \\ 0.26999 & 0.31999 & 0.31999 \end{bmatrix},$$

$$\mathbf{P}^4 = \mathbf{P}^2 \cdot \mathbf{P}^2 = \begin{bmatrix} 0.23149 & 0.21149 & 0.21899 \\ 0.46449 & 0.47699 & 0.47199 \\ 0.30399 & 0.31149 & 0.30899 \end{bmatrix},$$

$$\mathbf{P}^8 = \mathbf{P}^4 \cdot \mathbf{P}^4 = \begin{bmatrix} 0.21841 & 0.21807 & 0.21819 \\ 0.47259 & 0.47279 & 0.47272 \\ 0.30900 & 0.30913 & 0.30908 \end{bmatrix},$$

$$\mathbf{P}^{16} = \mathbf{P}^8 \cdot \mathbf{P}^8 = \begin{bmatrix} 0.21818 & 0.21818 & 0.21818 \\ 0.47273 & 0.47273 & 0.47273 \\ 0.30909 & 0.30909 & 0.30909 \end{bmatrix},$$

$$\mathbf{P}^{32} = \mathbf{P}^{16} \cdot \mathbf{P}^{16} = \begin{bmatrix} 0.21818 & 0.21818 & 0.21818 \\ 0.47272 & 0.47272 & 0.47272 \\ 0.30909 & 0.30909 & 0.30909 \end{bmatrix},$$

The matrices are continually squared, as convergence is faster. The compositional terms $P(A)$, $P(B)$, and $P(C)$ are the elements of any column of the matrix. For this example $P(A)=0.21818$, $P(B)=0.47272$, and $P(C)=0.30909$. This procedure is simple, requiring almost no mathematical manipulation and very little programming, since matrix multiplication subroutines are available. Furthermore, the method is completely general.

Another general method of seeking $\mathbf{P}^s$ for large matrices is to obtain the eigenvalues and the eigenfunctions of $\mathbf{P}$ and perform a similarity transformation (6). If $\mathbf{P}$ is of the order m,

$$|\mathbf{P}-\lambda I|=0 \qquad (2.36)$$

yields $\lambda_1, \lambda_2, \ldots, \lambda_m$ eigenvalues of $\mathbf{P}$. In our case $\mathbf{P}$ is a stochastic matrix; that is the sum of every row:

$$\sum_{\text{all } j} P_{ij} = 1.$$

For this type of matrix it is possible to show that (1) there is one eigenvalue equal to unity, and (2) that all other eigenvalues must be less than unity. Since for the eigenvalue equal to unity

$$|A|=|\mathbf{P}-I|=0, \qquad (2.37)$$

the solutions $\mathbf{P}_j^s$ that we desire are given by

$$\mathbf{P}_j^s=CA^{jm}, \qquad (2.38)$$

where A^{jm} is the cofactor of the element A_{jm} in the determinant $|A|$ and C is a constant. In later chapters this method is applied to systems to allow generation of results for complex models where the kinetic or conditional probability approach becomes cumbersome.

REFERENCES

1. W. Feller, *An Introduction to Probability Theory and Its Applications*, 2nd ed., Vol. 1, John Wiley & Sons, New York, 1957.
2. M. L. Miller, *Structure of Polymers*, Reinhold, New York, 1964.
3. P. J. Flory, *Principles of Polymer Chemistry*, Cornell University Press, Ithaca, New York, 1953.
4. L. H. Peebles, *Molecular Weight Distribution in Polymers*, Wiley-Interscience Publishers, New York, 1971.
5. G. G. Lowry, *Markov Chains and Monte Carlo Calculations in Polymer Science*, Marcel Dekker, New York, 1970.
6. F. P. Price, *J. Chem. Phys.*, **36**, 209 (1962).
7. H. J. Harwood, Y. Kodaira, and Daniel L. Newman in *Computors in Polymer Science*, J. S. Mattson, Harry B. Mark Jr., and H. C. MacDonald Jr., Eds., Marcel Dekker, New York, 1977, Chapter 2.

3

Theory of the Characterization of Polymer Microstructure

3.1 INTRODUCTION

In our efforts to characterize the polymer chain microstructure, we will be able to determine such structural components as the composition, the concentration of AA, AB, and BB dyads, the concentration of AAA, ABB, and ABA triads, and, in rare cases, higher order sequence lengths, where A and B represent different structural components. We seek to relate these experimental results to the polymerization parameters. For simplicity, we consider only two different structures, A and B, although these symbols could represent differences in comonomer, stereoregularity, monomer isomerism, or branching. First, we derive general relationships for the various experimental sequence fractions that are independent of the polymerization mechanism. Then we derive specific relations characteristic of model polymerization mechanisms. Comparison of the relationships between the experimental quantities and the calculated results for specific mechanisms permits us to determine the polymerization mechanism. By knowing the mechanism and using the experimental data to determine the parameters, we can calculate the complete structural distribution function for the polymers.

In the following we assume a statistically stationary process and a sufficiently high degree of polymerization to neglect end effects.

3.2 GENERAL STATISTICAL RELATIONS (1)

Let us define $P_n(x_1 x_2 x_3 \cdots x_n)$ as the *measured* number fraction of a particular sequence of $x_1 x_2 x_3 x_4 \cdots x_n$ taken n at a time, where $x_1 x_2 x_3 \cdots x_n$

represent the various possible structural variations of the repeating units. Thus $P_3(ABB)$ is the number fraction of ABB triads. The subscript n is not usually necessary for short sequences, since the number of units is obvious. It may therefore be dropped without causing serious difficulty, and in this text is often not noted.

With this notation $P_1(A)$ is the mole fraction composition of A in the polymer. The measured number fraction of units taken one at a time must be related:

$$P_1(A) + P_1(B) = 1, \tag{3.1}$$

since a unit must be A or B.

Similarly, when pairs of units are considered, the definition requires

$$P_2(AA) + P_2(AB) + P_2(BA) + P_2(BB) = 1. \tag{3.2}$$

Since any sequence has a successor or a predecessor unit of A or B, the concentrations of lower order placements can be expressed as sums of the two appropriate higher order sequences: for example,

$$P_1(A) = P_2(AA) + P_2(AB) = P_2(AA) + P_2(BA),$$
$$P_1(B) = P_2(BB) + P_2(AB) = P_2(BB) + P_2(BA). \tag{3.3}$$

These dyads represent all the possible ways of arranging units before and after the single A and B units, respectively. By inspection, one observes that

$$P_2(AB) = P_2(BA). \tag{3.4}$$

Relationships of this type are called the *reversibility* relationships and are a result of the assumption of statistical stationariness. The measured results are independent of the direction of counting of the sequences. This is fortunate, since the two units AB and BA are indistinguishable experimentally.

Similarly for the dyads in terms of the triads, one can write, by adding before and after each dyad, the A and B elements:

$$P_2(AA) = P_3(AAA) + P_3(AAB) = P_3(AAA) + P_3(BAA),$$
$$P_2(AB) = P_3(AAB) + P_3(BAB) = P_3(ABA) + P_3(ABB),$$
$$P_2(BA) = P_3(BAA) + P_3(BAB) = P_3(ABA) + P_3(BBA),$$
$$P_2(BB) = P_3(BBB) + P_3(BBA) = P_3(BBB) + P_3(ABB), \tag{3.5}$$

which requires

$$P_3(AAB) = P_3(BAA),$$
$$P_3(BBA) = P_3(ABB), \tag{3.6}$$

yielding the reversibility relations for the triads. The extension of this process to longer sequences should be obvious.

However, in our consideration of the polymerization process, the addition of a unit to a chain may depend on whether the end unit is A or B or any additional terminal sequence. We utilize the following nomenclature.

Let $P(x_{n+1}|x_1 \cdots x_n)$ define the *conditional* probability that a polymer chain with terminal sequence $x_1 x_2 \cdots x_n$ adds the particular monomer x_{n+1} as the next unit. For example, $P(A|B)$ is the probability of adding an A unit given the previous unit is a B. Using this definition and Bayes' theorem yields the relative probability of any measurable sequence, that is,

$$P_2(\text{BA}) = P(\text{B})P(\text{A}/\text{B}),$$

concluding that the dyad fraction is given by the probability of finding a B times the conditional probability of adding an A given a B unit.

Likewise,

$$P(\text{AA}) = P(\text{A})P(\text{A}/\text{A}),$$
$$P(\text{AAA}) = P(\text{A})P(\text{A}/\text{A})P(\text{A}/\text{AA}),$$
$$P(\text{BAA}) = P(\text{B})P(\text{A}/\text{B})P(\text{A}/\text{BA}), \tag{3.7}$$

where $P(A/A)$, $P(A/B)$, $P(A/AA)$, $P(A/AB)$ represent the conditional probabilities that the added unit is A, given that the preceding placement (or placement pair) is A, B, AA, or AB, respectively. By rearranging these equations, the conditional probabilities can be computed as ratios of the unconditional probabilities or the measured number fractions as follows:

$$P(\text{A}/\text{A}) = \frac{P(\text{AA})}{P(\text{A})}, \qquad P(\text{B}/\text{A}) = \frac{P(\text{AB})}{P(\text{A})},$$

$$P(\text{A}/\text{B}) = \frac{P(\text{BA})}{P(\text{B})}, \qquad P(\text{B}/\text{B}) = \frac{P(\text{BB})}{P(\text{B})},$$

$$P(\text{A}/\text{AA}) = \frac{P(\text{AAA})}{P(\text{AA})}, \qquad P(\text{B}/\text{AA}) = \frac{P(\text{AAB})}{P(\text{AA})},$$

$$P(\text{A}/\text{BA}) = \frac{P(\text{BAA})}{P(\text{BA})}, \qquad P(\text{B}/\text{BA}) = \frac{P(\text{BAB})}{P(\text{BA})},$$

$$P(\text{A}/\text{AB}) = \frac{P(\text{ABA})}{P(\text{AB})}, \qquad P(\text{B}/\text{AB}) = \frac{P(\text{ABB})}{P(\text{AB})},$$

$$P(\text{A}/\text{BB}) = \frac{P(\text{BBA})}{P(\text{BB})}, \qquad P(\text{B}/\text{BB}) = \frac{P(\text{BBB})}{P(\text{BB})}. \tag{3.8}$$

An additional relationship exists, since by the law of total probability

$$P(\text{A}/\text{A}) + P(\text{B}/\text{A}) = 1,$$
$$P(\text{A}/\text{B}) + P(\text{B}/\text{B}) = 1, \tag{3.9}$$

which are the statements of completeness indicating that these conditional probabilities represent all of the possible ways of adding to A and B, respectively. Similarly,

$$P(A/AA) + P(B/AA) = 1,$$
$$P(A/BA) + P(B/BA) = 1,$$
$$P(A/AB) + P(B/AB) = 1,$$
$$P(A/BB) + P(B/BB) = 1. \tag{3.10}$$

In calculating the compound events like $P(AAB)$,

$$P(AAB) = P(A)P(A/A)P(B/AA),$$

one must use conditional probabilities of different order, whereas for a given polymerization system only one type of conditional probability holds.

In general, the number fraction or probability of a sequence of n consecutive units being all A's is

$$P_n(A^n) = P(A)P(A/A)P(A/AA) \cdots P(A/A^{n-1}).$$

Written in terms of a particular set of conditional probabilities, this is

$$P_n(A^n) = P(A^k)P(A^{n-k}/A^k).$$

If k is the smallest integer that we have, for any sequence $T^{(j)}$ of any length j

$$P(X/T^{(j)}S^{(k)}) = P(X/S^k) \qquad \text{where } X, S = A \text{ or } B$$

We call the comonomer distribution kth order Markovian, and substitution above yields

$$P_n(A^n) = P(A)P(A/A) \cdots P(A/A^{k-1})[P(A/A^k)]^{n-k}$$
$$= P(A^k)[P(A/A^k)]^{n-k}.$$

For example, if $k = 2$, we have a second order Markovian chain and

$$P_n(A^n) = P(A)P(A/A)[P(A/AA)]^{n-2}$$
$$= P(AA)[P(A/AA)]^{n-2}.$$

Similarly, for a first order (or single) Markovian comonomer distribution we have $k = 1$ and

$$P_n(A^n) = P(A)[P(A/A)]^{n-1}.$$

Finally, if $k = 0$, we have a zeroth order Markovian or Bernoullian distribution, for which

$$P_n(A^n) = [P(A)]^n.$$

If the distribution is not of finite order Markovian, it is called *non-Markovian*.

One must work out relationships between the conditional probabilities of different orders. We can determine the relationships between $P_1(A)$ and the conditional probabilities $P(A/B)$ and $P(B/A)$ starting with the reversibility relation

$$P_2(AB) = P_2(BA),$$

which becomes, by definition,

$$P_1(A)P(B/A) = P_1(B)P(A/B).$$

Rearranging, we obtain the composition in terms of the conditional probabilities,

$$\frac{P_1(A)}{P_1(B)} = \frac{P(A/B)}{P(B/A)}. \tag{3.11}$$

Also it is necessary that

$$P_1(A) + P_1(B) = 1.$$

Substituting for $P_1(B)$ from above, we obtain

$$P_1(A) + \frac{P_1(A)P(B/A)}{P(A/B)} = 1,$$

so

$$P_1(A) = \frac{P(A/B)}{P(A/B) + P(B/A)},$$

$$P_1(B) = \frac{P(B/A)}{P(A/B) + P(B/A)}, \tag{3.12}$$

which are the desired relationships.

In similar fashion the relationship between $P(A/B)$ and $P(B/A)$ and the next higher conditional probabilities $P(A/AB)$, and so on, can be found by starting with

$$P_3(AAB) = P_3(BAA),$$

which, written in terms of the conditional probabilities, becomes

$$P_2(AA)P(B/AA) = P_2(BA)P(A/BA).$$

Further, we can use eq. 3.4,

$$P_2(BA) = P_2(AB) = P_1(A)P(B/A),$$

and write

$$P_1(A)P(A/A)P(B/AA) = P_1(A)P(B/A)P(A/BA)$$

and, since

$$P(A/A) = 1 - P(B/A),$$

$$(1 - P(B/A))P(B/AA) = P(B/A)P(A/BA),$$

the first order conditional probabilities in terms of the second order conditional probabilities are

$$P(B/A) = \frac{P(B/AA)}{P(B/AA) + P(A/BA)}. \tag{3.13}$$

Similarly, using

$$P_3(BBA) = P_3(ABB),$$

one obtains

$$P(A/B) = \frac{P(A/BB)}{P(B/AB) + P(A/BB)}. \tag{3.14}$$

The conversion to lower order is accomplished thusly:

$$\frac{P_1(A)}{P_1(B)} = \frac{P(A/B)}{P(B/A)} = \frac{1 + P(A/BA)/P(B/AA)}{1 + P(B/AB)/P(A/BB)}. \tag{3.15}$$

It should be noted that this process cannot be simply extended to conditional probabilities of higher order, as an insufficient number of reversibility relationships are available.

We are interested in the number $N_A(n)$ fractions of the sequences of A units which are defined as

$$N_A(n) = \frac{P_{n+2}(BA_n B)}{\sum\limits_{n=1}^{\infty} P_{n+2}(BA_n B)}.$$

It should also be noted that

$$P_2(AB) = \sum\limits_{n=1}^{\infty} P_{n+2}(BA_n B),$$

$$P_2(BA) = \sum\limits_{n=1}^{\infty} P_{n+2}(AB_n B),$$

so

$$N_A(n) = \frac{P_{n+2}(BA_n B)}{P_2(BA)}.$$

The number average sequence length, l_A, has the definition

$$l_A = \frac{\sum\limits_{n=1}^{\infty} n N_A(n)}{\sum\limits_{n=1}^{\infty} N_A(n)} = \sum\limits_{n=1}^{\infty} n N_A(n).$$

Upon substitution we obtain

$$\bar{l}_A = \frac{\sum\limits_{n} n P_{n+2}(BA_n B)}{P_2(BA)}$$

and, since

$$\sum\limits_{n} n P_{n+2}(BA_n B) = P_1(A),$$

we obtain the number average sequence length in terms of simple measurable sequences:

$$\bar{l}_A = \frac{P_1(A)}{P_2(BA)}.$$

The weight average fraction of A runs with length n is given by

$$W_A(n) = \frac{nN_A(n)}{\sum\limits_{n=1} nN_A(n)},$$

and the weight-average length of A runs is

$$\bar{l}_{w_A} = \sum_{n=1}^{\infty} \frac{nW_A(n)}{\sum\limits_{n=1}^{\infty} W_A(n)} = \sum_{n=1}^{\infty} nW_A(n)$$

$$= \frac{\sum\limits_{n=1}^{\infty} n^2 N_A(n)}{\sum\limits_{n=1}^{\infty} nN_A(n)}.$$

It follows that the basic requirement is the knowledge of number fractions of the various sequences of A's or B's.

Now we can define [absolute value of $P_2(AB)$ is used here] the following:

$$\bar{l}_A = \text{number average length of A runs} = \frac{P_1(A)}{P_2(AB)},$$

$$\bar{l}_B = \text{number average length of B runs} = \frac{P_1(B)}{P_2(BA)},$$

$$\alpha_A = \text{fraction of A—A bonds in A—X bonds} = \frac{P_2(AA)}{P_1(A)},$$

$$\alpha_B = \text{fraction of B—B bonds in B—X bonds} = \frac{P_2(BB)}{P_1(B)},$$

where X = A and B.

We also want to define a measure of departure from random statistics:

$$\chi = \frac{P_2(AB)}{P_1(A)P_1(B)}. \tag{3.16}$$

Since for a completely random process

$$P_2(AB) = P_1(B)P_1(A),$$

then when

$$\chi = 1 \quad \text{random statistics,}$$

$$\chi > 1 \quad \text{more alternating tendency process,}$$

$$\chi < 1 \quad \text{more block character,}$$

$$\chi = 2 \quad \text{completely alternating,}$$

$$\chi = 0 \quad \text{completely block.}$$

These relationships for the conditional probabilities in terms of the measured number fractions of units are important, since the conditional probabilities can be calculated for specific polymerization models. Thus the experimental test of the mechanism consists of a measurement of the microstructure and a comparison of the experimental microstructure with the calculated microstructure for the model being tested.

We need to be able to relate our experimental measurements to the general statistical relations described above.

3.3 MICROSTRUCTURE OF INSTANTANEOUS BINARY COPOLYMERIZATION MODELS

3.3.1 Terminal Copolymerization Model

For the model in which the rate of addition of monomer depends only on the nature of the terminal group, the system obeys first order Markov statistics and a general solution can be developed for the resulting polymer microstructure.

This model is represented by the following reactions and their corresponding rates.

Terminal Group	Added Group	Rate	Final
~A*	[A]	$K_{AA}[A^*][A]$	~AA*
~B*	[A]	$K_{BA}[B^*][A]$	~BA*
~A*	[B]	$K_{AB}[A^*][B]$	~AB*
~B*	[B]	$K_{BB}[B^*][B]$	~BB* (3.17)

The terminal conditional probabilities can be written as the ratio of the rate of formation of the desired product (AA) to the rates of the reaction of the starting unit (A*):

$$P(A/A) = \frac{K_{AA}[A][A^*]}{K_{AA}[A][A^*] + K_{AB}[B][A^*]} = \frac{K_{AA}[A]/K_{AB}[B]}{\{K_{AA}[A]/K_{AB}[B]\} + 1} = \frac{r_A x}{1 + r_A x},$$

$$P(B/A) = \frac{1}{1 + r_A x} \tag{3.18}$$

Likewise, a treatment of the reactions of B* yields

$$P(B/B) = \frac{r_B/x}{1 + r_B/x}, \qquad P(A/B) = \frac{1}{1 + r_B/x}, \qquad (3.19)$$

where

$$r_A = \frac{k_{AA}}{k_{AB}} = \text{reactivity ratio for } [A^*],$$

$$r_B = \frac{k_{BB}}{k_{BA}} = \text{reactivity ratio for } [B^*],$$

and

$$x = \frac{[A]}{[B]} = \text{monomer feed ratio.}$$

There are only two independent conditional probabilities, since eq. 3.9 holds:

$$P(A/A) + P(B/A) = 1,$$

$$P(B/B) + P(A/B) = 1.$$

We use $P(A/B)$ and $P(B/A)$.

The composition of a copolymer in terms of the conditional probabilities is given by eq. 3.11:

$$\frac{P_1(A)}{P_2(B)} = \frac{P(A/B)}{P(B/A)},$$

and substitution for the terminal model yields

$$\frac{P_1(A)}{P_1(B)} = \frac{1 + r_A x}{1 + r_B/x}, \qquad (3.20)$$

which is the familiar copolymer composition equation first derived by Mayo and Lewis (2).

The dyad concentrations for the terminal model are easily calculated:

$$P_2(AA) = P_1(A)P(A/A) = \frac{P(A/B)[(1 - P(B/A)]}{P(B/A) + P(A/B)},$$

$$P_2(AB) = P_2(AB) = P_1(A)P(B/A) = \frac{P(A/B)P(B/A)}{P(B/A) + P(A/B)},$$

$$P_2(BB) = 1 - 2P(AB) - P_2(AA).$$

The triad fraction can also be written

$$F_{AAA} = \frac{P_3(AAA)}{P_1(A)} = [(1 - P(B/A)]^2,$$

$$F_{BAB} = [P(B/A)]^2,$$

$$F_{BAA} = F_{AAB} = P(B/A)[(1 - P(B/A)].$$

Table 3.1 Microstructure of Terminal Model Copolymerization

Experimental	General	Terminal Conditional Probabilities	Reactivity Ratios and Monomer Feed
$P(BA) + P(AB)$	$2P_2(BA)$	$\dfrac{2}{1/P(B/A) + 1/P(A/B)}$	$\dfrac{2}{r_A x + 2 + r_B/x}$
χ	$\dfrac{P_2(BA)}{P_1(A)P(B)}$	$P(A/B) + P(B/A)$	$\dfrac{r_A x + 2 + r_B/x}{r_A x + 1 + r_A r_B + r_B/x}$
l_A	$\dfrac{P_1(A)}{P_2(AB)}$	$1/P(B/A)$	$1 + r_A x$
l_B	$\dfrac{P_1(B)}{P_2(BA)}$	$1/P(A/B)$	$1 + r_B/x$
$1/\alpha_A$	$\dfrac{P_1(A)}{P_2(AA)}$	$\dfrac{1}{1 - P(B/A)}$	$1 + 1/r_A x$
$1/\alpha_B$	$\dfrac{P_1(B)}{P_2(BB)}$	$\dfrac{1}{1 - P(A/B)}$	$1 + x/r_B$

We can now calculate all the experimental parameters in terms of the two independent terminal conditional probabilities $P(B/A)$ and $P(A/B)$. The results for the terminal model are shown in Table 3.1.

It is apparent from this model that the value of the $r_A r_B$ product determines whether the copolymer is alternating, random, or block. When $r_A r_B < 1$, then $\chi > 1$ and the polymer has an alternating tendency; if $r_A r_B = 1$, then $\chi = 1$ and the polymer is random; if $r_A r_B > 1$, then $\chi < 1$ and the polymer has a block character.

Since the number fraction of A sequences for the terminal model is

$$N_A(n) = [P(A/A)]^{n-1}[1 - P(A/A)], \qquad (3.21)$$

this distribution function allows the calculation of the amounts of the different sequences. Tosi (3) has made these calculations for various values of the product $r_A r_B$. In Tables 3.2a through 3.2e the distribution of sequences of various lengths are tabulated for different mole-percentages of the comonomer in the polymer.

These tables give values of $r_A r_A = 0.03$, corresponding to highly alternating, to $r_A r_B = 3$, corresponding to a block tendency.

Another experimental factor that can be calculated is the weight fraction of monomer in the copolymer found in each of these various sequences. This is the probability that a monomer unit belongs to a sequence of n units:

$$W_A(n) = n[P(A/A)]^{n-1}[1 - P(A/A)]^2. \qquad (3.22)$$

Tables 3.3a through 3.3e show the tabulation of the weight distribution of monomer among these various sequences. It is useful to use these tables to

Table 3.2a Distribution of $P_n(A_n)$ Sequences of Various Length, for $r_1 r_2 = 0.03$

A (mole-%)	$n = 1$	$n = 2$	$n = 3$	$n = 4$	$n = 5$	$n = 6$	$n = 7$	$n = 8$	$n = 9$	$n = 10$	$n = 11$	$n = 12$
10	99.63	0.37										
20	99.02	0.97	0.01									
30	97.88	2.08	0.04									
40	95.06	4.69	0.23	0.01								
50	85.24	12.58	1.86	0.27	0.04							
60	63.38	23.21	8.50	3.11	1.14	0.42	0.15	0.06	0.02			
70	41.95	24.35	14.14	8.21	4.76	2.77	1.61	0.93	0.54	0.31	0.18	0.11
80	24.76	18.68	14.02	10.55	7.94	5.97	4.49	3.38	2.54	1.91	1.44	1.13
90	11.07	9.84	8.75	7.79	6.92	6.16	5.48	4.87	4.33	3.85	3.43	3.05

Table 3.2b Distribution of $P_n(A_n)$ Sequences of Various Length, for $r_1 r_2 = 0.1$

A (mole-%)	$n = 1$	$n = 2$	$n = 3$	$n = 4$	$n = 5$	$n = 6$	$n = 7$	$n = 8$	$n = 9$	$n = 10$	$n = 11$	$n = 12$
10	98.78	1.20	0.01									
20	96.90	3.00	0.09									
30	93.71	5.89	0.37	0.02								
40	87.67	10.81	1.33	0.16	0.02							
50	75.97	18.25	4.39	1.05	0.25	0.06	0.01					
60	58.45	24.29	10.09	4.19	1.74	0.72	0.30	0.13	0.05	0.02	0.01	
70	40.16	24.03	14.38	8.61	5.15	3.08	1.84	1.10	0.66	0.40	0.24	0.14
80	24.22	18.36	13.91	10.54	7.99	6.05	4.59	3.47	2.63	2.00	1.51	1.15
90	10.98	9.77	8.70	7.74	6.89	6.14	5.46	4.86	4.33	3.85	3.43	3.06

Table 3.2c Distribution of $P_n(A_n)$ Sequences of Various Length, for $r_1 r_2 = 0.3$

A (mole-%)	$n = 1$	$n = 2$	$n = 3$	$n = 4$	$n = 5$	$n = 6$	$n = 7$	$n = 8$	$n = 9$	$n = 10$	$n = 11$	$n = 12$
10	96.52	3.36	0.12									
20	91.80	7.53	0.62	0.05								
30	85.27	12.56	1.85	0.27	0.04							
40	76.30	18.08	4.29	1.02	0.24	0.06	0.01					
50	64.61	22.86	8.09	2.86	1.01	0.36	0.13	0.04	0.02			
60	50.87	24.99	12.27	6.03	2.96	1.46	0.72	0.35	0.17	0.08	0.04	0.02
70	36.54	23.19	14.72	9.34	5.93	3.76	2.39	1.51	0.96	0.61	0.39	0.25
80	22.95	17.68	13.62	10.50	8.09	6.23	4.80	3.70	2.85	2.20	1.69	1.30
90	10.72	9.57	8.55	7.63	6.81	6.08	5.43	4.85	4.33	3.86	3.45	3.08

Table 3.2d   Distribution of $P_n(A_n)$ Sequences of Various Length, for $r_1r_2 = 1$

A (mole-%)	$n=1$	$n=2$	$n=3$	$n=4$	$n=5$	$n=6$	$n=7$	$n=8$	$n=9$	$n=10$	$n=11$	$n=12$
10	90.00	9.00	0.90	0.09								
20	80.00	16.00	3.20	0.64	0.13							
30	70.00	21.00	6.30	1.89	0.57	0.17	0.05					
40	60.00	24.00	9.60	3.84	1.54	0.62	0.25	0.10	0.04			
50	50.00	25.00	12.50	6.25	3.13	1.56	0.78	0.39	0.20	0.10	0.05	
60	40.00	24.00	14.40	8.64	5.18	3.11	1.87	1.12	0.67	0.40	0.24	0.14
70	30.00	21.00	14.70	10.29	7.20	5.04	3.53	2.47	1.73	1.21	0.85	0.59
80	20.00	16.00	12.80	10.24	8.19	6.55	5.24	4.19	3.36	2.68	2.15	1.72
90	10.00	9.00	8.10	7.29	6.56	5.90	5.31	4.78	4.30	3.87	3.59	3.23

Table 3.2e   Distribution of $P_n(A_n)$ Sequences of Various Length, for $r_1r_2 = 3$

A (mole-%)	$n=1$	$n=2$	$n=3$	$n=4$	$n=5$	$n=6$	$n=7$	$n=8$	$n=9$	$n=10$	$n=11$	$n=12$
10	77.88	17.23	3.81	0.84	0.19	0.04	0.01					
20	63.73	23.11	8.38	3.04	1.10	0.40	0.14	0.05	0.02			
30	53.08	24.91	11.69	5.48	2.57	1.21	0.57	0.27	0.12	0.06	0.03	0.01
40	44.30	24.68	13.74	7.66	4.26	2.38	1.33	0.74	0.41	0.23	0.13	0.07
50	36.60	23.21	14.71	9.33	5.91	3.75	2.38	1.51	0.96	0.61	0.38	0.24
60	29.53	20.81	14.66	10.33	7.28	5.13	3.62	2.55	1.80	1.27	0.89	0.63
70	22.75	17.58	13.58	10.49	8.10	6.26	4.83	3.73	2.89	2.23	1.72	1.33
80	15.94	13.40	11.26	9.47	7.96	6.69	5.62	4.73	3.97	3.34	2.81	2.36
90	8.65	7.90	7.22	6.60	6.02	5.50	5.03	4.59	4.19	3.83	3.50	3.20

Table 3.3a   Weight Distribution of A_n Units in Sequences of Various Length, for $r_1r_2 = 0.03$

A (mole-%)	$n=1$	$n=2$	$n=3$	$n=4$	$n=5$	$n=6$	$n=7$	$n=8$	$n=9$	$n=10$	$n=11$	$n=12$
10	99.26	0.74										
20	98.05	1.92	0.03									
30	95.80	4.07	0.13									
40	90.37	8.93	0.66	0.04								
50	72.65	21.45	4.75	0.94	0.17	0.03						
60	40.17	29.42	16.16	7.89	3.61	1.59	0.68	0.28	0.12	0.05	0.02	
70	17.60	20.43	17.79	13.77	9.99	6.96	4.72	3.13	2.05	1.32	0.84	0.53
80	6.13	9.22	10.41	10.44	9.82	8.87	7.79	6.70	5.67	4.74	3.92	3.36
90	1.23	2.18	2.91	3.45	3.83	4.09	4.24	4.31	4.32	4.26	4.17	4.05

Table 3.3b Weight Distribution of A_n Units in Sequences of Various Length, for $r_1 r_2 = 0.1$

A (mole-%)	$n=1$	$n=2$	$n=3$	$n=4$	$n=5$	$n=6$	$n=7$	$n=8$	$n=9$	$n=10$	$n=11$	$n=12$
10	97.58	2.38	0.04									
20	93.90	5.82	0.27	0.01								
30	87.82	11.05	1.04	0.09								
40	76.86	18.95	3.51	0.58	0.09	0.01						
50	57.72	27.74	10.00	3.20	0.96	0.28	0.08	0.02				
60	34.16	28.39	17.70	9.80	5.09	2.54	1.23	0.58	0.27	0.13	0.06	0.03
70	16.13	19.30	17.33	13.82	10.34	7.42	5.18	3.54	2.39	1.59	1.04	0.68
80	5.87	8.89	10.11	10.21	9.67	8.80	7.78	6.73	5.74	4.83	4.03	3.33
90	1.20	2.14	2.86	3.40	3.78	4.04	4.20	4.27	4.28	4.23	4.14	4.02

Table 3.3c Weight Distribution of A_n Units in Sequences of Various Length, for $r_1 r_2 = 0.3$

A (mole-%)	$n=1$	$n=2$	$n=3$	$n=4$	$n=5$	$n=6$	$n=7$	$n=8$	$n=9$	$n=10$	$n=11$	$n=12$
10	93.17	6.48	0.34	0.01								
20	84.27	13.82	1.70	0.19	0.02							
30	72.71	21.42	4.73	0.93	0.17	0.03						
40	58.22	27.59	9.81	3.10	0.92	0.26	0.07	0.02				
50	41.75	29.55	15.69	7.40	3.27	1.39	0.57	0.23	0.09	0.04	0.01	
60	25.88	25.43	18.74	12.28	7.54	4.44	2.55	1.43	0.79	0.43	0.23	0.13
70	13.35	16.95	16.13	13.65	10.83	8.24	6.10	4.43	3.16	2.23	1.56	1.08
80	5.27	8.12	9.38	9.64	9.28	8.58	7.71	6.79	5.89	5.04	4.27	3.59
90	1.15	2.05	2.75	3.27	3.65	3.91	4.08	4.16	4.18	4.14	4.07	3.96

Table 3.3d Weight Distribution of A_n Units in Sequences of Various Length, for $r_1 r_2 = 1$

A (mole-%)	$n=1$	$n=2$	$n=3$	$n=4$	$n=5$	$n=6$	$n=7$	$n=8$	$n=9$	$n=10$	$n=11$	$n=12$
10	81.00	16.20	2.43	0.32	0.04							
20	64.00	25.60	7.68	2.05	0.51	0.12	0.03					
30	49.00	29.40	13.23	5.29	1.98	0.71	0.25	0.09	0.03			
40	36.00	28.80	17.28	9.22	4.61	2.21	1.03	0.47	0.21	0.09	0.04	0.02
50	25.00	25.00	18.75	12.50	7.81	4.69	2.73	1.56	0.88	0.49	0.27	0.15
60	16.00	19.20	17.28	13.82	10.37	7.46	5.23	3.58	2.42	1.61	1.06	0.70
70	9.00	12.60	13.23	12.35	10.80	9.08	7.41	5.93	4.67	3.63	2.80	2.14
80	4.00	6.40	7.68	8.19	8.19	7.86	7.34	6.71	6.04	5.37	4.73	4.12
90	1.00	1.80	2.43	2.92	3.28	3.54	3.72	3.83	3.87	3.87	3.84	3.77

Table 3.3e   Weight Distribution of A_n Units in Sequences of Various Length, for $r_1r_2 = 3$

A (mole-%)	$n = 1$	$n = 2$	$n = 3$	$n = 4$	$n = 5$	$n = 6$	$n = 7$	$n = 8$	$n = 9$	$n = 10$	$n = 11$	$n = 12$
10	60.66	26.83	8.90	2.63	0.73	0.19	0.05	0.01				
20	40.62	29.46	16.02	7.75	3.51	1.53	0.65	0.27	0.11	0.04	0.02	
30	28.17	26.44	18.61	11.64	6.83	3.84	2.10	1.13	0.60	0.31	0.16	0.08
40	19.62	21.86	18.27	13.57	9.44	6.31	4.11	2.62	1.64	1.01	0.62	0.38
50	13.40	16.99	16.15	13.66	10.82	8.23	6.09	4.41	3.15	2.22	1.55	1.07
60	8.72	12.29	12.99	12.21	10.75	9.09	7.48	6.02	4.77	3.74	2.90	2.23
70	5.18	8.00	9.27	9.54	9.22	8.54	7.70	6.80	5.91	5.07	4.31	3.63
80	2.54	4.27	5.38	6.03	6.34	6.40	6.27	6.03	5.70	5.32	4.92	4.51
90	0.75	1.37	1.87	2.28	2.61	2.86	3.05	3.18	3.27	3.32	3.33	3.32

determine the fraction of units present in sequences of 5 or more units $U_5(A)$ and 10 or more units, $U_{10}(A)$, respectively. These results are tabulated in Table 3.4.

Additionally, the average sequence length can be calculated for each of the reactivity ratio products; these are tabulated in Table 3.5.

3.3.2   Penultimate Copolymerization Model

When the nature of the penultimate unit has a significant effect on the absolute rate constant in copolymerization, eight reactions yielding the penultimate conditional probabilities can be derived in terms of these absolute rate constants and the monomer feed by the following equation:

Penultimate Group	Added Group	Rate	Final	
$\sim AA^*$	[A]	$k_{AAA}[AA^*][A]$	$\sim AAA^*$	
$\sim AA^*$	[B]	$k_{AAB}[AA^*][B]$	$\sim AAB^*$	
$\sim BA^*$	[A]	$k_{BAA}[BA^*][A]$	$\sim BAA^*$	
$\sim BA^*$	[B]	$k_{BAB}[BA^*][B]$	$\sim BAB^*$	
$\sim AB^*$	[A]	$k_{ABA}[AB^*][A]$	$\sim ABA^*$	
$\sim AB^*$	[B]	$k_{ABB}[AB^*][B]$	$\sim ABB^*$	
$\sim BB^*$	[A]	$k_{BBA}[BB^*][A]$	$\sim BBA^*$	
$\sim BB^*$	[B]	$k_{BBB}[BB^*][B]$	$\sim BBB^*$	(3.23)

$$P(A/AA) = \frac{k_{AAA}[A][AA^*]}{k_{AAA}[A][AA^*] + k_{AAB}[B][AA^*]} = \frac{k_{AAA}[A]/k_{AAB}[B]}{\{k_{AAB}[A]/k_{AAB}[B]\} + 1} = \frac{r_A x}{1 + r_A x}$$

Four independent conditional probabilities can be written, with four

Table 3.4 Fraction of A_n Units Present in Sequences of 5 or More Units and 10 or More Units, Respectively

$r_1 r_2$	A (mole-%)	$U_5(A)$	$U_{10}(A)$
0.03	50	0.21	
	60	6.36	0.08
	70	30.41	3.56
	80	63.80	24.95
	90	90.23	69.44
0.1	40	0.10	
	50	1.34	
	60	9.95	0.24
	70	33.42	4.55
	80	64.92	26.20
	90	90.40	69.83
0.3	30	0.02	
	40	1.28	
	50	5.61	0.06
	60	17.67	0.92
	70	39.92	7.16
	80	67.59	29.34
	90	90.78	70.80
1	10	0.05	
	20	0.67	
	30	3.08	0.02
	40	8.70	0.17
	50	18.75	1.08
	60	33.70	4.64
	70	52.82	14.93
	80	73.73	37.59
	90	91.85	73.61
3	10	0.98	
	20	6.15	0.08
	30	15.14	0.64
	40	26.68	2.56
	50	39.80	7.10
	60	53.79	15.68
	70	68.01	29.84
	80	81.78	51.04
	90	93.73	78.76

Table 3.5 Average Length of A Sequences

A (mole-%)	$r_1r_2 = 0.03$	$r_1r_2 = 0.1$	$r_1r_2 = 0.3$	$r_1r_2 = 1$	$r_1r_2 = 3$
10	1.004	1.012	1.036	1.111	1.284
20	1.010	1.032	1.089	1.250	1.569
30	1.022	1.067	1.173	1.429	1.884
40	1.052	1.141	1.311	1.667	2.257
50	1.173	1.316	1.548	2.000	2.732
60	1.578	1.711	1.966	2.500	3.386
70	2.384	2.490	2.736	3.333	4.395
80	4.039	4.128	4.357	5.000	6.275
90	9.034	9.111	9.324	10.000	11.557

monomer reactivity ratios, in a manner similar to the terminal model:

$$P(B/AA) = \frac{1}{1 + r_A x}, \qquad P(A/AA) = 1 - P(B/AA),$$

$$P(A/BA) = \frac{r'_A x}{1 + r'_A x}, \qquad P(B/BA) = 1 - P(A/BA),$$

$$P(B/AB) = \frac{r'_B/x}{1 + r'_B/x}, \qquad P(A/AB) = 1 - P(B/AB),$$

$$P(A/BB) = \frac{1}{1 + r_B/x}, \qquad P(B/BB) = 1 - P(A/BB), \qquad (3.24)$$

where

$$r_A = \frac{k_{AAA}}{k_{AAB}}, \qquad r'_A = \frac{k_{BAA}}{k_{BAB}},$$

$$r_B = \frac{k_{BBB}}{k_{BBA}}, \qquad r'_B = \frac{k_{ABB}}{k_{ABA}}.$$

The composition of the copolymer in terms of the conditional penultimate probabilities is found by substitution in eq. 3.15:

$$\frac{P_1(A)}{P_1(B)} = \frac{P(A/B)}{P(B/A)} = \frac{1 + P(A/BA)/P(B/AA)}{1 + P(B/AB)/P(A/BB)}. \qquad (3.25)$$

The above results can be expressed in terms of the reactivity coefficients and monomer feed:

$$\frac{P_1(A)}{P_1(B)} = \frac{1 + r'_A x(1 + r_A x)/(1 + r'_A x)}{1 + (r'_B/x)(1 + r_B/x)/(1 + r'_B/x)}, \qquad (3.26)$$

which is the copolymer composition equation for the penultimate model derived by Mertz, Alfrey, and Goldfinger (4).

The other experimental parameters of interest for the penultimate system are shown in Table 3.6. These results are easily derivable from the

Table 3.6 Microstructure for Penultimate Model Copolymerization

Experimental	General	Conditional Probabilities	Reactivity Ratios and Monomer Feed
$P_2(AB)$	$2P_2(BA)$	$\dfrac{2}{P(A\|BA)/P(B\|AA) + 2 + \dfrac{P(B\|AB)}{P(A\|BB)}}$	$\dfrac{2}{\dfrac{r_A x(1+r_A x)}{1+r'_A x} + 2 + \dfrac{(r_B/x)(r_B+x)}{r'_B+x}}$
x	$\dfrac{P_2(BA)}{P_1(B)P_1(A)}$	$\dfrac{P(A\|BA)/P(B\|AA) + 2 + P(B\|AB)/P(A\|BB)}{P(A\|BA)/P(B\|AA) + 1 + P(A\|BA)/P(B\|AA)P(B\|AB)/P(A\|BB) + P(B\|AB)/P(A\|BB)}$	$\dfrac{r'_A x\left(\dfrac{1+r_A x}{1+r'_A x}\right) + 2 + \dfrac{r'_B}{x}\left(\dfrac{r_B+x}{r'_B+x}\right)}{r'_A\left(\dfrac{1+r_A x}{1+r'_A x}\right) + 1 + r'_A r'_B\left(\dfrac{1+r_A x}{1+r'_A x}\right)\left(\dfrac{r_B+x}{r'_B+x}\right) + \dfrac{r'_B}{x}\left(\dfrac{r_B+x}{r'_B+x}\right)}$
$\bar{l}_A$	$\dfrac{P_1(A)}{P_2(AB)}$	$1+\dfrac{P(A\|BA)}{P(B\|AA)}$	$1+\left[r'_A x\left(\dfrac{1+r_A x}{1+r'_A x}\right)\right]x$
$\bar{l}_B$	$\dfrac{P_1(B)}{P_2(BA)}$	$1+\dfrac{P(B\|AB)}{P(A\|BB)}$	$1+\dfrac{r'_B\left(\dfrac{r_B+x}{r'_B+x}\right)}{x}$
$1/\alpha_A$	$\dfrac{P_1(A)}{P_2(BA)}$	$1+\dfrac{P(B\|AA)}{P(A\|BA)}$	$1+\left[\dfrac{1}{r'_A}\left(\dfrac{1+r_A x}{1+r'_A x}\right)\right]x$
$1/\alpha_B$	$\dfrac{P_1(B)}{P_2(BB)}$	$1+\dfrac{P(A\|BB)}{P(B\|AB)}$	$1+\dfrac{x}{r_B\left(\dfrac{r_B+x}{r'_B+x}\right)}$

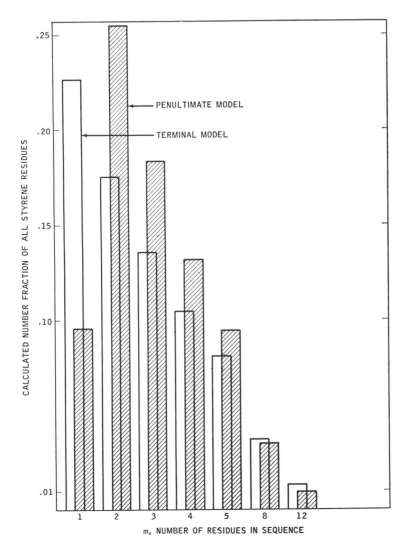

Figure 3.1 Number distribution of monomer sequences calculated from terminal and penultimate models. For the terminal model $r_A = 0.1$, $r_B = 1.0$; for the penultimate model $r_A = 0.99$, $r'_A = 0.01$, $r_B = 0.9$, $r'_B = 0.01$. (Reprinted by permission of Ref. 5.)

substitution of the conditional probabilities into the general relationships. The relationships are cumbersome. Note that whether $\chi < 1$ or $\chi > 1$ generally depends not only on the monomer reactivity ratios but also on the monomer feed ratio.

The number distribution function for the penultimate model can be written

$$P(BA_mB) = P(A/BA)P(A/AA)^{n-2}P(B/AA),$$

with

$$P_3(BAB) = P(B/BA) = 1 - P(A/BA).$$

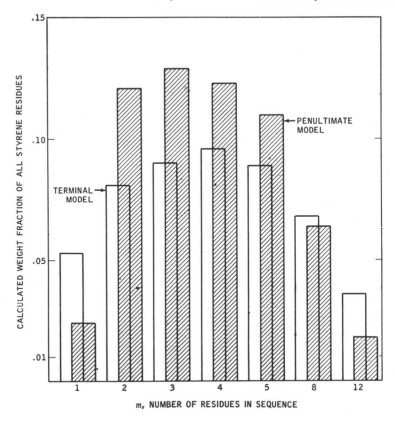

Figure 3.2 Weight distribution of monomer sequences calculated from terminal and penultimate models. For the terminal model $r_A = 0.1$, $r_B = 1.0$; for the penultimate model $r_A = 0.99$, $r'_A = 0.01$, $r_B = 0.9$, $r''_B = 0.01$. (Reprinted by permission of Ref. 5.)

In Figure 3.1 the calculated number fraction for a penultimate model polymerization is compared with a terminal model (5).

The weight fraction of monomer sequences can be calculated by using the relationship

$$W_A(m) = \frac{mP(BA_mB)P(B/AA)}{P(B/AA) + P(A/BA)}.$$

The calculated weight fraction of sequence lengths for a penultimate model is compared with the terminal model in Figure 3.2 (5).

Inspection of these two figures indicates that, at least for the reactivity ratios used for this model calculation, a considerable difference exists between the microstructure of the terminal and the penultimate model.

3.3.3 Generalized Approach to Calculation of Microstructure for Effects of Remote Units

Development of Method. The methods of deriving the relationships between the microstructure parameters and the conditional probabilities that are used for the terminal and penultimate models cannot be generalized for Markov chains with $k \geq 3$, as the number of reversibility relationships is insufficient. The method can be worked out, but it is extremely cumbersome and complex. Price (6) has shown that the mathematics of Markovian processes are applicable to this problem. This method can be used in a variety of circumstances useful to us later and therefore is presented at this time. It is demonstrated that this treatment gives results mathematically identical to those of the kinetic or steady-state probability methods that were used previously.

We have seen that a given copolymer sequence can be described by a number of expressions that contain (1) the probability of finding a given sequence and (2) the conditional probabilities. These conditional probabilities depend only on the initial and final state of the chain resulting from the addition of a single unit to the end of the growing chain and do not depend on the state of the end of the chain at any prior time. For this reason the chain is Markovian.

The bookkeeping process is considerably simplified if we adopt a binary notation. The numbers 0 and 1 will be substituted for the comonomers of the copolymer. In this notation the copolymerization for the penultimate model is illustrated first for simplicity.

In this case we have the possibility of four terminal sequences, 00, 01, 10, and 11, which as binary numbers correspond to 0, 1, 2, and 3 in the decimal system. To each of these terminal sequences we add a 0 and 1 and form a new terminal sequence. The conditional probabilities are defined as follows.

Initial Sequence			Final Sequence		
Binary	Decimal	Add	Binary	Decimal	Conditional Probability
00	0	0	00	0	P_{00}
00	0	1	01	1	P_{01}
01	1	0	10	2	P_{12}
01	1	1	11	3	P_{13}
10	2	0	00	0	P_{20}
10	2	1	01	1	P_{21}
11	3	0	10	2	P_{32}
11	3	1	11	3	P_{33}

Since a given terminal sequence must add a 0 or a 1 and become transformed into another state, the conditional probabilities obey the condition

$$\sum_s P_{rs} = 1,$$

where r refers to the initial state and s to the final state. We define $P_0^{(2)}$, $P_1^{(2)}$, $P_2^{(2)}$, $P_3^{(2)}$ as the probability that any given pair will be, respectively, 00, 01, 10, or 11. In this notation the sequence

$$
\begin{array}{cccccc}
a & b & c & d & e & f \\
\ldots 0 & 1 & 0 & 0 & 0 & 1 \ldots
\end{array}
$$

has the probability

$$P^{(2)}_{a-j} = P_1^{(2)} P_{12} P_{20} P_{00} P_{01}.$$

Of course, since a given singlet, doublet, triplet, and so on, must be in the same state,

$$\sum_j P_j^{(k)} = 1,$$

where k is the number of units in the terminal sequence being studied. For the above penultimate model, $k = 2$.

If one neglects the initiation and termination stages in the reaction, which is a good approximation for a high molecular weight chain, the Markov chain has a stationary distribution.

To find the steady-state concentration of 0's and 1's, we seek solution of the equation

$$V_n P_n = \lambda V_n,$$

where V_n is the row vector with the components $V_n(0)$ and $V_n(1)$ representing the relative population in the completed chain, and λ is a constant. The value of n denotes the number of units in the terminal sequence of the growing chain. The absolute probability that a given unit is a 0, is

$$P_1(0) = \frac{V_n(0)}{V_n(1) + V_n(0)}.$$

We are interested in the eigenvector of this equation which is determined from the secular equation

$$|A| = |P - I| = 0,$$

where I is the identity matrix. The solutions V_j that we desire are given by

$$V_j = CA^{js},$$

where A^{js} is the cofactor of the element A_{js} in the determinant $|A|$ and C is a constant.

Thus to obtain the relative number of the various states V_j, it is only necessary to subtract unity from each element along the main diagonal of the array $(\mathbf{P})$ and find the cofactors of the elements of one row of the resulting determinant. The composition can be obtained by proper summation of the appropriate states, V_j's.

For the penultimate model, we arrange the conditional probabilities into a matrix $(\mathbf{P})$ whose elements are P_{rs}, where r (initial) refers to the column and s (final state) to the row:

Initial	0	1	2	3
Final				
0	P_{00}	0	$1 - P_{21}$	0
1	$1 - P_{00}$	0	P_{21}	0
2	0	P_{12}	0	$1 - P_{33}$
3	0	$1 - P_{12}$	0	P_{33}

So

$$\mathbf{P}_{(2)} = \left\{ \begin{array}{cccc} P_{00} & 0 & 1 - P_{21} & 0 \\ 1 - P_{00} & 0 & P_{21} & 0 \\ 0 & P_{12} & 0 & 1 - P_{33} \\ 0 & 1 - P_{12} & 0 & P_{33} \end{array} \right\},$$

and the determinant $|P - I|$ is

$$\begin{vmatrix} P_{00} - 1 & 0 & 1 - P_{21} & 0 \\ 1 - P_{00} & -1 & P_{21} & 0 \\ 0 & P_{12} & -1 & 1 - P_{33} \\ 0 & 1 - P_{12} & 0 & P_{33} - 1 \end{vmatrix} = 0.$$

The relative concentrations for the final steady state are given by the appropriate cofactors. Thus

$$V_2(0) = C(P_{33} - 1)(1 - P_{21}),$$

$$V_2(1) = V_2(2) = C(1 - P_{00})(P_{33} - 1),$$

$$V_2(3) = C(1 - P_{00})(P_{12} - 1).$$

If we let N_0 equal the number of 0's in the completed chain and N_1 the number of 1's in the completed chain,

$$\frac{N_2(0)}{N_2(1)} = \frac{V_2(0) + V_2(2)}{V_2(1) + V_2(3)}.$$

This equation, upon insertion of the proper reactivity ratios and monomer feed, yields the composition equation for the penultimate case (eq. 3.26).

Application to Antepenultimate Model. General Case. For this model, we assume that the terminal radical, the penultimate monomer, and the next preceding monomer all affect the probability of addition of a new monomer to the polymer chain ($k = 3$). Here we recognize eight states, 000, 001, 010, 011, 100, 101, 110, 111, which we designate 0, 1, 2, 3, 4, 5, 6, and 7, respectively. These states are transformed according to the following addition scheme:

Initial Sequence			Final Sequence		Conditional
Binary	Decimal	Add	Binary	Decimal	Probability
000	0	0	000	0	P_{00}
000	0	1	001	1	P_{01}
001	1	0	010	2	P_{12}
001	1	1	011	3	P_{13}
010	2	0	100	4	P_{24}
010	2	1	101	5	P_{25}
011	3	0	110	6	P_{36}
011	3	1	111	7	P_{37}
100	4	0	000	0	P_{40}
100	4	1	001	1	P_{41}
101	5	0	010	2	P_{52}
101	5	1	011	3	P_{53}
110	6	0	100	4	P_{64}
110	6	1	101	5	P_{65}
111	7	0	110	6	P_{76}
111	7	1	111	7	P_{77}

The transition matrix ($\mathbf{P}_{(3)}$) then becomes

$$
(\mathbf{P}_{(3)}) = \begin{Bmatrix}
P_{00} & 0 & 0 & 0 & 1-P_{41} & 0 & 0 & 0 \\
1-P_{00} & 0 & 0 & 0 & P_{41} & 0 & 0 & 0 \\
0 & P_{12} & 0 & 0 & 0 & 1-P_{53} & 0 & 0 \\
0 & 1-P_{12} & 0 & 0 & 0 & P_{53} & 0 & 0 \\
0 & 0 & P_{24} & 0 & 0 & 0 & 1-P_{65} & 0 \\
0 & 0 & 1-P_{24} & 0 & 0 & 0 & P_{65} & 0 \\
0 & 0 & 0 & P_{36} & 0 & 0 & 0 & 1-P_{77} \\
0 & 0 & 0 & 1-P_{36} & 0 & 0 & 0 & P_{77}
\end{Bmatrix}.
$$

After unity is subtracted from the diagonal elements, the appropriate cofactors are

$$V_0^{(3)} = c(1 - P_{41})(P_{77} - 1)(P_{24} + P_{53} - P_{24}P_{53} - P_{53}P_{65}),$$

$$V_2^{(3)} = c(1 - P_{00})(P_{77} - 1)(P_{12} + P_{65} - P_{12}P_{65} - P_{53}P_{65}),$$

$$V_4^{(3)} = c(1 - P_{00})(P_{77} - 1)(P_{24} + P_{53} - P_{24}P_{53} - P_{53}P_{65}),$$

$$V_6^{(3)} = c(1 - P_{00})(P_{77} - 1)(P_{24} + P_{53} - P_{24}P_{53} - P_{12}P_{24}),$$

$$V_1^{(3)} = c(1 - P_{00})(P_{77} - 1)(P_{24} + P_{53} - P_{24}P_{53} - P_{53}P_{65}),$$

$$V_3^{(3)} = c(1 - P_{00})(P_{77} - 1)(P_{24} + P_{53} - P_{24}P_{53} - P_{12}P_{24}),$$

$$V_5^{(3)} = c(1 - P_{00})(P_{77} - 1)(P_{12} + P_{65} - P_{12}P_{24} - P_{12}P_{65}),$$

$$V_7^{(3)} = c(P_{00} - 1)(1 - P_{36})(P_{24} + P_{53} - P_{24}P_{12} - P_{24}P_{53}).$$

We note that

$$V_1^{(3)} = V_4^{(3)},$$

$$V_3^{(3)} = V_6^{(3)},$$

and

$$V_2^{(3)} + V_3^{(3)} = V_1^{(3)} + V_5^{(3)}.$$

Then the composition becomes

$$\frac{P_1(0)}{P_1(1)} = \frac{V_0^{(3)} + V_2^{(3)} + V_4^{(3)} + V_6^{(3)}}{V_1^{(3)} + V_3^{(3)} + V_5^{(3)} + V_7^{(3)}},$$

or

$$\frac{P_1(0)}{P_1(1)} = \frac{1 + \dfrac{1 + (1 - P_{41})/(1 - P_{00})}{1 + \left[\dfrac{1 - P_{24}}{P_{24}} + \left(\dfrac{1 - P_{12}}{P_{12}}\right)\left(\dfrac{P_{65}}{P_{24}}\right)\right] \Big/ \left[\left(\dfrac{P_{53}}{P_{12}}\right)\left(\dfrac{1 - P_{53}}{P_{53}} + \dfrac{1 - P_{65}}{P_{24}}\right)\right]}}{1 + \dfrac{1 + (1 - P_{36})/(1 - P_{77})}{1 + \left[\left(\dfrac{P_{65}}{P_{24}}\right)\left(\dfrac{1 - P_{65}}{P_{65}} + \dfrac{1 - P_{53}}{P_{12}}\right)\right] \Big/ \left[\dfrac{1 - P_{12}}{P_{12}} + \left(\dfrac{1 - P_{24}}{P_{24}}\right)\left(\dfrac{P_{53}}{P_{12}}\right)\right]}} .$$

The definitions of the conditional probabilities are

$$P_{00} = \frac{r_0 x}{1 + r_0 x}, \qquad P_{12} = \frac{x}{r_1 + x},$$

$$P_{24} = \frac{r_2 x}{1 + r_2 x}, \qquad P_{36} = \frac{x}{r_3 + x},$$

$$P_{41} = \frac{1}{1 + r_4 x}, \qquad P_{53} = \frac{r_5}{r_5 + x},$$

$$P_{65} = \frac{1}{1 + r_6 x}, \qquad P_{77} = \frac{r_7}{r_7 + x}.$$

These definitions give

$$
\frac{P_2(0)}{P_1(1)} = \frac{1 + \dfrac{1 + (r_4 x)[(1 + r_0 x)/(1 + r_4 x)]}{1 + \left\{\dfrac{1}{r_2 x}\left[1 + \dfrac{r_1}{x}\left(\dfrac{1 + r_2 x}{1 + r_6 x}\right)\right]\right\} \Big/ \left\{\left(\dfrac{r_1 + x}{r_5 + x}\right)\left(\dfrac{r_5}{x}\right)\left[\dfrac{x}{r_5} + \dfrac{r_6}{r_2}\left(\dfrac{1 + r_2 x}{1 + r_6 x}\right)\right]\right\}}}{1 + \dfrac{1 + (r_3/x)[(r_7 + x)/(r_3 + x)]}{1 + \left\{\left(\dfrac{1}{r_2 x}\right)\left(\dfrac{1 + r_2 x}{1 + r_6 x}\right)\left[r_6 x + \dfrac{r_1 + x}{r_5 + x}\right]\right\} \Big/ \left\{\dfrac{1}{x}\left[r_1 + \left(\dfrac{r_5}{r_2 x}\right)\left(\dfrac{r_1 + x}{r_5 + x}\right)\right]\right\}}} .
$$

For the composition of a copolymer whose radical reactivity depends on the antepenultimate unit, we have eight independent parameters. If the monomer preceding the penultimate group has no effect, then $r_0 = r_4$, $r_1 = r_5$, $r_2 = r_6$, and $r_3 = r_7$. If these restrictions hold, this equation reduces to the preceding equation derived for the penultimate case. If only the terminal radical affects the reactivity, then

$$r_0 = r_2 = r_4 = r_6,$$

$$r_1 = r_2 = r_6 = r_7,$$

and the above equation reduces to the familiar composition equation for the terminal model.

  Antepenultimate Model Where One Monomer Cannot Homopolymerize.   When monomer 1 cannot add to a terminal sequence ending in 1, then

$$r_1 = r_3 = r_5 = r_7 = 0,$$

and the conditional probabilities become

$$P_{12} = P_{36} = 1,$$

$$P_{53} = P_{77} = 0,$$

which yields

$$V_0^{(3)} = -CP_{24}(1 - P_{41}),$$

$$V_1^{(3)} = -CP_{24}(1 - P_{00}),$$

$$V_2^{(3)} = -C(1 - P_{00}),$$

$$V_4^{(3)} = -CP_{24}(1 - P_{00}),$$

$$V_5^{(3)} = -C(1 - P_{00})(1 - P_{24}),$$

and

$$V_3^{(3)} = V_6^{(3)} = V_7^{(3)} = 0.$$

In this case

$$
\frac{P_1(0)}{P_1(1)} = 1 + P_{24}\left(1 + \frac{1 - P_{41}}{1 - P_{00}}\right)
$$

$$
= 1 + \left(\frac{r_2 x}{1 + r_2 x}\right)\left[1 + \frac{r_4 x}{r_0}\left(\frac{1 + r_0 x}{1 + r_4 x}\right)\right].
$$

Reactivity Effects of Four Comonomer Units Where One Monomer Cannot Homopolymerize. Ham (7) has extended this method to the copolymer systems where significant influence on the addition of the monomers can be effected by units as far back from the growing end as four including influences of less distant units. The notation for the system is given below.

Initial Sequence			Final Sequence		Conditional
Binary	Decimal	Add	Binary	Decimal	Probability
0000	0	0	0000	0	P_{00}
0000	0	1	0001	1	$1 - P_{00}$
0001	1	0	0010	2	1
0010	2	0	0100	4	P_{24}
0010	2	1	0101	5	$1 - P_{24}$
0100	4	0	1000	8	P_{48}
0100	4	1	1001	9	$1 - P_{48}$
0101	5	0	1010	10	1
1000	8	0	0000	0	P_{80}
1000	8	1	0001	1	$1 - P_{80}$
1010	10	0	0100	4	P_{10-4}
1010	10	1	0101	5	$1 - P_{10-4}$

The transition matrix becomes

$$
\text{Final}\ \begin{matrix} \\ 0\\1\\2\\3\\4\\5\\6\\7\\8\\9\\10 \end{matrix}
\overset{\text{Initial}}{\begin{bmatrix}
0 & 1 & 2 & 3 & 4 & 5 & 6 & 7 & 8 & 9 & 10 \\
P_{00} & 0 & 0 & 0 & 0 & 0 & 0 & 0 & P_{80} & 0 & 0 \\
1-P_{00} & 0 & 0 & 0 & 0 & 0 & 0 & 0 & 1-P_{80} & 0 & 0 \\
0 & 1 & 0 & 0 & 0 & 0 & 0 & 0 & 0 & 1 & 0 \\
0 & 0 & 0 & 0 & 0 & 0 & 0 & 0 & 0 & 0 & 0 \\
0 & 0 & P_{24} & 0 & 0 & 0 & 0 & 0 & 0 & 0 & P_{10-4} \\
0 & 0 & 1-P_{24} & 0 & 0 & 0 & 0 & 0 & 0 & 0 & 1-P_{10-4} \\
0 & 0 & 0 & 0 & 0 & 0 & 0 & 0 & 0 & 0 & 0 \\
0 & 0 & 0 & 0 & 0 & 0 & 0 & 0 & 0 & 0 & 0 \\
0 & 0 & 0 & 0 & P_{48} & 0 & 0 & 0 & 0 & 0 & 0 \\
0 & 0 & 0 & 0 & 1-P_{48} & 0 & 0 & 0 & 0 & 0 & 0 \\
0 & 0 & 0 & 0 & 0 & 1 & 0 & 0 & 0 & 0 & 0
\end{bmatrix}}.
$$

After the appropriate algebra, one finds

$$
\frac{P_1(0)}{P_1(1)} = 1 + \frac{[(1-P_{00})P_{48} + P_{80}P_{48} + 1 - P_{00}]P_{10-4}}{[P_{10-4}P_{48} + (1-P_{24})(P_{48} + (1-P_{24})(1-P_{48}) + (1-P_{48})P_{10-4}](1-P_{00})}.
$$

Written in terms of the appropriate sequence condition probabilities, where the last unit is the added monomer, this equation becomes

$$\frac{P_1(0)}{P_1(1)} = 1 + \frac{P_{10100}(P_{00001} \cdot P_{01000} + P_{1000} \cdot P_{01000} + P_{00001})}{P_{00001}[P_{10100}P_{01000} + (1 - P_{00100})P_{01000} + (1 - P_{00100})P_{01001} + P_{01001}P_{10100}]}.$$

With substitution of the reactivity ratios and monomer feed, this equation is appropriate for any system (1 not adding to 1) where the effects exist within the last four units of the growing chain. This equation reduces to the form of the penultimate model with the appropriate reduction of the conditional probabilities. For highly polar monomers which exhibit enhanced repulsion from more distant units or sterically hindered monomers, these equations may find application. Particular forms of this model have been found useful (7).

In the case of the styrene (0)–fumaronitrile (1) system, substantial effects arise from fumaronitrile as the fourth unit as well as the penultimate unit from the chain ends, but no effects arise beyond two terminal adjacent styrene units. The appropriate composition equation becomes

$$\frac{P_1(0)}{P_1(1)} = 1 + \frac{P_{10100}(P_{001} \cdot P_{000} + P_{000} \cdot P_{111} + P_{001})}{P_{001}[P_{10100}P_{000} + (1 - P_{00100})P_{000} + (1 - P_{00100})P_{001} + P_{001}P_{10100}]},$$

which reduces to

$$\frac{P_1(0)}{P_1(1)} = 1 + \frac{P_{10100}}{P_{001}[P_{00101} + P_{10100}]}.$$

This equation holds for the styrene–fumaronitrile system (7).

For the case where addition to the sequence ~AAA (000) is substantially different from ~BAA (100) (no addition of 1 to 1),

$$P_{00001} = P_{0001},$$
$$P_{01000} = P_{1000},$$
$$P_{10000} = P_{0000},$$
$$P_{01001} = P_{1001}.$$

The equation becomes

$$\frac{P_1(0)}{P_1(1)} = 1 + \frac{P_{10100}[P_{1000} + P_{0001}]}{P_{0001}[P_{00101} + P_{10100}]}.$$

This equation exhibits close correlation with the data for the system α-methylstyrene–fumaronitrile.

Finally, an equation can be derived for the case where the effects on reactivity arise only from an isolated monomer (1) unit up to four units from the growing chain. For this case

$$P_{01000} = P_{1000},$$
$$P_{01001} = P_{1001},$$
$$P_{10100} = P_{100},$$
$$P_{00100} = P_{100},$$

and the composition equation is

$$\frac{P_1(0)}{P_1(1)} = 1 + \frac{P_{100}(P_{00001} \cdot P_{1000} + P_{10000} \cdot P_{1000} + P_{00001})}{P_{00001}}.$$

This equation applies to the styrene–maleic anhydride system (7).

3.3.4 Terminal Complex Copolymerization Model (8)

One can consider the possibility of the formation of a charge-transfer complex during copolymerization, and such a complex would influence the polymerization. This process can be treated by assuming that monomers A and B form a reversible AB complex. Simple dissociative rules can be used. Thus this equilibrium appears as

$$A + B \rightleftharpoons [AB] \equiv [C],$$

$$K = \frac{[AB]}{([A_0] - [AB])([B_0] - [AB])}, \tag{3.27}$$

where $[A_0]$ and $[B_0]$ are the concentrations of each monomer if there is no complex formation, while $[A]$ and $[B]$ are the true monomer concentrations.

A generalized scheme must include the possibilities of either monomer or either side of the AB complex reacting with chains ending in A· or B· radicals. When the complex reacts with the chain radical, both monomer units add to the polymer chain. For illustration, if the A side of the complex is attacked, it adds into the chain as ~AB·. Once the complex adds, the radical formed must be that of one of the monomers. If it is attacked on the A side, the result is ~B·. Attack on the B side of the complex must produce ~A·.

The eight equations required to describe the terminal complex model are:

	Added Group	Rate	Final	
~A*	A	$k_{AA}[A^*][A]$	~AA*	
~A*	B	$k_{AB}[A^*][B]$	~AB*	
~B*	A	$k_{BA}[B^*][A]$	~BA*	
~B*	B	$k_{BB}[B^*][B]$	~BB*	
~A*	[AB]	$k_{ACA}[A^*][C]$	~AAB*	
~A*	[BA]	$k_{ACB}[A^*][C]$	~ABA*	
~B*	[AB]	$k_{BCA}[B^*][C]$	~BAB*	
~B*	[BA]	$k_{BCB}[B^*][C]$	~BBA*	(3.28)

In this notation k_{AA} is the rate constant for addition of radical ~A· to monomer A, and so forth, while k_{ACA} is the rate constant for attack of radical ~A· on the A side of the complex, and so forth. Note that the first four equations are identical to terminal mode. The remaining equations show radical attack on the A side of the AB complex and on the B side.

The complex has been postulated to be more reactive than either of the two monomers separately. This can be rationalized on the basis of polarizability of the complex versus polarizability of the individual monomers. The charge-transfer complex has a larger π electron system (two monomers), making the system as a whole highly polarizable. As such, it can interact more readily with an approaching radical of the correct polarity than with an individual monomer.

Introducing the reaction steps and factoring out [A·] and [B·], respectively, we obtain the complex formation conditional probability:

$$P(B/A) = \frac{k_{AB}[B] + (k_{ACA} + k_{ACB})[C] + k_{BCA}[C]([B\cdot]/[A\cdot])}{k_{AA}[A] + k_{AB}[B] + [2k_{ACA} + k_{ACB}][C] + k_{BCA}[C]([B\cdot]/[A\cdot])},$$
(3.29)

$$P(A/B) = \frac{k_{BA}[A] + (k_{BCB} + k_{BCA})[C] + k_{ACB}[C]([A\cdot]/[B\cdot])}{k_{BB}[B] + k_{BA}[A] + [2k_{BCB} + k_{BCA}][C] + k_{ACB}[C]([A\cdot]/[B\cdot])},$$
(3.30)

where [C] is the molar concentration of complex, CA refers to attack at the A side of the complex, and CB refers to attack at the B side. Note that in eq. 3.29 a term must be included to represent attack of a chain ending in radical B· on the complex, since addition of this radical to the A side of the complex generates one AB link. Likewise, in eq. 3.30 radical A· terms must be considered. Also, addition of ~A· to the A side of the complex produces one AA sequence and one AB sequence in the polymer chain, hence the inclusion of the factor of 2 for this process.

To simplify these equations the radical ratios in the last term of each must be eliminated. Utilizing the steady-state approximation of equal radical turnovers, we can write the four turnover equations as

$$[A\cdot] + [B] \rightarrow [AB\cdot],$$

$$[A\cdot] + [C] \rightarrow [AAB\cdot],$$

$$[B\cdot] + [A] \rightarrow [BA\cdot],$$

$$[B\cdot] + [C] \rightarrow [BBA\cdot].$$

Now, equating turnovers, we obtain

$$[A\cdot] \rightarrow [B\cdot] = [B\cdot] \rightarrow [A\cdot],$$

$$k_{AB}[A\cdot][B] + k_{AA}[A\cdot][C] = k_{BA}[B\cdot][A] + k_{BCB}[B\cdot][C],$$

$$\frac{[A\cdot]}{[B\cdot]} = \frac{k_{BA}[A] + k_{BCB}[C]}{k_{AB}[B] + k_{ACA}[C]}.$$

If we substitute, multiply eq. 3.29 and 3.30 by $(1/k_{AB}[B])/(1/k_{AB}[B])$ and $(1/k_{BA}[A])/(1/k_{BA}[A])$ where appropriate, and substitute, we obtain the completely general terminal complex equation. Note that in this case we define $r_{AB} = k_{AA}/k_{AB}$, $r_{ACA} = k_{AA}/k_{ACA}$, $r_{ACB} = k_{AA}/k_{ACB}$, $r_{BA} = k_{BB}/k_{BA}$, $r_{BCA} = k_{BB}/k_{BCA}$, and $r_{BCB} = k_{BB}/k_{BCB}$. We arrive at the following general terminal complex equation:

$$\frac{P_1(A)}{P_1(B)} = \frac{P(A/B)}{P(B/A)} = \frac{\dfrac{1 + \left(\dfrac{r_{BA}}{r_{BCB}} + \dfrac{r_{BA}}{r_{BCA}}\right)\dfrac{[C]}{[A]} + \dfrac{r_{AB}}{r_{ACB}}\dfrac{[C]}{[A]}\left[\dfrac{1 + \dfrac{r_{BA}[C]}{r_{BCB}[A]}}{1 + \dfrac{r_{AB}[C]}{r_{ACA}[B]}}\right]}{r_{BA}\dfrac{[B]}{[A]} + 1 + \left(\dfrac{2r_{BA}}{2r_{BCB}} + \dfrac{r_{BA}}{r_{BCA}}\right)\dfrac{[C]}{[A]} + \dfrac{r_{AB}}{r_{ACB}}\dfrac{[C]}{[B]}\left[\dfrac{1 + \dfrac{r_{BA}[C]}{r_{BCB}[A]}}{1 + \dfrac{r_{AB}[C]}{r_{ACA}[B]}}\right]}}{\dfrac{1 + \left(\dfrac{r_{AB}}{r_{ACA}} + \dfrac{r_{AB}}{r_{ACB}}\right)\dfrac{[C]}{[B]} + \dfrac{r_{BA}}{r_{BCA}}\dfrac{[C]}{[A]}\left[\dfrac{1 + \dfrac{r_{AB}[C]}{r_{ACA}[B]}}{1 + \dfrac{r_{BA}[C]}{r_{BCB}[B]}}\right]}{r_{AB}\dfrac{[A]}{[B]} + 1 + \left(\dfrac{2r_{AB}}{r_{ACA}} + \dfrac{r_{AB}}{r_{ACB}}\right)\dfrac{[C]}{[B]} + \dfrac{r_{BA}}{r_{BCA}}\dfrac{[C]}{[A]}\left[\dfrac{1 + \dfrac{r_{AB}[C]}{r_{ACA}[B]}}{1 + \dfrac{r_{BA}[C]}{r_{BCB}[A]}}\right]}}.$$

$$(3.31)$$

In addition define $k_{AC} = k_{ACA} + k_{ACB}$, $k_{BC} = k_{BCA} + k_{BCB}$, $r_{AC} = k_{AA}/(k_{ACA} + k_{ACB}) = k_{AA}/k_{AC}$, $r_{BC} = k_{BB}/(k_{BCA} + k_{BCB}) = k_{BB}/k_{BC}$. Depending on the assumptions one is willing to make, eq. 3.31 can be cast into several very useful forms. If one assumes weak complex formation, where $K \leq 0.03$, then the concentration of complex must be much lower than that of either uncomplexed monomer: $[A] \gg [C]$ and $[B] \gg [C]$ and $[A_0] \cong [A]$ and $[B_0] \cong [B]$. Therefore

$$K \cong \frac{[C]}{[A][B]},$$

which can be written

$$[C] \cong K[A][B].$$

Substituting yields the weak complex equation:

$$\frac{P_1(A)}{P_1(B)} = \frac{\dfrac{1 + (r_{BA}/r_{BCB} + r_{BA}/r_{BCA})K[B] + \dfrac{Kr_{AB}}{r_{ACB}}[A]\dfrac{1 + Kr_{BA}[B]/r_{BCB}}{1 + Kr_{AB}[A]/r_{ACA}}}{r_{BA}[B]/[A] + 1 + (2r_{BA}/r_{BCB} + r_{BA}/r_{BCA})K[B] + \dfrac{Kr_{12}}{r_{ACB}}[A]\dfrac{1 + Kr_{BA}[B]/r_{BCB}}{1 + Kr_{AB}[A]/r_{ACA}}}}{\dfrac{1 + (r_{AB}/r_{ACA} + r_{AB}/r_{ACB})K[A] + \dfrac{Kr_{BA}}{r_{BCA}}[B]\dfrac{1 + Kr_{AB}[A]/r_{ACA}}{1 + Kr_{BA}[B]/r_{BCB}}}{r_{AB}[A]/[B] + 1 + (2r_{AB}/r_{ACA} + r_{AB}/r_{ACB}K[A] + \dfrac{Kr_{BA}}{r_{BCA}}[B]\dfrac{1 + Kr_{AB}[A]/r_{ACA}}{1 + Kr_{BA}[B]/r_{BCB}}}}.$$

$$(3.32)$$

It can be seen that each complex reactivity ratio is associated with K. This weak-complex form of the equation can now be specialized still further for application to specific systems. An example is the special case where radical B· does not add to monomer B or to the B side of the complex. In this case $P(A/B) = 1.0$ and the numerator of eq. 3.32 can be rearranged to the form

$$\left(\frac{P_A}{P_B} - 1\right)\left[1 + \frac{Kr_{BA}[B]}{r_{BCA}}(1 + Kr_{AB}[A]/r_{ACA}) + Kr_{AB}[A](1/r_{ACA} + 1/r_{ACB})\right]$$

$$= r_{AB}[A]/[B] + Kr_{AB}[A]/r_{ACA}. \qquad (3.33)$$

Note that for high dilution eq. 3.32 reduces to the classical terminal composition:

$$\frac{P_1(A)}{P_1(B)} = \frac{1 + r_{AB}[A]/[B]}{1 + r_{BA}[B]/[A]}. \qquad (3.34)$$

This is expected in light of the fact that at high dilution complex formation becomes negligible. Therefore the model reverts to the classical terminal case.

3.3.5 Copolymerization Model with Depropagation

Case Where Only One Monomer Pair Depolymerizes (10). The copolymerization models that have been considered have assumed that there is no depropagation. However, since some comonomers exhibit depropagation at the temperatures of polymerization, the impact of this effect on the copolymer structure should be examined. First we consider the depropagation model where terminal addition of A and B occurs and monomer B depolymerizes whenever it is attached to another B unit but not when attached to an A unit. The A comonomer is assumed not to depolymerize. These assumptions lead to the following copolymerization reactions:

Terminal Unit	Added	Final	Rate	
Copolymerization				
$\sim A_n^*$	A	$\sim A_{n+1}^*$	$k_{AA}[A^*][A]$	
$\sim A_n^*$	B	$\sim AB^*$	$k_{AB}[A^*][B]$	
$\sim B_n^*$	A	$\sim BA^*$	$k_{BA}[B^*][A]$	
$\sim B_n^*$	B	$\sim B_{n+1}^*$	$k_{BB}[B^*][B^*]$	(3.35)
Depropagation				
$\sim[\sim B_{n+1}^*]$	$(-)B$	$\sim B^*$	$k_{BB}'[\sim B_n^*]$	

The steady-state condition for the B* radical is

$$k_{AB}[A^*][B] + k_{BB}'[BB^*] = k_{BB}[B^*][B] + k_{BA}[B^*][A],$$

while for the [BB*] radical the steady state is

$$k_{BB}[B^*][B] + k'_{BB}[BBB^*] = k'_{BB}[BB^*] + k_{BB}[BB^*][B] + k_{AB}[BB^*][A].$$

We define

$$d = \frac{k_{BB}}{k'_{BB}},$$

$$\alpha_B^* = \frac{[B^*]_{n+1}}{[B^*]_n}.$$

Substitution of these symbols in the equation for [BB*] yields

$$(\alpha_B^*)^2 - \{1 + d[B] + (d/r_B)[A]\}\alpha_B^* + d[B] = 0.$$

Solving for α_B^*, we obtain

$$2\alpha_B^* = (1 + [B]d + (d/r_B)[A]) - \{(1 + [B]d + (d/r_B)[A])^2 - 4[B]d^{1/2}\}.$$

The steady-state equations for [A*] have the form

$$k_{BA}[A] \sum_{n=1}^{\infty} [(B_n^*)] - k_{AB}[A^*][B] = 0.$$

We can find that

$$\frac{[B^*]}{[A^*]} = \frac{k_{AB}[B](1 - \alpha_B^*)}{k_{BA}[A]}.$$

Using this relationship we can write

$$[B_n^*] = \frac{k_{AB}[B](1 - \alpha)(\alpha^{n-1})[A^*]}{k_{BA}[A]}.$$

The composition of the copolymer depends on the relative rates of disappearance of the monomers. With depropagation, the rate of disappearance of B is the rate at which B units are trapped in the chain by addition of A units:

$$\frac{-d[B]}{dt} = \sum_{n=1}^{\infty} nk_{BA}[B_n^*][A],$$

where the integer n accounts for the number of trapped B units.

The disappearance of A is given by

$$\frac{-d[A]}{dt} = k_{AA}[A^*][A] + \sum_{n=1}^{\infty} k_{BA}[B_n^*][A].$$

Dividing these two equations and substituting for $[B_n^*]$ in terms of [A*] give

$$\frac{P_1(B)}{P_1(A)} = \frac{k_{AB}[A^*][B](1 - \alpha_B^*)(1 + 2\alpha_B^* + \cdots + n\alpha_B^{*n-1})}{k_{AA}[A^*][A] + k_{AB}[A^*][B](1 - \alpha_B^*)(1 + \alpha_B^{*2} + \cdots + \alpha_B^{*n})},$$

which becomes

$$\frac{P_1(B)}{P_1(A)} = \frac{[B][1/(1 - \alpha_B^*)]}{r_A[A] + [B]}. \tag{3.36}$$

This composition equation is complicated by the presence of the term α_B^*, which involves the radical species, hence the equation has little value experimentally. However, as the temperature of the polymerization is increased, one approaches the ceiling temperature, the temperature at which the rate of propagation equals the rate of depropagation. Thus temperature has a marked effect on the copolymer composition cures near the ceiling temperature. Lowry (10) has calculated the composition curves for temperatures both above and below the ceiling temperature. His results are shown in Figure 3.3. For $r_A = r_B = 1$ for all temperatures with a ceiling temperature of 50°C for ΔE^* for k_{BB}, he selected 2000 cal/mole and the ΔE^* for k'_{BB} was arbitrarily chosen as 12,000 cal/mole. A lesser amount of B in the polymer for the same mole fraction of monomer B is obtained at 0°C, with considerably less at the ceiling temperature of 50°C. The polymer only gains the B monomer with great difficulty, whereas above the ceiling temperature at 100°C only an alternating sequence of AB can be prepared even at high mole fraction of B because the BB units depolymerize.

Since the monomer concentrations appear individually rather than as ratios, dilution of the monomers with an inert solvent has an effect on the shape of the copolymer composition curve near the ceiling temperature. Lowry's calculated results are shown in Figure 3.4.

Case Where all Four Comonomer Dyads Undergo Depolymerization (11). Let us now consider the copolymerization of monomers A and B to occur such that all four of the possible reactions can undergo depropagation. The possible reactions and associated rate constants are

$$\sim A_n^* + A \underset{k_2}{\overset{k_1}{\rightleftarrows}} \sim A_n A_{n+1}^*,$$

$$\sim A_n^* + B \underset{k_4}{\overset{k_3}{\rightleftarrows}} \sim A_n B_{n+1}^*,$$

$$\sim B_n^* + A \underset{k_6}{\overset{k_5}{\rightleftarrows}} \sim B_n A_{n+1}^*,$$

$$\sim B_n^* + B \underset{k_8}{\overset{k_7}{\rightleftarrows}} \sim B_n B_{n+1}^*. \tag{3.37}$$

Since the depropagation depends on the chance of the chain ending in A or B at a given time, it is necessary to define the probability of the dyads as well.

In unit time the number of chains of n units ending in A is

$$\frac{d(A_n)}{dt} = [(A_{n-1}^*)k_1 + (B_{n-1}^*)k_5 - (B_n A_{n+1}^*)k_6 - (A_n A_{n+1}^*)k_2] = N(A_n).$$

Similarly, the number of chains of n units ending in B is

$$\frac{d(B_n)}{dt} = [(B_{n-1}^*)k_7 + (A_{n-1}^*)k_3 - (A_n B_{n+1}^*)k_4 + (B_n B_{n+1}^*)k_8] = N(B_n).$$

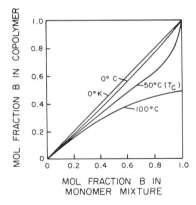

Figure 3.3 Effect of temperature on composition for a copolymer with one depolymerizing monomer pair. (Adapted from Ref. 10.)

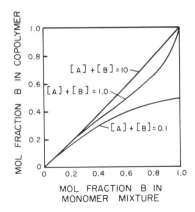

Figure 3.4 Effect of monomer concentration on the composition of a copolymer with one monomer pair depolymerizing at its ceiling temperature. (Adapted from Ref. 10.)

The propagation constants that are written as pseudo-first order are more properly written as $k_1(A)$, $k_3(B)$, and so on, but this is done for simplicity here. For the dyads the rate of imbibing into the chain is given by

$$\frac{d(A_n A_{n+1})}{dt} = (A_n^*)k_1 - (A_{n+1}A_{n+2}^*)k_2 = N(A_n A_{n+1}),$$

$$\frac{d(A_n B_{n+1})}{dt} = (A_n^*)k_3 - (A_{n+1}B_{n+2}^*)k_4 = N(A_n B_{n+1}),$$

$$\frac{d(B_n A_{n+1})}{dt} = (B_n^*)k_5 - (B_{n+1}A_{n+2}^*)k_6 = N(B_n A_{n+1}),$$

$$\frac{d(B_n B_{n+1})}{dt} = (B_n^*)k_7 - (B_{n+1}B_{n+2}^*)k_8 = N(B_n B_{n+1}),$$

and the probability of an $A_n A_{n+1}$ dyad is

$$P(A_n A_{n+1}) = \frac{N(A_n A_{n+1})}{N(A_n A_{n+1}) + N(A_n B_{n+1})},$$

and so forth.

From probability we know that

$$P(A_n A_{n+1}) = P(A_n)P(A_{n+1}/A_n),$$

$$P(B_n B_{n+1}) = P(B_n)P(B_{n+1}/B_n).$$

If we divide,

$$\frac{P(A_n)}{P(B_n)} = \frac{P(A_n A_{n+1})/P(A_{n+1}/A_n)}{P(B_n B_{n+1})/P(B_{n+1}/B_n)}.$$

Substitution yields

$$\frac{P(A_n)}{P(B_n)} = \frac{[(A_n^*)k_1 - (A_{n+1}A_{n+2}^*)k_2]/P(A_{n+1}/A_n)}{[(B_n^*)k_7 - (B_{n+1}B_{n+2}^*)k_8]/P(B_{n+1}/B_n)}.$$

This is the ratio of the relative probabilities of the nth unit being an A or B. We must sum over all values of n and let

$$\sum_n (A_n^*) = a,$$

$$\sum_n (B_n^*) = b,$$

$$P(A_{n+1}/A_n) = \epsilon,$$

$$P(B_{n+1}/B_n) = \eta.$$

We obtain

$$\frac{P_1(A)}{P_1(B)} = \frac{a\eta(k_1 - \epsilon k_2)}{b\epsilon(k_7 - \eta k_8)}. \tag{3.38}$$

This is the copolymer composition first derived by Howell, Izu, and O'Driscoll (11). By letting $k_2 = k_4 = k_6 = k_8 = 0$, we get the composition equation for the terminal model; with $k_2 = k_4 = k_6 = 0$ we get eq. 3.36. Application of eq. 3.38 for depropagation is difficult, as it utilizes eight independent rate constants. Izu and O'Driscoll (12) have applied this equation to their study of the system α-methylstyrene–methyl methacrylate, which have low ceiling temperatures of 61 and 164°C, respectively. At a polymerization temperature of 60°C, the results were consistent with a terminal model copolymerization, but at higher temperatures considerable deviations were observed. The composition of the copolymer could be fit by using as a model the reversible copolymerization of the dyads.

With depropagation the general trend of the behavior of the composition-feed curve is that the shape of the curves becomes flatter with increasing temperature. The copolymer formed at the higher temperature is nearly a homopolymer of the monomer having a higher ceiling temperature, even under the condition of a large excess of the other monomer in the feed. This model can also be used to calculate the microstructure (12).

3.3.6 Other Polymerization Mechanisms

From the above it is clear that for even relatively simple polymerization mechanisms, the algebra involved in reducing the mechanism to microstructural information is arduous but not necessarily difficult. The results quickly become complex and in many cases impractical to use. However, computer techniques can be used to simulate the polymerization, and in many cases they can yield sufficient information to gain some insight into the applicability of the mechanism. It must be remembered, however, that as the mechanism becomes more complex, the number of parameters increases, making more demands on the quality of the experimental measurements. But when trends are useful, these techniques represent the only practical approach (9).

In the most simple case, in a binary system, two radicals and two monomers are considered. Accordingly, the composition and microstructure are determined by the relative rate of the four propagation reactions. Any intramolecular or intermolecular interaction will lead to considering kinetically more than two radicals, more than two monomers, and certainly more than four chain propagation reactions in the system. Consequently, equations describing the microstructure of the copolymer are more complex with more than two parameters. The most important case we have considered is that of the penultimate and other chain reactivity effects. We have also considered the case where the monomers are capable of association and more than two monomer types must be treated. Similar treatments can be carried out for monomer association with growing radicals and/or solvent molecules. The case where the monomers show a tendency for depolymerization has also been treated, but variations on this process may also occur and these require specific evaluation.

Further complications can arise when the monomer can undergo isomerism and the monomer molecules can present themselves to the active ends in several configurations. If one of the monomers is a diene, it may participate in chain propagation in various ways (1, 2 or 3, 4, or 1, 4 addition or cycloaddition). It is also possible that the individual radicals may react in two different energetic states. Accordingly, at least eight chain propagation steps must be considered.

Finally, it must be clearly understood that the treatments apply to "instantaneous" polymerization, that is, polymerizations that are carried out under isothermal conditions with minimal change in conversion. Any changes in polymerization conditions or comonomer concentration must be accounted for.

3.4 MICROSTRUCTURE FOR INTEGRAL COPOLYMERIZATION (13)

3.4.1 Composition for Integral Copolymerization

In the preceding discussions the various equations are given for instantaneous copolymer polymerization. It has been assumed that the conversion is sufficiently low that the comonomer feed is relatively unchanged from its initial value. For nearly all copolymerizations the comonomer feed and copolymer composition change with conversion. The comonomer feed changes in composition, since one of the monomers is reacting preferentially and more enters the polymer chain.

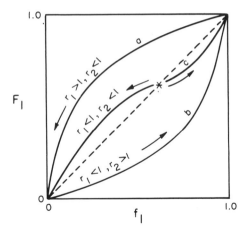

Figure 3.5 The mole fraction of monomer F_1 in polymer as a function of mole fraction f_1 in monomer feed for an ideal copolymer (dashed line), $r_1 = r_2 = 1$; (a) $r_1 > 1$; $r_2 < 1$; (b) $r_1 < 1$, $r_2 > 1$; (c) $r_1 < 1$, $r_2 < 1$. At the cross-over point (*) $F_1 = f_1 = (1 - r_2)/(2 - r_1 - r_2)$. (Reprinted by permission of Ref. 13.)

The change in copolymer composition with feed can most easily be demonstrated by expressing the copolymer composition equation in terms of mole fractions instead of concentrations. Define the monomer mole fractions as

$$f_1 = 1 - f_2 = \frac{[M_1]}{[M_1] + [M_2]}$$

and the polymer mole fractions as

$$F_1 = 1 - F_2 = \frac{d[M_1]}{d[M_1] + d[M_2]}.$$

Then, using the copolymer composition equation in the form

$$\frac{d[M_1]}{d[M_2]} = \frac{[M_1]}{[M_2]} \frac{(r_1[M_1] + [M_2])}{([M_1] + r_2[M_2])},$$

we can substitute to obtain

$$F_1 = \frac{r_1 f_1^2 + f_1 f_2}{r_1 f_1^2 + 2 f_1 f_2 + r_2 f_2^2}.$$

Since $f_2 = 1 - f_1$, we can write this equation as

$$F_1 = \frac{(r_1 - 1)f_1^2 + f_1}{(r_1 + r_2 - 2)f_1^2 + 2(1 - r_2)f_1 + r_2}.$$

In general, three different types of copolymerization behavior are observed, and the changes in composition with conversion are shown in Figure 3.5. The direction of drift is indicated by the arrows.

The changes in composition and sequence distribution must be considered for batch polymerizations. The simplest approach is from a material balance approach. Consider a system initially containing a total of $M(= [M_1] + [M_2])$ moles of the two monomers and in which the copolymer formed is richer in monomer M_1 than is the feed $(F_1 > f_1)$. When dM moles of monomers have been polymerized, the polymer will contain $F_1 dM$ moles of monomer 1 and the feed will contain $\Delta M_1 = \Delta M \Delta f_1 = (M - dM)(f_1 - df_1)$ moles of monomer 1. A material balance for monomer 1 requires that the moles of M_1 copolymerized equal the difference in the moles of M_1 in the feed before and after reaction, or

$$Mf_1 - (M - dM)(f_1 - df_1) = F_1 dM.$$

This can be rearranged (neglecting the $df_1 dM$ term) to give

$$\frac{dM}{M} = \frac{df_1}{(F_1 - f_1)}.$$

Letting M_0 and f_1^0 be the initial values of M and f_1, we need to integrate

$$\int_{M_0}^{M} \frac{dM}{M} = \ln \frac{M}{M_0} = \int_{f_1^0}^{f_1} \frac{df_1}{(F_1 - f_1)}. \tag{3.39}$$

For given values of the reactivity ratios, graphical or numerical methods may be used to calculate the expected change in monomer feed and copolymer composition corresponding to the mole conversion, $1 - (M/M_0)$. Computer techniques have been described for this purpose (15). We can substitute for F_1 in eq. 3.39 and obtain

$$\ln \frac{M}{M_0} = \frac{1}{(2 - r_1 - r_2)} \int_{f_1^0}^{f_1} \frac{(r_1 + r_2 - 2)f_1^2 + 2(1 - r_2)f_1 + r_2}{f_1(f_1 - 1)\left(f_1 - \dfrac{1 - r_2}{2 - r_1 - r_2}\right)} df_1. \tag{3.40}$$

If we let $\delta = (1 - r_2)/(2 - r_1 - r_2)$, we can expand and integrate each term and, after collecting terms and rearranging, we obtain

$$\frac{M}{M_0} = \left(\frac{f_1}{f_1^0}\right)^{\alpha} \left(\frac{f_2}{f_2^0}\right)^{\beta} \left(\frac{f_1^0 - \delta}{f_1 - \delta}\right)^{\gamma}, \tag{3.41}$$

where $\alpha = r_2/(1 - r_2)$, $\beta = r_1/(1 - r_1)$, and $\gamma = (1 - r_1 r_2)/(1 - r_1)(1 - r_2)$.

The slope of the mole conversion–instantaneous copolymer composition curve is

$$\frac{d}{dF_1}\left(1 - \frac{M}{M_0}\right) = -\frac{1}{M^0}\frac{dM}{dF_1},$$

and differentiation of the preceding equation and substitution yield

$$\frac{1}{M^0}\frac{dM}{df_1} = \left(\frac{f_1}{f_1^0}\right)^{\alpha} \left(\frac{f_1 - 1}{f_1^0 - 1}\right)^{\beta} \left(\frac{f_1^0 - \delta}{f_1 - \delta}\right)^{\gamma} \left(\frac{\alpha}{f_1} + \frac{\beta}{f_1 - 1} - \frac{\gamma}{f_1 - \delta}\right),$$

which may be simplified to

$$\frac{1}{M^0}\frac{dM}{df_1} = \frac{M}{M^0}\left(\frac{\alpha}{f_1} - \frac{\beta}{f_2} - \frac{\gamma}{f_1-\delta}\right).$$

The desired differential copolymer composition equation for binary copolymerization is

$$-\frac{1}{M^0}\frac{dM}{dF_1} = -\frac{M}{M^0}\left(\frac{\alpha}{f_1} - \frac{\beta}{f_2} - \frac{\gamma}{f_1-\delta}\right)\frac{[(r_1+r_2-2)f_1^2+2(1-r_2)f_1+r_2]^2}{[(r_1+r_2-2r_1r_2)f_1^2+2r_2(r_1-1)f_1-r_2]}.$$
$$(3.42)$$

Let us consider an example of the copolymer of styrene ($r_1 = 55$) and vinyl acetate ($r_2 = 0.01$). In Figure 3.6 it is seen that over a fairly wide starting composition range (0.1–1.0) the copolymer is over 90% styrene. In Figure 3.6 the expected change in monomer mixture composition as a

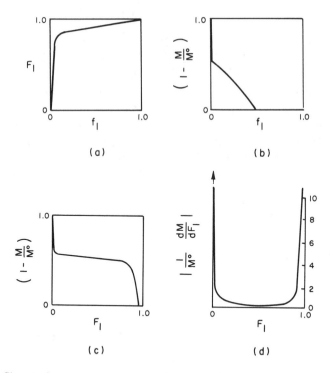

Figure 3.6 Changes in copolymer as a function of conversion: M_1 = styrene; M_2 = vinyl acetate (M_1) = (M_2); $r_1 = 55$, $r_2 = 0.01$. (a) Composition of polymer as a function of monomer feed ratio; (b) change in monomer composition with conversion for an initially equal molar monomer mixture; (c) change in copolymer composition with conversion for an initially equal molar monomer mixture; (d) slope of the mole conversion–instantaneous copolymer composition curve with conversion. (Reprinted by permission of Ref. 13.)

function of conversion for an equimolar mixture of styrene–vinyl acetate has been calculated. Figure 3.6 shows the corresponding change in instantaneous copolymer composition. From Figure 3.6 one observes that during the first half of the polymerization a copolymer of predominately styrene is found. The last half of the copolymerization is essentially vinyl acetate. The absolute value of the slope of the mole conversion–instantaneous copolymer composition curve is shown in Figure 3.6. This type of U-shaped curve is considered characteristic of "incompatible" copolymerizations, that is, copolymerizations in which the homopolymerization of one of the species is favored initially and the other species is favored during the final stages of the polymerization.

3.4.2 Copolymer Microstructures for Integral Copolymerization

A similar theory for the calculation of copolymer microstructure has been developed (14) as a function of conversion. In simplest terms

$$P_n(\ldots) = \frac{1}{M} \int_{M_1}^{M} P_n(\ldots) dM, \tag{3.43}$$

and the previously presented formalism has been used to calculate changes in M. A plot of the instantaneous concentration of the sequence versus

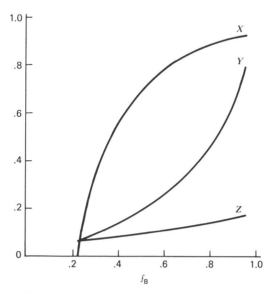

Figure 3.7 Calculated mole fraction converted (X), instantaneous composition (Y), and average copolymer composition (Z) as a function of mole fraction of vinyl chloride in unreacted monomer. Initial monomer feed composition 22% vinyl chloride. (Reprinted by permission of Ref. 15.)

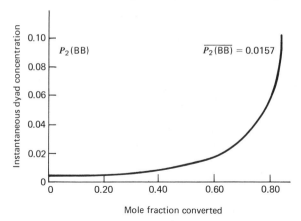

Figure 3.8 Instantaneous dyad concentration as a function of mole fraction converted ($f_B^0 = 0.22$). (Reprinted by permission of Ref. 15.)

mole fraction converted, when integrated and divided by the total monomer converted, gives the total sequence concentration $\{P_n(...)\}$.

We demonstrate the results for the vinylidene chloride ($r_1 = 4.0$) and vinyl chloride ($r_2 = 0.20$) system (16). Figure 3.7 shows the percent conversion, the instantaneous composition, and the average composition as a function of mole fraction of vinyl chloride. Figure 3.8 shows the change in the dyad $P_2(BB)$ as a function of conversion with an initial concentration of 22% vinyl chloride. Figures 3.9 and 3.10 show the change for an initial concentration of 61% and 82.5% vinyl chloride. The integrated composition

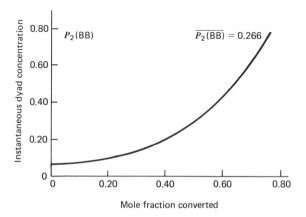

Figure 3.9 Instantaneous dyad concentration as a function of mole fraction converted ($f_B = 0.61$). (Reprinted by permission Ref. 15.)

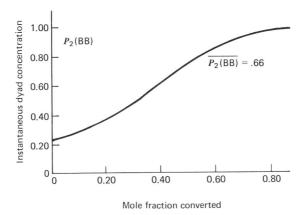

Figure 3.10 Instantaneous dyad concentration as a function of mole fraction converted ($f_B = 0.825$). (Reprinted by permission of Ref. 15.)

changes are

f_B^0	$\overline{P_2(BB)}$
0.22	0.0157
0.61	0.266
0.825	0.66

Table 3.7 summarizes the results for various sequences as a function of f_B^0 and conversion.

It is possible to calculate the concentration of any sequence as a function of f and conversion from a knowledge of the reactivity ratios.

Table 3.7 Calculated Comonomer Sequence Concentrations and Initial Vinyl Chloride Monomer Content (16)

f_B^0	Conversion	$\overline{P_1\{B\}}$	$\overline{P_2\{BB\}}$	$\overline{P_3\{BBB\}}$	$\overline{P_2\{BA\}}$	$\overline{P_2\{AA\}}$	$\overline{P_4\{AAAA\}}$
0		0	0	0	0	1.00	1.00
0.22	0.83	0.13	0.0015	0.0035	0.22	0.77	0.60
0.46	0.59	0.26	0.0062	0.016	0.39	0.55	0.31
0.61	0.75	0.49	0.27	0.16	0.47	0.27	0.10
0.74	0.85	0.69	0.51	0.41	0.36	0.13	0.028
0.83	0.87	0.80	0.66	0.56	0.28	0.057	0.020
0.90	0.91	0.89	0.80	0.72	0.19	0.024	0.010
1.00		1.00	1.00	1.00	0	0	0

REFERENCES

1. K. Ito and Y. Yamashita, *J. Polym. Sci.*, **3A**, 2165 (1965).
2. F. R. Mayo and F. M. Lewis, *J. Am. Chem. Soc.*, **66**, 1594 (1944).
3. C. Tosi, *Adv. Polym. Sci.*, **5**, 451 (1968).
4. E. Merz, T. Alfrey, and G. Goldfinger, *J. Polym. Sci.*, **1**, 75 (1946).
5. E. Berger and I. Kuntz, *J. Polym. Sci.*, **2A**, 1687 (1964).
6. F. Price, *J. Chem. Phys.*, **36**, 209 (1962).
7. G. E. Ham, *Copolymerization*, Interscience Publishers, New York, 1974.
8. J. A. Seiner and M. H. Litt, *Macromolecules*, **4**, 308 (1971).
9. H. L. Frisch, M. Bishop, and J. Roth, *J. Chem. Phys.*, **67**, 1082 (1977).
10. G. C. Lowry, *J. Polym. Sci.*, **42**, 463 (1960).
11. J. A. Howell, M. Izu, and K. F. O'Driscoll, *J. Polym. Sci.*, *A-1*, **8**, 699 (1970).
12. M. Izu and K. F. O'Driscoll, *J. Polym. Sci.*, *A-1*, **8**, 1687 (1970).
13. V. Meyer and G. Lowry, *J. Polym. Sci.*, **3A**, 2843 (1965).
14. Y. Yamashita, K. Ito, H. Ishii, S. Hashino, and M. Kai, *Macromolecules*, **1**, 529 (1968).
15. H. J. Harwood, *J. Polym. Sci.*, **25C**, 37 (1968).
16. M. Meeks and J. L. Koenig, *J. Polym. Sci.*, *A-2*, **9**, 717 (1971).

4

Multicomponent
Copolymerization
Theory

In spite of the expected complexity of multicomponent copolymers, they have found considerable utilization in industry. The addition of a small amount of a third component to a binary copolymer can give desirable chemical, physical, and mechanical properties that differ from those of the copolymer consisting of only two monomers. It therefore behooves us to consider multicomponent systems from both the theoretical and experimental point of view.

4.1 COMPOSITION OF TERPOLYMERS WITH TERMINAL PROPAGATION

4.1.1 General Case

For a terpolymer there are nine propagation steps for the terminal mechanism of copolymerization:

Terminal Group	Added Group	Rate	Final
~A*	A	$k_{AA}[A^*][A]$	~AA*
~A*	B	$k_{AB}[A^*][B]$	~AB*
~A*	C	$k_{AC}[A^*][C]$	~AC*
~B*	A	$k_{BA}[B^*][A]$	~BA*
~B*	B	$k_{BB}[B^*][B]$	~BB*
~B*	C	$k_{BC}[B^*][C]$	~BC*
~C*	A	$k_{CA}[C^*][A]$	~CA*
~C*	B	$k_{CB}[C^*][B]$	~CB*
~C*	C	$k_{CC}[C^*][C]$	~CC*

$$(4.1)$$

We adopt the approach of Alfrey and Goldfinger (1), who first proposed the relationships between the composition of a terpolymer and that of the corresponding monomer mixture.

The rates of monomer disappearance are

$$\frac{-d[A]}{dt} = k_{AA}[A^*][A] + k_{BA}[B^*][A] + k_{CA}[C^*][A],$$

$$\frac{-d[B]}{dt} = k_{AB}[A^*][B] + k_{BB}[B^*][B] + k_{CB}[C^*][B],$$

$$\frac{-d[C]}{dt} = k_{AC}[A^*][C] + k_{BC}[B^*][C] + k_{CC}[C^*][C]. \tag{4.2}$$

We make the steady-state approximation, which means that the number of growing chains of type A^* (or B^*, C^*) that disappear in unit time is equal to the number of A^*'s (or B^*, C^*) that form in unit time. This relationship requires

$$k_{AB}[A^*][B] + k_{AC}[A^*][C] = k_{BA}[B^*][A] + k_{CA}[C^*][A],$$

$$k_{BA}[B^*][A] + k_{BC}[B^*][C] = k_{AB}[A^*][B] + k_{CB}[C^*][B],$$

$$k_{CA}[C^*][A] + k_{CB}[C^*][B] = k_{AC}[A^*][C] + k_{BC}[B^*][C]. \tag{4.3}$$

Combining and substitution yield the following terpolymerization equations:

$$d[A]:d[B]:d[C] = P_1(A):P_1(B):P_1(C)$$

$$= [A]\left\{\frac{[A]}{r_{CA}r_{BA}} + \frac{[B]}{r_{BA}r_{CB}} + \frac{[C]}{r_{CA}r_{BC}}\right\}\left\{[A] + \frac{[B]}{r_{AB}} + \frac{[C]}{r_{AC}}\right\}$$

$$:[B]\left\{\frac{[A]}{r_{AB}r_{CA}} + \frac{[B]}{r_{AB}r_{CB}} + \frac{[C]}{r_{CB}r_{AC}}\right\}\left\{\frac{[A]}{r_{BA}} + [B] + \frac{[C]}{r_{BC}}\right\}$$

$$:[C]\left\{\frac{[A]}{r_{AC}r_{BA}} + \frac{[B]}{r_{BC}r_{AB}} + \frac{[C]}{r_{AC}r_{BC}}\right\}\left\{\frac{[A]}{r_{CA}} + \frac{[B]}{r_{CB}} + [C]\right\}, \tag{4.4}$$

where the reactivities have the usual definitions

$$r_{XY} = \frac{k_{XX}}{k_{XY}},$$

where X, Y = A, B, or C.

For brevity let us define

$$A' = [A] + \frac{[B]}{r_{AB}} + \frac{[C]}{r_{AC}},$$

$$B' = \frac{[A]}{r_{BA}} + [B] + \frac{[C]}{r_{BC}},$$

$$C' = \frac{[A]}{r_{CA}} + \frac{[B]}{r_{CA}} + [C], \tag{4.5}$$

which leads to

$$P_1(A):P_1(B):P_1(C) = [A]\left\{\frac{[A]}{r_{CA}r_{BA}} + \frac{[B]}{r_{BA}r_{CB}} + \frac{[C]}{r_{CA}r_{BC}}\right\}A'$$

$$:[B]\left\{\frac{[A]}{r_{AB}r_{CA}} + \frac{[B]}{r_{AB}r_{CB}} + \frac{[C]}{r_{CB}r_{AC}}\right\}B'$$

$$:[C]\left\{\frac{[A]}{r_{AC}r_{BA}} + \frac{[B]}{r_{BC}r_{AB}} + \frac{[C]}{r_{AC}r_{BC}}\right\}C' \qquad (4.6)$$

A comparison between the experimental and calculated terpolymer composition is shown in Table 4.1 for four different monomer feeds (2).

A simpler version of the terpolymer composition equation has been utilized (3) based on the steady-rate relations

$$k_{AB}[A^*][B] = k_{BA}[B^*][A],$$

$$k_{BC}[B^*][C] = k_{CB}[C^*][B],$$

$$k_{CA}[C^*][A] = k_{AC}[A^*][C]. \qquad (4.7)$$

Table 4.1 Terpolymer Compositions for Various Monomer Feed Ratios

			Mole Fractions in the Terpolymer		
No.	Monomer	Mole Fractions in the Feed	Found	Calculated by Eq. 4.4	Calculated by Eq. 4.8
I	m_1	0.359	0.447	0.436	0.452
	m_2	0.360	0.261	0.292	0.338
	m_3	0.281	0.292	0.262	0.210
II	m_1	0.532	0.526	0.529	0.542
	m_2	0.265	0.202	0.232	0.260
	m_3	0.203	0.272	0.239	0.198
III	m_1	0.283	0.384	0.414	0.422
	m_2	0.282	0.230	0.227	0.267
	m_3	0.435	0.386	0.359	0.311
IV	m_1	0.278	0.364	0.368	0.370
	m_2	0.520	0.406	0.438	0.476
	m_3	0.202	0.230	0.194	0.154

The reactivity ratios of the binary copolymerizations are (3)

$$r_{12} = 0.50 \pm 0.02, \qquad r_{31} = 0.04 \pm 0.04, \qquad r_{23} = 1.20 \pm 0.14,$$

$$r_{21} = 0.50 + 0.02, \qquad r_{13} = 0.41 \pm 0.08, \qquad r_{32} = 0.15 \pm 0.07$$

where m_1 is styrene, m_2 is methyl methacrylate, and m_3 is acrylonitrile.

These equations are a consequence of the assumption that the number of A—B bonds is equal to that of B—A bonds for sufficiently long polymer chains. The simplified terpolymer equation becomes

$$P_1(A):P_1(B):P_1(C) = [A]A':[B]B'\frac{r_{BA}}{r_{AB}}:[C]C'\frac{r_{BA}}{r_{AC}}. \tag{4.8}$$

This simplified equation is valid if

$$r_{AB}r_{BC}r_{CA} = r_{BA}r_{AC}r_{CB}, \tag{4.9}$$

which is obtained by multiplying eq. 4.7 member by member, which yields

$$k_{AB}k_{BC}k_{CA} = k_{BA}k_{AC}k_{CB}.$$

Division by $k_{AA}k_{BB}k_{CC}$ yields eq. 4.8. A comparison of the calculated results for these two equations is shown in Table 4.1.

A further simplification occurs for special relationships between the reactivity ratios. Thus when $r_{XY}r_{YX} = 1$,

$$r_{BC} = \frac{r_{AC}}{r_{AB}}, \tag{4.10}$$

which is often the case in ionic terpolymerizations. The result is

$$P_1(A):P_1(B):P_1(C) = [A]:\frac{[B]}{r_{AB}}:\frac{[C]}{r_{AC}}. \tag{4.11}$$

This equation has been used with success for the terpolymerization of ethylene–propylene–butene-1 by an anionic coordination mechanism.

These relationships may also be expressed using the conditional probability notation:

$$\begin{aligned}
P_1(A):P_1(B):P_1(C) = \ &P(A/B)P(A/C) + P(B/C)P(B/A) \\
&+ P(C/B)P(A/C):P(B/A)P(B/C) + P(A/C)P(A/B) \\
&+ P(C/A)P(B/C):P(C/A)P(C/B) + P(A/B)P(C/A) \\
&+ P(B/A)P(C/B),
\end{aligned} \tag{4.12}$$

where

$$\begin{aligned}
P(B/A) &= \frac{k_{AB}[A^*][B]}{k_{AA}[A^*][A] + k_{AB}[A^*][B] + k_{AC}[A^*][C]} \\
&= \frac{[B]/r_{AB}}{[A] + [B]/r_{AB} + [C]/r_{AC}}.
\end{aligned} \tag{4.13}$$

This form of the equation demonstrates the ability to calculate the composition of a terpolymer from the reactivity coefficients obtained from binary reactions. The above relations require knowledge of six reactivity ratios, where none is infinite or equal to 0.

4.1.2 One of Three Comonomers Cannot Homopolymerize

When one of the monomers cannot homopolymerize ($k_{CC} = 0$), the terpolymer equation becomes

$$d[A] : d[B] : d[C] = P_1(A) : P_1(B) : P_1(C)$$

$$= [A]\left\{\frac{R_C[A]}{r_{BA}} + \frac{[B]}{r_{BA}} + \frac{R[C]}{r_{BC}}\right\}\left\{[A] + \frac{[B]}{r_{AB}} + \frac{[C]}{r_{AC}}\right\} : [B]\left\{\frac{R_C[A]}{r_{AB}}\right.$$

$$\left. + \frac{[B]}{r_{AB}} + \frac{[C]}{r_{AC}}\right\}$$

$$\times \left\{[B] + \frac{[A]}{r_{BA}} + \frac{[C]}{r_{BC}}\right\} : [C]\left\{\frac{[A]}{r_{AC}r_{BA}} + \frac{[B]}{r_{AB}r_{BC}} + \frac{[C]}{r_{AC}r_{BC}}\right\}\{R_C[B]$$

$$+ [B]\},$$

where $R_C = k_{CA}/k_{CB}$. A single terpolymerization experiment determines R_C.

4.1.3 Two of the Termonomers Cannot Homopolymerize

When two of the monomers have vanishing rates of self-propagation ($k_{BB} = k_{CC} = 0$), but can add to each other, the proper equations are

$$\frac{P_1(A)}{P_1(B)} = \frac{[A]\{R_B[B] + R_B R_C[A] + R_B[C]\}}{[B]\left\{\dfrac{R_C[A]}{r_{AB}} + \dfrac{[B]}{r_{AB}} + \dfrac{[C]}{r_{AC}}\right\}}\left[\frac{\left\{[A] + \dfrac{[B]}{r_{AB}} + \dfrac{[C]}{r_{AC}}\right\}}{\{R_B[A] + [C]\}}\right],$$

$$\frac{P_1(A)}{P_1(B)} = \frac{[A]\{R_B[B] + R_B R_C[A] + R_B[C]\}}{[C]\left\{\dfrac{R_B[A]}{r_{AB}} + \dfrac{[B]}{r_{AB}} + \dfrac{[C]}{r_{AC}}\right\}}\left[\frac{\left\{[A] + \dfrac{[B]}{r_{AB}} + \dfrac{[C]}{r_{AC}}\right\}}{\{R_B[A] + [B]\}}\right],$$

$$\frac{P_1(B)}{P_1(C)} = \frac{[B]\left\{\dfrac{R_B[A]}{r_{AB}} + \dfrac{[B]}{r_{AB}} + \dfrac{[C]}{r_{AC}}\right\}\{R_B[A] + [C]\}}{[C]\left\{\dfrac{R_B[A]}{r_{AC}} + \dfrac{[B]}{r_{AB}} + \dfrac{[C]}{r_{AC}}\right\}\{R_C[A] + [B]\}},$$

where

$$R_B = \frac{k_{BA}}{k_{BC}},$$

$$R_C = \frac{k_{CA}}{k_{CB}}.$$

Values of R_B and R_C must be determined by means of a terpolymerization experiment.

4.1.4   Two Termonomers Cannot Add to Themselves or to Each Other

If B and C cannot add to themselves ($k_{BB} = 0$, $k_{CC} = 0$) or to each other ($k_{BC} = 0$, $k_{CB} = 0$) then

$$\frac{P_1(A)}{P_1(B)} = 1 + \frac{r_{AB}[A]}{[B]} + \frac{r_{AB}[C]}{r_{AC}[B]},$$

$$\frac{P_1(A)}{P_1(C)} = 1 + \frac{r_{AC}[A]}{[C]} + \frac{r_{AC}[B]}{r_{AB}[C]},$$

$$\frac{P(B)}{P(C)} = \frac{r_{AC}[B]}{r_{AB}[C]}. \tag{4.16}$$

4.1.5   Two Termonomers Can Only Add to Third Monomer

The final special case of terpolymerization is the case where only cross addition of A and B to C can occur but not addition to each other. Then

$$\frac{P_1(A)}{P_1(B)} = \frac{k_{CA}[A]}{k_{CB}[B]}. \tag{4.17}$$

This special case has been observed for the terpolymerization of maleic anhydride with substituted α-methylstyrenes. In the chains maleic anhydrides alternate with one or the other of the substituted α-methyl styrenes.

The relative reactivities toward the maleic anhydride radical, k_{CA}/k_{CB}, can easily be determined, as integration of eq. 4.17 yields

$$\log \frac{[A]_0}{[A]} = \frac{k_{CA}}{k_{CB}} \log \frac{[B]_0}{[B]},$$

where the zero subscripts indicate starting concentrations.

4.2   SEQUENCE DISTRIBUTION IN TERPOLYMERS (2)

For a terpolymer the probability that a sequence of each of the monomers X contains n units is

$$P_n(A^n) = \left(\frac{[A]}{A'}\right)^{n-1} \left(1 - \frac{[A]}{A'}\right),$$

$$P_n(B^n) = \left(\frac{[B]}{B'}\right)^{n-1} \left(1 - \frac{[B]}{B'}\right),$$

$$P_n(C^n) = \left(\frac{[C]}{C'}\right)^{n-1} \left(1 - \frac{[B]}{B'}\right). \tag{4.18}$$

Table 4.2 (2) shows the calculated sequence distribution for the four particular terpolymers whose composition is given in Table 4.1. The

Table 4.2   Distribution of X Sequences of Various Length, $P_n(X^n)$

No.	Monomer	$n = 1$	$n = 2$	$n = 3$	$n = 4$	$n = 5$	$n = 6$
I	m_1	0.796	0.162	0.033	0.007	0.001	
	m_2	0.725	0.199	0.055	0.015	0.004	0.001
	m_3	0.976	0.023	0.001			
	m_1	0.658	0.225	0.077	0.026	0.009	0.003
II	m_2	0.801	0.160	0.032	0.006	0.001	
	m_3	0.987	0.013				
	m_1	0.852	0.126	0.019	0.003		
III	m_2	0.767	0.179	0.042	0.010	0.002	
	m_3	0.954	0.044	0.002			
	m_1	0.847	0.130	0.020	0.003		
IV	m_2	0.582	0.243	0.102	0.043	0.018	0.008
	m_3	0.981	0.019				

average length of the sequence $\bar{l}_x$ is obtained using

$$\bar{l}_x = \frac{\sum_{n=1}^{\infty} nP_n(X)^n}{\sum_{n=1}^{\infty} P_n(X)^n}.$$

Evaluation of the sums yields

$$\bar{l}_A(t) = 1 + \frac{[A]}{A' - [A]},$$

$$\bar{l}_B(t) = 1 + \frac{[B]}{B' - [B]},$$

$$\bar{l}_C(t) = 1 + \frac{[C]}{C' - [C]}. \qquad (4.19)$$

The average sequence lengths are shown in Table 4.3.

Table 4.3   Average   Length   of m_i Sequences, $\bar{l}_i$, for Terpolymer Systems

No.	$\bar{l}_1$	$\bar{l}_2$	$\bar{l}_3$
I	1.256	1.378	1.025
II	1.519	1.249	1.013
III	1.174	1.304	1.048
IV	1.181	1.721	1.019

It is interesting to note that the average sequence length in ter-polymers is shorter than in the corresponding copolymers. This can be demonstrated in the following manner. The average sequence length for A $[\bar{l}_A(C)]$ in a copolymer is

$$\bar{l}_A(C) = 1 + r_{AB}\frac{[A]}{[B]} = 1 + \frac{[A]}{[B]/r_{AB}},$$

whereas the equation for the terpolymer $[l_A(t)]$ *is*

$$\bar{l}_A(t) = 1 + \frac{[A]}{A' - [A]} = \frac{[A]}{([B]/r_{AB}) + ([C]/r_{AC})},$$

and obviously

$$\bar{l}_{(A)}(t) < \bar{l}_A(C).$$

This result is particularly important in the field of elastomers where a third monomer makes a substantial reduction in the longer sequences, making the possibility of crystallization less likely.

The probability of $U(X, n)$ that a monomeric unit of X belongs to a sequence of n members is given by

$$U(X, n) = \frac{nP(X)^n}{\sum\limits_{n=1}^{\infty} nP(X)^n},$$

so

$$U(A, n) = n\left(\frac{[A]}{A'}\right)^{n-1}\left(1 - \frac{[A]}{A'}\right)^2,$$

$$U(B, n) = n\left(\frac{[B]}{B'}\right)^{n-1}\left(1 - \frac{[B]}{B'}\right)^2.$$

$$(4.20)$$

We can also calculate the bond fractions α_{AA}, α_{BB}, α_{CC}, α_{AC}, and so on. We need the total number of units

$$P_1(A) + P_1(B) + P_1(C) = 1,$$

and the total number of sequences of each type (S_X) can be obtained by noting

$$S_A = \frac{1}{\bar{l}_A} = \frac{A' - [A]}{A'},$$

$$S_B = \frac{1}{\bar{l}_B} = \frac{B' - [B]}{B'},$$

$$S_C = \frac{1}{\bar{l}_C} = \frac{C' - [C]}{C}.$$

The number of sequences contained in a terpolymer is

$$S = P_1(A)S_A + P_1(B)S_B + P_1(C)S_C.$$

From these equations one can calculate the fraction of X—X bonds for a terpolymer (α_{XX}^t):

$$\alpha_{AA}^{(t)} = P_1(A)\frac{A' - [A]}{A'} \sum_{n=2}^{\infty} (n-1)\left(\frac{[A]}{A'}\right)^{n-1}\left(1 - \frac{[A]}{A'}\right)$$

$$= P_1(A)\frac{[A]}{A'}. \tag{4.20}$$

Similarly,

$$\alpha_{BB}^t = P_1(B)\frac{[B]}{B'},$$

$$\alpha_{CC}^t = P_1(C)\frac{[C]}{C'}.$$

On the other hand, the number of heterobonds $(X - Y)$ is equal to the number of contact points between the sequences of the three monomers:

$$\alpha_{AB}^t = \alpha_{BA}^t = \tfrac{1}{2}[P_1(A)S_A + P_2(B)S_B - P_2(C)S_C],$$

$$\alpha_{AC}^t = \alpha_{CA}^t = \tfrac{1}{2}[P_1(A)S_A + P_1(C)S_C - P_1(B)S_B],$$

$$\alpha_{BC}^t = \alpha_{CB}^t = \tfrac{1}{2}[P_1(B)S_B + P_1(C)S_C - P_1(A)S_A]. \tag{4.21}$$

A tabulation of the bond fractions for the four terpolymers is given in Table 4.4 (2). The probability of M_C occurring in a sequence of n length is given by

$$U(C, n) = n\left(\frac{[C]}{C'}\right)^{n-1}\left(1 - \frac{[C]}{C'}\right)^2 \tag{4.22}$$

These probabilities have been calculated and are tabulated for the four terpolymers in Table 4.5 (2).

Table 4.4   Bond Fractions α_{ij}, for Terpolymer System

No.	α_{11}	α_{12}	α_{13}	α_{21}	α_{22}	α_{23}	α_{31}	α_{32}	α_{33}
I	0.092	0.200	0.160	0.200	0.093	0.045	0.160	0.045	0.005
II	0.185	0.185	0.172	0.185	0.052	0.023	0.172	0.023	0.003
III	0.063	0.1335	0.2255	0.1335	0.062	0.0715	0.2255	0.0715	0.014
IV	0.057	0.2195	0.935	0.2195	0.199	0.0575	0.0935	0.0575	0.003

Table 4.5 Distribution of m_1 Monomer Units in Sequences of Various Length, $U(m_1, n)$

No.	Monomer	$n = 1$	$n = 2$	$n = 3$	$n = 4$	$n = 5$	$n = 6$
I	m_1	0.634	0.258	0.079	0.022	0.005	0.001
	m_2	0.526	0.289	0.119	0.044	0.015	0.005
	m_3	0.952	0.046	0.002			
II	m_1	0.433	0.296	0.152	0.069	0.030	0.012
	m_2	0.641	0.256	0.076	0.020	0.005	0.001
	m_3	0.974	0.026				
III	m_1	0.725	0.215	0.048	0.009	0.002	
	m_2	0.588	0.274	0.096	0.030	0.009	0.002
	m_3	0.910	0.084	0.006			
IV	m_1	0.717	0.220	0.051	0.010	0.002	
	m_2	0.338	0.283	0.178	0.099	0.052	0.026
	m_3	0.962	0.037	0.001			

The number-average sequence lengths in terpolymers can also be calculated in terms of the dyad and triad monomer distributions (5). The relationship between monomer composition $P_1(A)$ and the dyad concentration is

$$P_1(A) = P_2(AA) + \tfrac{1}{2}P_2(AB) + \tfrac{1}{2}P_2(AC),$$

$$P_1(B) = P_2(BB) + \tfrac{1}{2}P_2(AB) + \tfrac{1}{2}P_2(BC),$$

$$P_2(C) = P_2(CC) + \tfrac{1}{2}P_2(AC) + \tfrac{1}{2}P_2(BC).$$

In this equation the heterodyads $P_2(AB)$, and so on, represent the sum of the concentrations of the two indistinguishable dyads (AB and BA). The dyad fraction can be counted from runs of the following:

$$P_2(AA) = \sum_{n=0} nP(BA[A]_n B) + \sum_{n=0} nP(BA[A]_n C) + \sum_{n=0} nP(BA[A]_n C).$$

$$(4.24)$$

The upper limits of the sums are the longest runs in the system. Similar expressions are easily written for the other possible dyads. The number-average sequence of runs of A units is given by

$$\bar{l}_B = \frac{\displaystyle\sum_{n=0} nP(B[A]_n B) + \sum_{n=0} nP(B[A]_n C) + \sum_{n=0} nP(C[A]_n C)}{\displaystyle\sum_{n=1} P(B[A]_n B) + \sum_{n=1} P(B[A]_n C) + \sum_{n=1} P(C[A]_n C)}.$$

$$(4.25)$$

Corresponding expressions can be written for number-average sequences of bond C units. Expanding the above sum (eq. 4.25) to the same form as expressed for the dyads (eq. 4.24) and substituting, we obtain

$$\bar{l}_A = \frac{P_2(AA) + \frac{1}{2}P_2(AB) + \frac{1}{2}P_2(AC)}{\frac{1}{2}P_2(AB) + \frac{1}{2}P_2(AC)},$$

$$\bar{l}_B = \frac{P_2(BB) + \frac{1}{2}P_2(AB) + \frac{1}{2}P_2(BC)}{\frac{1}{2}P_2(AB) + \frac{1}{2}P_2(BC)},$$

$$\bar{l}_C = \frac{P_2(CC) + \frac{1}{2}P_2(BC) + \frac{1}{2}P_2(AC)}{\frac{1}{2}P_2(BC) + \frac{1}{2}P_2(AC)}.$$

For a triad distribution the number-average sequence lengths can be developed in a similar manner, and the results are

$$\bar{l}_A = \frac{P_3(BAB) + P_3(BAC) + P_3(CAC) + P_3(AAA) + P_3(AAB) + P_3(AAC)}{P_3(BAB) + P_3(BAC) + P_3(CAC) + \frac{1}{2}P_3(AAB) + \frac{1}{2}P_3(AAC)},$$

$$\bar{l}_B = \frac{P_3(ABA) + P_3(ABC) + P_3(CBC) + P_3(BBB) + P_3(BBA) + P_3(BBC)}{P_3(ABA) + P_3(ABC) + P_3(CAC) + \frac{1}{2}P_3(BBA) + \frac{1}{2}P_3(BBC)},$$

$$\bar{l}_C = \frac{P_3(ACA) + P_3(ACB) + P_3(BCB) + P_3(CCC) + P_3(CCA) + P_3(CCB)}{P_3(ACA) + P_3(ACB) + P_3(BCB) + \frac{1}{2}P_3(CCA) + \frac{1}{2}P_3(CCB)}.$$

4.3 MULTICOMPONENT COPOLYMERIZATION WITH n MONOMERS (4)

For the copolymerization of n monomers the rates of disappearance are given by

$$-\frac{d[M_1]}{dt} = K_{11}[M_1^{\cdot}][M_1] + K_{21}[M_2^{\cdot}][M_1] + \cdots K_{n1}[M_n^{\cdot}][M_1],$$

$$-\frac{d[M_2]}{dt} = K_{12}[M_1^{\cdot}][M_2] + K_{22}[M_2^{\cdot}][M_2] + \cdots + K_{n2}[M_n^{\cdot}][M_2],$$

$$-\frac{d[M_n]}{dt} = K_{1n}[M_1^{\cdot}][M_n] + K_{2n}[M_2^{\cdot}][M_n] + \cdots + K_{nn}[M_n^{\cdot}][M_n].$$

The steady-state approximation is invoked, which supposes that the number of growing chain ends of type $M_n^{\cdot}$ that disappear in a unit of time is equal to the number of $M_n^{\cdot}$'s that form in unit time. This leads to the following equations:

$$K_{12}[M_1^{\cdot}][M_2] + K_{13}[M_1^{\cdot}][M_3] + \cdots + K_{1n}[M_1^{\cdot}][M_n]$$
$$= K_{21}[M_2^{\cdot}][M_1] + K_{31}[M_3^{\cdot}][M_1] + \cdots + K_{n1}[M_n^{\cdot}][M_1],$$

$$K_{21}[M_2^{\cdot}][M_1] + K_{23}[M_2^{\cdot}][M_3] + \cdots + K_{2n}[M_2^{\cdot}][M_n]$$
$$= K_{12}[M_1^{\cdot}][M_2] + K_{32}[M_3^{\cdot}][M_2] + \cdots + K_{n2}[M_n^{\cdot}][M_2],$$

$$K_{n1}[M_n^{\cdot}][M_1] + K_{n2}[M_n^{\cdot}][M_2] + \cdots + K_{n,n-1}[M_n^{\cdot}][M_{n-1}]$$
$$= K_{1n}[M_1^{\cdot}][M_n] + K_{2n}[M_2^{\cdot}][M_n] + \cdots + K_{n-1,n}[M_{n-1}^{\cdot}][M_n].$$

Substituting and performing the proper series of algebraic transformations, we obtain the following multicomponent copolymerizations:

$$d[M_1]:d[M_2]:\cdots:d[M_n] = m_1:m_2:\cdots:m_n$$

$$= [M_1]D_1\left\{[M_1] + \frac{[M_2]}{r_{12}} + \frac{[M_3]}{r_{13}} + \cdots + \frac{[M_n]}{r_{1n}}\right\}:$$

$$[M_2]D_2\left\{\frac{[M_1]}{r_{21}} + [M_2] + \frac{[M_3]}{r_{23}} + \cdots + \frac{[M_n]}{r_{2n}}\right\}:$$

$$\vdots$$

$$[M_n]D_n\left\{\frac{[M_1]}{r_{n1}} + \frac{[M_2]}{r_{n2}} + \cdots + \frac{[M_{n-1}]}{r_{n,n-1}} + [M_n]\right\},$$

where

$$D_1 =$$

$$\begin{vmatrix} -\dfrac{1}{r_{n1}} & \dfrac{1}{r_{21}} & \dfrac{1}{r_{31}} & \cdots & \dfrac{1}{r_{n-1,1}} \\[2ex] -\dfrac{[M_2]}{r_{n2}} & -\left(\dfrac{[M_1]}{r_{21}} + \dfrac{[M_3]}{r_{23}} + \cdots + \dfrac{[M_n]}{r_{2n}}\right) & \dfrac{[M_2]}{r_{32}} & \cdots & \dfrac{[M_2]}{r_{n-1,2}} \\[2ex] -\dfrac{[M_3]}{r_{n3}} & \dfrac{[M_3]}{r_{23}} & -\left(\dfrac{[M_1]}{r_{31}} + \dfrac{[M_2]}{r_{32}} + \cdots + \dfrac{[M_n]}{r_{3n}}\right) & \cdots \\[2ex] \vdots \\[1ex] -\dfrac{[M_{n-1}]}{r_{n,n-1}} & \cdots & & & -\left(\dfrac{[M_1]}{r_{n-1,1}} + \dfrac{[M_2]}{r_{n-1,2}} + \cdots + \dfrac{[M_n]}{r_{n-1,n}}\right) \end{vmatrix},$$

and analogous relationships give $D_2 \ldots D_n$. This equation, called the Wallings-Briggs equation (4), allows the calculation of the composition of a copolymer formed by n monomers if the monomer concentrations and the $n(n-1)$ reactivity ratios are known. The reactivity ratios are determinable from data of binary copolymerizations. A computer program has been written using these equations to calculate the composition of multicomponent copolymer systems from a catalog of reactivity ratios. In this manner the expected composition of a variety of systems can be generated without the laborious task of carrying out the synthesis of all of the possible combinations. From the computer results an estimation of the thermal, physical, and mechanical properties of a particular multicomponent system can be made.

Experimentally, the multicomponent copolymer composition equation has been tested for the tetrapolymer system styrene–methyl metacrylate–acrylonitrile–vinyl chloride (1). The comparison between the experimental values of the composition and the calculated values using the Wallings-Briggs equation is given in Table 4.6. The reactivity ratios used are given in Table 4.7.

Table 4.6 Comparison Between the Experimental Values of the Composition of the Styrene–Methyl Methacrylate–Acrylonitrile–Vinylidene Chloride Tetrapolymer and Those Calculated by Means of the Walling-Briggs Equation

Feed		Mole-% in the Copolymer	
Monomer	Mole-%	Found	Calculated
1-Styrene	25.21	40.7	41.0
2-Methyl methacrylate	25.48	25.5	27.3
3-Acrylonitrile	25.40	25.8	24.8
4-Vinylidene chloride	23.91	8.0	6.9

Table 4.7 Monomer Reactivity Ratios

M_1	M_2	r_1	r_2
Styrene	Methyl methacrylate	0.50 ± 0.02	0.50 ± 0.02
Styrene	Vinylidene chloride	2.0 ± 0.1	0.14 ± 0.05
Methyl methacrylate	Vinylidene chloride	2.53 ± 0.1	0.24 ± 0.05
Methyl methacrylate	Acrylonitrile	1.20 ± 0.14	0.15 ± 0.07
Acrylonitrile	Vinylidene chloride	0.91 ± 0.1	0.37 ± 0.1
Styrene	Acrylonitrile	0.41 ± 0.08	0.04 ± 0.04

REFERENCES

1. T. Alfrey and G. Goldfinger, *J. Chem. Phys.*, **12**, 322 (1944).
2. C. Tosi, *Eur. Polym. J.*, **6**, 161 (1970).
3. A. Valvassori and G. Sartori, *Adv. Polym. Sic.*, **5**, 28 (1967).
4. C. Wallings and E. R. Briggs, *J. Am. Chem. Soc.*, **67**, 1774 (1945).
5. J. C. Randall, *J. Polym. Sci., Phys. Ed.*, **15**, 1451 (1977).

5

Copolymer
Microstructure and
Its Experimental
Treatment

5.1 INTRODUCTION

As the experimental research proceeds, it becomes desirable to represent
the system under study by a polymerization model. We would like to
predict the copolymer microstructure for any starting feed and to understand the kinetic features of the copolymerization. We have described a
series of models that can possibly be used in analyzing the experimental
results. The first requirement of the experimental data is a determination of
the model applicable to the particular polymerizing system. The methods of
evaluating the potential models should be simple and rapid. Later, before
one can use the model to predict results, it will be necessary to determine
the unknown parameters of the model. But with a knowledge of the model
early in the research, we can use the equations to design appropriate
experiments that provide not only further tests of the model but also the
most reliable results for calculation of the parameters.

In the principle of minimum hypothesis the simplest model (terminal)
should be tested first and, if found unsatisfactory, other models should be
pursued. One must also be conscious of the differential or integral nature
of the copolymerization. We demonstrate this approach for some systems
and later in the chapter discuss the methods of determining the parameters
of the polymerization once the proper model has been selected.

5.2 EXPERIMENTAL METHODS OF DISTINGUISHING POLYMERIZATION MODELS

Experimentally, we may be able to make measurements of the composition, dyads, and triad fractions and, in rare cases, longer sequences, so we desire relations between these sequences and the conditional probabilities for the terminal and penultimate models. These relations are shown in Table 5.1. We desire simple, fast, graphical methods so that we can visually ascertain the utility of a particular model. For more complex models like the terminal complex and depropagation models, other independent evidence for complex formation or depropagation must be available.

Table 5.1 Experimental Variables for Terminal and Penultimate Models

	Terminal	Penultimate
Copolymer composition		
$P_1\{A\}$	$P\{A/B\}/D_t$	$\left[1 + \dfrac{P\{A/BA\}}{P\{B/AA\}}\right]\Big/ D_p$
$P_1\{B\}$	$1 - P_1\{A\}$	
Dyad concentration		
$P_2\{AA\}$	$P\{A/B\}[1 - P\{B/A\}]/D_t$	$\left[\dfrac{P\{A/BA\}}{P\{B/AA\}}\right]\Big/ D_p$
$P_2\{AB\}$	$P\{A/B\}P\{B/A\}/D_t$	$1/D_p$
$P_2\{BB\}$	$1 - 2P_2\{AB\} - P_2\{AA\}$	$1 - 2P_2\{AB\} - P_2\{AA\}$

where $D_t = P\{B/A\} + P\{A/B\}$

$$D_p = \frac{P\{A/BA\}}{P\{B/AA\}} + 2 + \frac{P\{B/AB\}}{P\{A/BB\}}$$

Triad fraction

$$F_{AAA} = \frac{P_3\{AAA\}}{P_1\{A\}} = \frac{P_2\{AA\}^2}{P_1\{A\}^2} \quad [(1 - P\{B/A\})]^2 \quad P\{A/BA\}[1 - P\{A/AA\}]/D$$

$$F_{BAB} = \frac{P_3\{BAB\}}{P_1\{A\}} = \frac{P_2\{AB\}^2}{P_1\{A\}^2} \quad P\{B/A\}^2 \quad P\{B/AA\}[1 - P\{A/BA\}]/D$$

$$F_{BAA} = F_{AAB} = \frac{P_3\{AAB\}}{P_1\{A\}} = \quad P\{B/A\}(1 - P\{B/A\}) \quad P\{A/BA\}P\{B/AA\}/D$$

$$= \frac{P_2\{AA\}P_2\{AB\}}{P_1\{A\}^2}$$

where $D = P\{A/BA\} + P\{B/AA\}$

5.2.1 Copolymer Composition Measurements

In principal, the composition of copolymers is different for each of the polymerization models, but the question is whether these compositional differences are sufficiently large to ascertain whether a terminal or penultimate mechanism is operating.

Berger and Kuntz (1) have pointed out that very precise determinations of composition, particularly at low and high monomer feed ratio, are required to distinguish between the terminal and penultimate models using only measurements of composition. Their results are illustrated in Figure 5.1 for hypothetical copolymerizations. When reactivity ratios for penultimate and terminal models are used, the calculated compositions show very slight differences; only very accurate measurements of composition allow one to differentiate between the two models.

The copolymer composition equation may be written

$$\frac{P_1\{A\}}{P_1\{B\}} = \frac{1 + (r_A)x}{1 + (r_B)/x},$$

where for the terminal model the (r_A) and (r_B) are independent of x and

$$(r_A) = r_A,$$

$$(r_B) = r_B.$$

For the penultimate model

$$(r_A) = \frac{r'_A(1 + r_A x)}{(1 + r'_A x)},$$

$$(r_B) = \frac{r'_B(1 + r_B/x)}{(1 + r'_B/x)},$$

and these are dependent on the monomer feed ratio. Observe that

$$\frac{P_1\{A\}}{P_1\{B\}} = \frac{x}{r_B} \qquad \text{as } x \to 0,$$

$$\frac{P_1\{A\}}{P_1\{B\}} = r_A x \qquad \text{as } x \to \infty.$$

The penultimate effect produces only slight deviation from a straight line except at high and low values of x. So in either mode r_A and r_B can be estimated from the slopes of the $P_1\{A\}/P_1\{B\}$ versus x plot as $x \to \infty$ and $x \to 0$. Then it is possible to test whether r_A and r_B thus estimated can (terminal model) or cannot (penultimate model) describe the copolymer composition over the whole range of x employed.

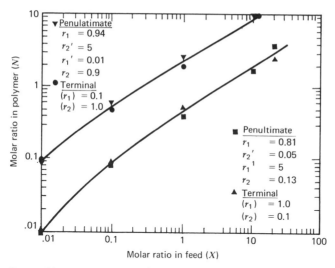

Figure 5.1 Composition versus monomer feed for two hypothetical copolymerizations using the terminal and penultimate models. (Reprinted by permission of Ref. 1.)

A useful graphical test of the polymer composition data can be obtained in linear form. If we let

$$y = \frac{P_1\{A\}}{P_1\{B\}},$$

we can arrange the composition equation into the form known as the Fineman–Ross equation (2):

$$\frac{x}{y}(y - 1) = r_A\frac{x^2}{y} - r_B,$$

where a plot of $(x/y)(y - 1)$ versus x^2/y linearizes the equation and allows a graphical test of the model. An example of a Fineman–Ross plot is shown in Figure 5.2 for the data on vinyl chloride (A) and methyl acrylate (B)

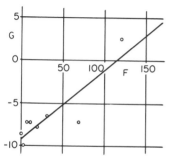

Figure 5.2 Data of the system vinyl chloride–methyl acrylate–benzoyl peroxide at 50°C, plotted according to the Fineman-Ross equation (eq. 5.1) with the data from Table 5.2. (Reprinted by permission of Ref. 4.)

given by Chapin, Ham, and Fordyce (3). Their data are given in Table 5.2.

There are several other linear forms that can be used. An obvious one is the inverted form of eq. 5.1. For this purpose let

$$G = \frac{x(y-1)}{y},$$

$$F = \frac{x^2}{y}.$$

The Fineman–Ross equation then becomes

$$G = r_A F - r_B, \tag{5.1}$$

the inverse of which is

$$\frac{G}{F} = -r_B \frac{1}{F} + r_A. \tag{5.2}$$

Figure 5.3 shows the plot of the data of Table 5.2 for eq. 5.2. Other forms include

$$F = \frac{1}{r_A} G + \frac{r_B}{r_A}, \tag{5.3}$$

which is graphically illustrated in Figure 5.4 for the data of Table 5.2. Finally, another linear form is

$$\frac{F}{G} = \frac{r_B}{r_A} \frac{1}{G} + \frac{1}{r_A} \tag{5.4}$$

which is shown graphically in Figure 5.5. None of these equations are invariant to the inversion of the data. By reindexing the monomers, for F we have $1/F$, for G we obtain $-G/F$, and for r_A we derive r_B and vice versa.

Table 5.2 Composition of Polymer for Various Feeds (3)

Vinyl Chloride, [A] (mole fraction)	Methyl Acrylate, [B] (mole fraction)	m_B in Polymer (mole fraction)
0.925	0.075	0.441
0.846	0.154	0.699
0.763	0.237	0.753
0.674	0.326	0.828
0.579	0.421	0.864
0.479	0.521	0.900
0.256	0.744	0.968
0.133	0.867	0.983

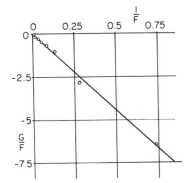

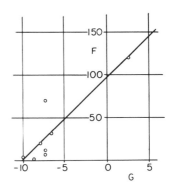

Figure 5.3 Data of the system vinyl chloride–methyl acrylate–benzoyl peroxide at 50°C, plotted according to eq. 5.2 with the data from Table 5.2. (Reprinted by permission of Ref. 4.)

Figure 5.4 Data of the system vinyl chloride–methyl acrylate–benzoyl peroxide at 50°C, plotted according to eq. 5.3 with the data from Table 5.2. (Reprinted by permission of Ref. 4.)

For eq. 5.3 the corresponding equations become

$$\frac{1}{F} = -\frac{1}{r_B}\frac{G}{F} + \frac{r_A}{r_B}, \tag{5.5}$$

and for eq. 5.4 we get

$$\frac{1}{G} = \frac{r_A}{r_B}\frac{F}{G} - \frac{1}{r_B}. \tag{5.6}$$

The data in Table 5.2 were used to calculate the corresponding plots, which are given in Figures 5.6 and 5.7, respectively. These figures (5.2 through 5.7) show considerable differences as a result of the various weightings given to the experimental data.

Although the Fineman–Ross equation and the corresponding linear forms are useful for a visual graphical testing of the polymer composition

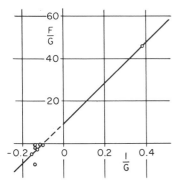

Figure 5.5 Data of the system vinyl chloride–methyl acrylate–benzoyl peroxide at 50°C, plotted according to eq. 5.5 with the data from Table 5.2. (Reprinted by permission of Ref. 4.)

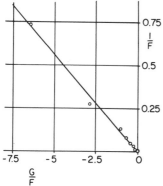

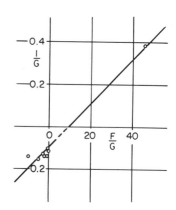

Figure 5.6 Data of the system vinyl chloride–methyl acrylate–benzoyl peroxide at 50°C, plotted according to eq. 5.4 with the data from Table 5.2. (Reprinted by permission of Ref. 4.)

Figure 5.7 Data of the system vinyl chloride–methyl acrylate–benzoyl peroxide at 50°C, plotted according to eq. 5.6 with the data from Table 5.2. (Reprinted by permission of Ref. 4.)

data, they should not be used to determine the reactivity coefficients except for a "guestimation" of the magnitude of the reactivity ratios. There are several reasons for not using these linear forms of the composition equations for the determination of the reactivity ratios. These problems include the fact that the equations are not symmetrical to the method of designating the monomers and as a result give two different solutions for the same data. This effect is illustrated in Table 5.3, where the reactivity ratios are determined. Secondly, in the linear form not all the experimental data are uniformly weighted in the determination of the slope. In particular the polymer data obtained at the lowest comonomer concentrations have the greatest influence on the slope of the line. Finally, and perhaps most important, no valid expression can be presumed for the errors involved in the estimation of the reactivity. Both the "independent" and dependent variables that are plotted contain the observed response (i.e., the polymer composition). The reactivity ratios obtained by these equations, given in Table 5.3, illustrate some of the problems. The method of least squares was used to obtain the best straight line through the data, yet still a considerable range of values is obtained for the reactivity coefficients. This result simply reinforces the caution that these linear, graphical equations should only be used for *visual* testing of the model and *not* for *evaluation* of parameters.

When the linear, graphical equations indicate that perhaps the terminal model does not apply, one turns to models of higher order.

In the case for the penultimate model the penultimate composition equation can be arranged in a linear graphical form only when $r_B = r_B'$,

Table 5.3 Graphical Equations for Reactivity Ratios

Equation	r_A	r_B	$\Delta_1 \times 10^{2a}$	$\Delta_2 \times 10^{2b}$
$G = r_A F - r_B$	0.08266	9.337	1.318	0.665
$\dfrac{G}{F} = -r_B \dfrac{1}{F} + r_A$	0.05273	8.931	1.694	0.897
$F = \dfrac{1}{r_A} G + \dfrac{r_B}{r_A}$	0.10484	10.077	1.181	0.600
$\dfrac{F}{G} = \dfrac{r_B}{r_A} \dfrac{1}{G} + \dfrac{1}{r_A}$	0.10614	10.147	1.190	0.608
$\dfrac{1}{F} = -\dfrac{1}{r_B} \dfrac{G}{F} + \dfrac{r_A}{r_B}$	0.06063	8.978	1.573	0.819
$\dfrac{1}{G} = \dfrac{r_A}{r_B} \dfrac{F}{G} - \dfrac{1}{r_B}$	0.10461	10.223	1.251	0.625

[a] Δ_1 is the difference (average absolute value of deviations) between calculated and measured values of mole fractions $y/(1 + y)$.
[b] Δ_2 is the deviation between calculated and measured chlorine content (average absolute value of deviations).

which is

$$\frac{1}{x}(y - 2) = r_A - \frac{1}{r_A}\left(\frac{1}{x}\right)^2 (y - 1), \qquad (5.7)$$

where

$$r_A = \frac{k_{AAA}}{k_{AAB}}, \qquad r'_A = \frac{k_{BAA}}{k_{BAB}}.$$

This equation requires extremely accurate values of y, and small errors in y produce substantial changes in the curve. An example of this type of data treatment is shown in Figure 5.8 for the data given in Table 5.4 (16). The differences in polymer composition as a function of monomer feed for an ideal copolymerization and for terminal reactivity coefficients of similar magnitude, with no penultimate effect, as well as the results obtained from eq. 5.7 from experiment, are compared with the curve for the calculated penultimate effect in Figure 5.9. The improvement for the penultimate model is obvious (16).

Composition measurements are not to be recommended for the determination of the copolymer polymerization model. The extensive use in the literature is generally a result of the composition being the only information available. With modern instrumentation sequence distributions can be measured and used more effectively.

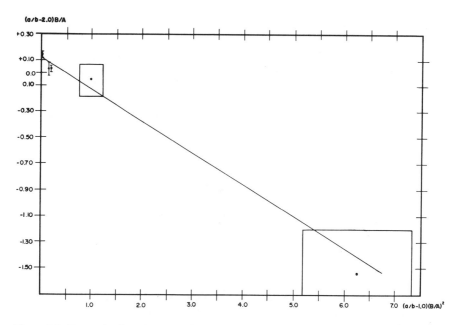

Figure 5.8 Determination of r and r' for the VAc–6FK system using eq. 5.7, $r = 0.12$, $r' = 4.1$. Boxed in points are experimental points including an estimate of the minimum error. (Reprinted by permission of Ref. 16.)

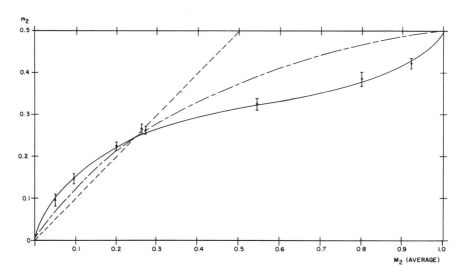

Figure 5.9 Copolymerization of vinyl acetate with 4FK (M_2); (points with bracketed lines) mole fraction of 4FK in polymer (M_2), minimum error shown; (unbroken line) theoretical curve for $r = 0.41$, $r' = 5.6$; (dashed line) line for $r_1 = r_2 = 1$; (longshort dashed line) theoretical curve for $r = r' = 0.7$ (no penultimate effect). (Reprinted by permission of Ref. 16.)

96

Table 5.4 Copolymerization of Vinyl Acetate and 4 FK at 60°C (16)

Run	Initial Monomer Feed (mole fraction)		W-% Conversion	Polymer Composition[a] (mole fraction)		Average Monomer Feed (mole fraction)	
	VAc	4 FK		VAc	4 FK	VAc	4 FK
5.1	0.950	0.050	9.3	0.912 ± 0.010	0.088 ± 0.010	0.952	0.0482
5.2	0.900	0.100	8.8	0.858 ± 0.011	0.142 ± 0.011	0.908	0.0925
5.3	0.800	0.200	10.4	0.779 ± 0.008	0.221 ± 0.008	0.801	0.199
5.4	0.750	0.250	4.9	0.752 ± 0.007	0.248 ± 0.007	0.750	0.250
3.2	0.750	0.250	70.	0.748 ± 0.011	0.252 ± 0.011	0.755	0.245
3.3	0.500	0.500	35.	0.675 ± 0.012	0.325 ± 0.012	0.458	0.542
3.4	0.250	0.750	19.4	0.613 ± 0.018	0.387 ± 0.018	0.201	0.799
3.5	0.100	0.900	6.0	0.576 ± 0.012	0.424 ± 0.012	0.079	0.921

[a] The composition was taken as the average of carbon, chlorine, and hydrogen analyses. Hydrogen analyses generally gave results 0.01–0.02 higher in mole fraction of vinyl acetate than the other two.

5.2.2 Copolymer Sequence Measurements

With chemical and physical techniques it is feasible to make quantitative measurements of various copolymer sequences. We desire to compare these experimental sequences with the predictions of the models. We want to test the copolymerization mechanism and ultimately determine the parameters controlling the copolymerization.

In general, as we have seen,

$$P_n(A^n) = P(A)P(A/A)P(A/AA)\ldots P(A/A^{n-1})$$
$$= P(A^k)P(A^{n-k}/A^k),$$

and we need to test whether

$k = 0$	random (Bernoullian),
$k = 1$	terminal (first order Markovian),
$k = 2$	penultimate (second order Markovian),
$k = 3$	antepenultimate (third order Markovian),
$\vdots$	$\vdots$
$k = n$	(nth order Markovian),

or whether the process is not a finite Markovian chain and is therefore a non-Markovian process (i.e., terminal complex, etc.). A Markovian sequence distribution of order k has the property that

$$\frac{P[B(A_{n+1})B]}{P[B(A_n)B]} = \text{constant} < 1$$

for all $n = k$. For sequences lower than A^k the data can be fit to a kth order Markovian chain. Ratios of other similarly related sequences must also obey this rule. From chromatographic measurements of the propylene oxide–maleic anhydride copolymer, the sequences $P[B(A_n)B]$ for n up to 11 can be measured from these data, and the conditional probabilities have been calculated (Table 5.5). Above $n \geq 3$ and up to $n = 9$, a constant within experimental error is obtained, indicating that the process is third order Markovian. The data for $n \geq 9$ are unreliable due to accumulation of error (19).

By *tested*, we mean that the data can be shown to be consistent or inconsistent with a given mechanism, at a given level of sequence discrimination. Thus dyads are required to *test* the Bernoullian model, triads the terminal, and tetrads the penultimate model. But the composition can be made to *fit* the Bernoullian composition; and the dyads, the terminal; and the composition, dyads, and the triads, the penultimate. If we can measure these sequences as a function of composition, the data allow the testing of the models that could only be fit at a given composition. But there is always the possibility that measurement of longer sequences will reveal inconsistencies between the predictions and actual sequence measurements.

Table 5.5 Conditional Probabilities for
Propylene Oxide–Maleic Anhydride
Copolymers (19)

N Sequence Length	$P(BA_n/B)$
0	0
1	0.025
2	0.22
3	0.30
4	0.25
5	0.26
6	0.28
7	0.31
8	0.30
9	0.43
10	0.6
11	1.0

From the measurement of composition $P(A)$ or $P(B)$ we can only write

$$P_1(A) = P(A) = P,$$

so from the composition we can fit the random process but no other models. If we make a series of composition measurements at different monomer feed ratios, we can *test* the random process through the relation

$$y = \frac{P(A)}{P(B)} = r\frac{[A]}{[B]} = rx,$$

and can *fit* the composition data to the terminal and penultimate models as seen in the preceding section.

From the measurement of the comonomer pairs and the composition, we can fit the terminal ($k = 1$) model using

$$P(A/A) = \frac{P_2(AA)}{P(A)},$$

$$P(B/A) = \frac{P_2(BA)}{2P(A)},$$

$$P(A/B) = \frac{P_2(BA)}{2P(B)},$$

$$P(B/B) = \frac{P_2(BB)}{P(B)},$$

and test the random or Bernoulli process if $P(A/B) = P(A/A)$ or $P(B/A) = P(B/B)$. The higher order Markov chains can also be made to fit the data for dyads and the composition.

From the measurement of the monomer dyad sequences in the copolymers, the reactivity ratios for the terminal model can be calculated directly from

$$r_A = \frac{2}{x} \frac{P_2(AA)}{P_2(AB)},$$

$$r_B = 2\frac{P_2(BB)}{P_2(AB)} x, \tag{5.8}$$

and the calculated reactivity ratios should be the same for samples prepared at different monomer feeds. The measured dyad fractions were measured by NMR spectroscopy for the vinylidene chloride–methacrylonitrile system in solution (18). The calculated reactivity ratios, plotted as a function of mole fraction VDC in the feed, are shown in Figure 5.10. The variation indicates that the terminal model does not apply over the entire composition range.

If the pair and composition measurements are made for a range of comonomer feed ratios, the terminal model can be *tested*, as it is required

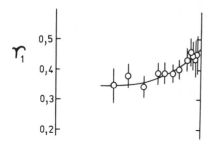

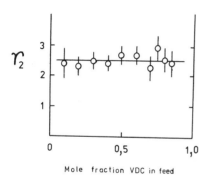

Mole fraction VDC in feed

Figure 5.10 Reactivity ratios calculated from measured dyad distributions as a function of mole fraction VDC in the field for copolymers prepared from vinylidene chloride and methacrylonitrile in solution. (Reprinted by permission of Ref. 18.)

that

$$\bar{l}_A = \frac{1}{1 - \alpha_A} = 1 + r_A x,$$

$$\bar{l}_B = \frac{1}{1 - \alpha_B} = 1 + \frac{r_B}{x}, \qquad (5.10)$$

where $\alpha_A = P_2(AA)/P(A)$ and $\alpha_B = P_2(BB)/P(B)$, which yields a linear relationship, whereas for the penultimate model

$$\bar{l}_A = 1 + R_A x, \qquad \text{where } R_A = \frac{1 + r_A x}{1 + r'_A x} r'_A,$$

and (5.11)

$$\bar{l}_B = 1 + R_B/x, \qquad \text{where } R_B = \frac{r_B + x}{r'_B + x} r'_B.$$

If $\bar{l}_A$ is plotted against x, it gives an intercept of unity. Then a straight line should be obtained with a slope r_A in the case of a terminal model, whereas a concave or a convex curve should be obtained in the penultimate case depending on whether $r_A > r'_A$ or $r_A < r'_A$.

For the penultimate case observe that

$$\frac{d\bar{l}_A}{dx} = r_A - \frac{r_A - r'_A}{(1 + r'_A x)^2}.$$

Thus as shown in Fig. 5.11 for the penultimate model the slope as $x \to 0$ will be r'_A, and similarly for l_B. Therefore r'_A and r_B can be obtained from the slopes of the curves.

A linear relationship for the determination of values for r can be made by noting that

$$\frac{l_A - 2}{x} = r_A - \left(\frac{1}{r'_A}\right)\left(\frac{l_A - 1}{x^2}\right),$$

$$\frac{l_B - 2}{x} = r_B - \left(\frac{1}{r'_B}\right)(l_B - 1)x^2. \qquad (5.12)$$

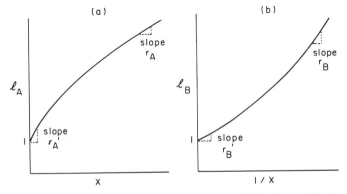

Figure 5.11 Plots of mean run length versus monomer feed ratio in the penultimate model: (a) l_A; (b) l_B. (Reprinted by permission of Ref. 17.)

Table 5.6   Data on Copolymer Composition and Microstructure (17)

No.	x	$P_1\{A\}$	$P_1\{B\}$	$P_2\{AA\}$	$P(AB)$	$\bar{l}_A$	$\bar{l}_B$	$P(AB)_{calc}$
1	3	0.168	0.832	0.0643	0.208	1.62	8.00	0.268
2	4	0.189	0.811	0.0563	0.266	1.42	6.10	0.322
3	9	0.388	0.612	0.225	0.326	2.38	3.76	0.472
4	19	0.592	0.408	0.425	0.334	3.54	2.44	0.528

Since each experiment gives values of l_A, l_B, and x, direct plots should approximately determine the four reactivity ratios.

Consequently, determination of copolymer composition and any one kind of dyad concentration over a range of x would permit estimation of four monomer reactivity ratios for the penultimate model.

An example of the use of these equations is given for the copolymerization of β-propialacetone and 3,3-bischloromethyloxetone (17). The composition and dyad data are given in Table 5.6 (17). The plots of average sequence length versus monomer feed are shown in Figure 5.12. The curvature expected for a penultimate effect at low monomer feed is observed. Finally, the appropriate functionality of the average sequence length is plotted in Figure 5.13, where straight lines are obtained that are suitable for determining the penultimate reactivity ratios. Table 5.7 shows a comparison of the experimental and calculated values based on the determined reactivity coefficients for each model (17). We can calculate the necessary terminal parameters from any one of the following sets of

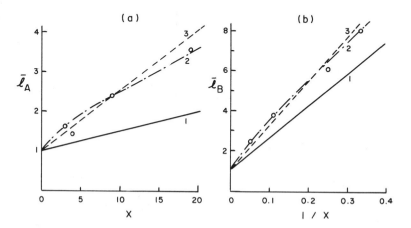

Figure 5.12   Plots of mean sequence length versus monomer feed ratio: (a) $\bar{l}_A$; (b) $\bar{l}_B$; (open circles) experimental data points. Plots were calculated with: (1) $r_A = 0.05$, $r_B = 16$; (2) $r_A = 0.10$, $r'_A = 0.30$, $r_B = 20$, $r'_B = 45$; (3) $(\bar{r}_A) = 0.152$, $(\bar{r}_B) = 22.5$. (Reprinted by permission of Ref. 17.)

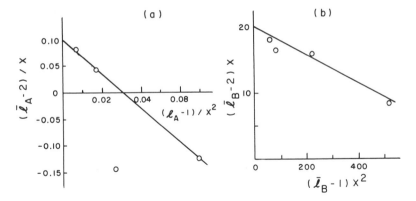

Figure 5.13 Plots according to eq. 5.12 to determine the four monomer reactivity ratios: (*a*) r_A and r'_A; (*b*) r_B and r'_B. (Reprinted by permission of Ref. 17.)

information:

(a) one $P(X)$ and one $P(XY)$,
(b) any two independent $P(XY)$,
(c) average sequence lengths, $\bar{l}_A$ and $\bar{l}_B$,
(d) $\bar{l}_A$ and any one of the $P(A)$, $P(AB)$, but not $\bar{l}_A$ and $P(BB)$,
(e) $\bar{l}_B$ and any one of the $P(X)$, $P(BX)$, but not $\bar{l}_B$ and $P(AA)$,
(f) $P_2(AB)$ and one of the $P(X)$, $P(XX)$.

From these experimental results we can calculate sequences of any length $P_n(X_n)$ for a first order Markovian distribution.

Table 5.7 Experimental and Theoretical Copolymer Composition and Dyad Concentration (17)

		$P_1\{A\}$			$P_2\{AA\}$		
			Calculated			Calculated	
No.	x	Ob-served	Termi-nal[a]	Penulti-mate[b]	Ob-served	Termi-nal	Penulti-mate
1	3	0.168	0.154	0.165	0.0643	0.0201	0.0627
2	4	0.189	0.194	0.213	0.0563	0.0323	0.0923
3	9	0.388	0.343	0.393	0.225	0.107	0.225
4	19	0.592	0.514	0.591	0.425	0.250	

[a] (terminal) with $r_A = 0.05$, $r_B = 16$.
[b] (penultimate) with $r_A = 0.10$, $r'_A = 0.30$, $r_B = 20$, $r'_B = 45$.

The dyad measurements allow one to determine the deviation from random statistics using the persistance ratio. The persistance ratio ρ is the ratio of the actual mean length of closed A (or B) sequences to the mean length that one could calculate for a random or Bernoullian distribution with the same values of $P(A)$ or $P(B)$. The persistance ratio is defined by

$$\rho = \bar{l}_A P(B) = \bar{l}_B P(A), \tag{5.13}$$

which may be expressed as

$$\rho = \frac{P(A)}{P(A/B)} = \frac{P(B)}{P(B/A)} = \frac{1 - P(A)}{1 - P(A/A)} = \frac{1 - P(B)}{1 - P(B/B)},$$

and has the range $\frac{1}{2} < \rho < \infty$. When $\rho = 1$, $P(A/A) = P(A/B) = P(A)$ and $P(B/A) = P(B/B) = P(B)$. (This is the random case.) When $\rho = \frac{1}{2}$, $P(A/B) = P(B/A) = 1$, $\bar{l}_A = \bar{l}_B = 1$, and $P_2(AB) = 1$. (This is the completely alternating case.) When $\rho = \infty$, $P(A/B)$ and $P(B/A) <<< 1$, $\bar{l}_A$ and $\bar{l}_B \to \infty$, and $P_2(AB) \to 0$. (This is the long blocks case.)

From the triad sequence (which also implies the dyad measurements) the penultimate model can be *fit*, and the Bernoulli or terminal model *tested*. The penultimate parameters are determined through

$$P(AA/A) = \frac{P_3(AAA)}{P_2(AA)},$$

and so on.

Deviation from the terminal or first order Markovian case may be discussed in terms of the following factors:

$$\Omega_A = \frac{P(A/AA)}{P(A/A)} = \frac{P_3(AAA)P(A)}{[P(AA)]^2},$$

$$\Omega_A' = \frac{P(B/AA)}{P(B/A)}, \tag{5.14}$$

$$\Omega_A'' = \frac{P(B/BA)}{P(B/A)},$$

with similar factors for the B monomer.

When $\Omega_A = \Omega_A' = \Omega_A'' = 1$, then all of the ratios are unity, signifying that the sequence distribution is a result of a terminal process up to $P_3(X_3)$ if $\rho \neq 1$, and random or Bernoullian if $\rho = 1$.

From measurements of the triad fractions as a function of monomer feed ratios, the penultimate model can be tested and the antepenultimate model fit. Extension of these results to measurement of longer sequences is obvious.

To summarize, we can draw the following conclusions about a sequence distribution:

1 If

$$\rho = \Omega_A = \Omega_B = 1,$$

then the system is random up to $P_3(X_3)$, most likely up to any $P_n(X_n)$.

2 If

$$\rho \neq 1, \qquad \Omega_A = \Omega_B = 1,$$

then the system is terminal at least up to $P(X_3)$ and most likely up to $P_n(X_n)$.

3 If

$$\rho \neq 1, \qquad \Omega_A \neq 1 \text{ and/or } \Omega_B = 1,$$

then the system is not terminal and is either penultimate or non-Markovian.

5.3 COPOLYMERIZATION REACTIVITY RATIOS FROM COMPOSITION

5.3.1 Preliminary Considerations

There are several purposes in determining the reactivity ratios with precision for any copolymerization system. First, we can predict copolymer composition for any starting feed. This goal obviously requires a good fit of the experimental points with the theoretical curve of copolymer composition versus monomer feed. Secondly, with proper reactivity ratios we can understand some aspects of the kinetic features of the copolymerization. Finally, copolymerization reactivity ratios are measured and tabulated for the purpose of describing and classifying the relative reactivities of various monomers toward other reactive chains for different types of polymerization including free radical, ionic, and coordination addition polymerization. This result allows the estimation of the composition of new copolymers without the necessity of experimentation.

According to Tidwell and Mortimer (6) a good estimation procedure for copolymerization reactivity ratios should have the following properties:

1. The method should give unbiased estimates of the parameters.
2. The method should utilize all (or nearly all) the information resident in the data with regard to the parameters to be estimated, thus providing precise estimates.
3. The parameter values calculated by the method should not depend on arbitrary factors (such as which monomer is subscripted 1 and the starting point of the calculation).

4. The method should supply a valid measure of the errors of the resulting estimates.
5. The method should be reasonably easy to use.

The method of least squares is a suitable method if the random errors associated with the observations are independent and normally distributed with constant variances. With the availability of digital computers and suitable computer programs, the criteria of ease of estimation using least squares techniques is also satisfied. By using least squares methods, the experimenter tacitly makes the following four assumptions:

1. The mathematical model adequately describes the data.
2. The random errors in the dependent variable (generally the polymer composition) are statistically independent from run to run.
3. The variance of the dependent variable is constant.
4. The random errors in the dependent variable are normally distributed.

Examination of the literature reveals that the question of the adequacy of the model is usually not tested, particularly the decision as to whether it is necessary to follow the more tedious route of integrating the model or whether it is sufficient merely to use the more convenient differential model. A useful test is to compare the residuals, the differences between the observed and computed results as a function of monomer feed. If these residuals exhibit a nonrandom pattern, there is reason to believe that the model is inadequate.

The other assumptions simply state that no mathematical analysis can hope to succeed if it uses imprecise polymer assay methods. The consequences are greatest when there is a bias in the polymer analysis. The possibility of bias arising because of some consistent error in the values for the polymer composition is particularly difficult to ferret out. The only suitable procedure is to compare two or more analytical methods of determining the polymer composition. This method is seldom, if ever, used due to the increased effort and the general lack of ability to do analysis by multiple techniques. The availability of primary standards, that is, copolymers with exactly known compositions, would be useful in properly calibrating polymer assay methods. The use of mass or material balance is a valuable method to discover bias in polymer composition and should be used whenever methods are available to determine the amount of unreacted monomer.

The replication of experimental runs made with proper care can be successful in detecting problems in analysis and should be a requirement of all sound experimental work in copolymerization.

Finally, it is often overlooked that the possibility of bias can arise because of consistent error in the values of the concentration of the starting monomer. It is universal practice to assume that no error whatever exists in the starting monomer concentrations. The possibility of bias is only compounded by the use of a common solution to prepare different initial molar ratios of monomer for the polymerizations. The use of the differential form of the copolymer equation, where the conversion is sufficient to cause appreciable change in the fraction of unreacted monomers, causes the reactivities to be biased. The extent of bias is dependent on the initial monomer concentration, the actual reaction rates, and the conversion.

5.3.2 Linear Least Squares Methods

If we adopt the least squares criterion, it can easily be applied to all the above equations and put in a form that can be evaluated numerically. For the Fineman–Ross equation (eq. 5.1) the following pairs of equations are used to determine the values of r_A and r_B:

$$r_A \sum_{i=1}^{n} F_i^2 - r_B \sum F_i = \sum_{i}^{n} F_i G_i, \tag{5.15}$$

$$r_A \sum_{i=1}^{n} F_i - r_B n = \sum_{i=1}^{n} G_i, \tag{5.16}$$

where n is the number of experimental data. If one reindexes the monomer, one obtains

$$r_A \sum \frac{1}{F} - r_B \sum \frac{1}{F^2} = \sum \frac{G}{F^2}, \tag{5.17}$$

$$r_A n - r_B \sum \frac{1}{F} = \sum \frac{G}{F}. \tag{5.18}$$

Combining eqs. 5.15 and 5.17, one finds

$$r_A = \frac{\sum FG \sum \frac{1}{F^2} - \sum \frac{G}{F^2} \sum F}{\sum F^2 \sum \frac{1}{F^2} - \sum \frac{1}{F} \sum F},$$

$$r_B = \frac{\sum FG \sum \frac{1}{F} - \sum \frac{G}{F^2} \sum F^2}{\sum F^2 \sum \frac{1}{F^2} - \sum \frac{1}{F} \sum F}. \tag{5.19}$$

Additional equations of this type can be obtained by combining the other linear equations. However, this method can lead to physically impossible (negative) copolymerization constants because the experimental data are unequally weighted, and it is only recommended that this procedure be used to obtain initial values for use in the nonlinear least squares method.

Kelen and Tudos (4) have suggested an alternative equation that corrects for most of the above-mentioned deficiencies. They recommend the equation

$$\frac{G}{\alpha + F} = \left(r_A + \frac{r_B}{\alpha}\right)\frac{F}{\alpha + F} - \frac{f_B}{\alpha},$$ (5.20)

where α denotes an arbitrary constant ($\alpha > 0$). This relationship is invariant to inversion of the data, and a uniform distribution of the experimental data in the interval $(0, 1)$ may be obtained by the proper choice of α. If the reactivity ratios are nearly equal, a value of $\alpha = 1$ is generally satisfactory. A recommended selection of α for markedly different reactivity ratios is

$$\alpha = \sqrt{F_m F_M},$$

where F_m stands for the lowest and F_M for the highest value. Applying the method of least squares yields

$$r_A = \frac{\sum \eta\xi\left(\eta - \sum \xi\right) - \sum \eta\left(\sum \xi - \sum \xi^2\right)}{\eta \sum \xi^2 - \left(\sum \xi\right)^2},$$

$$r_B = \alpha \frac{\sum \eta\xi \sum \xi - \sum \eta \sum \xi^2}{\eta \sum \xi^2 - \left(\sum \xi\right)^2},$$ (5.21)

where

$$\eta = \frac{G}{\alpha + F},$$

$$\xi = \frac{F}{\alpha + F}.$$

Another invariant linear equation that has been found to be useful is the equation suggested by Yezrielev, Brokhina, and Roskin (YBR) (7):

$$\frac{G}{\sqrt{F}} = r_A F - r_B \frac{1}{\sqrt{F}}.$$ (5.22)

This equation is the combination of the Finneman–Ross equation and its reversed monomer form. The equation pair for the YBR relation is

$$r_A \sum F - r_B n = \sum G,$$

$$r_A n - r_B \sum \frac{1}{F} = \sum \frac{G}{F}.$$ (5.23)

Since these linear methods are simpler than the nonlinear method to be described next, answers can be obtained more quickly and easily. However, all of the deficiencies of the lineariezed methods are not eliminated by the application of the method of least squares, so caution must be exercized.

5.3.3 Nonlinear Least Squares Method (9)

This method, based on the Gauss–Newton nonlinear least squares procedure of determining the reactivity ratios, has the virtue of allowing one to calculate through an iteration procedure the reactivity ratios and an estimation of whether the data are consistent with the assumption that the copolymer equation describes the relationship between the monomer and polymer composition. Given initial estimates of r_A and r_B, a set of computations is performed that, on repetition, gives a pair of values of the reactivity ratios, yielding a minimum value for the sum of the squares for the differences between the observed and computed polymer composition. Additionally, it provides the joint confidence limits after r_A and r_B are calculated.

One does a least squares refinement to reduce $S = \Sigma(d_i)^2$ to a minimum where

$$d = m_{Bi} - G_i^j \tag{5.24}$$

where m_{Bi} is the observed mole fraction of M_B in the polymer resulting from the experimental run and

$$G^j = \frac{r_B^j M_B^2 + M_B M_A}{f_B^j M_B^2 + 2M_A M_B + r_A^j M_A^2},$$

where M_A and M_B are the initial mole fractions of A and B, respectively, in the feed and r_B^j is the jth estimate of r_B. Then let G_i^j, $\partial G_i^j / \partial r_A$, and $\partial G_i^j / \partial r_B$ be the values for each of the n experimental runs for the jth set of estimates of r_A and r_B. Then

$$m_{Bi} = G_i^j + \frac{\partial G_i^j}{\partial r_A}(r_A^0 - r_1^j) + \frac{\partial G_i^j}{\partial r_B}(r_B^0 + r_B^j) + \epsilon_i, \tag{5.25}$$

where ϵ_i is a random variable, and r_A^0 and r_B^0 are the expectations of r_A^j and r_B^j, respectively. Then one computes the least squares estimates of β_A and β_B for the equation

$$d_i = m_{Bi} - G_i^j = \beta_A \frac{\partial G_i^j}{\partial r_A} + \beta_B \frac{\partial G_i^j}{\partial r_B} + \epsilon_i \tag{5.26}$$

Let $\hat{b}_A + \hat{b}_B$ be least squares estimates of β_A and β_B such that one can use

$$r_A^{j+1} = r_A^j + \hat{b}_a, \tag{5.27}$$

$$r_B^{j+1} = r_B^j + \hat{b}_B.$$

in eq. 5.26 to decrease the value of $\Sigma(d_i)^2$. One repeats the above calculations until $\Sigma(d_i)^2$ is reduced to its minimum value. The rate of convergence is increased if one calculates

$$S_h = \left[\sum (d_i)^2 \right]_h$$

for

$$r_A = r'_A + \left(\frac{h-1}{2} \right) \hat{b}_A, \tag{5.28}$$

$$r_B = r'_B + \left(\frac{h-1}{2} \right) \hat{b}_B,$$

where $h = 1, 2, 3$.

Let

$$V = \frac{1}{2} + \frac{S_1 - S_3}{4(S_1 - 2S_2 + S_3)}.$$

Then compute $S_4 = [\sum (d_i)^2]_4$ for

$$r_A = r'_A + V\hat{b}_A, \tag{5.29}$$

$$r_B = r'_B + V\hat{b}_B.$$

If $S_4 < S_1$, repeat the process, replacing r'_A and r'_B with new values calculated above. If $S_4 > S_1$, then reevaluate V first, halving $\hat{b}_A$ and $\hat{b}_B$. The method is illustrated in Table 5.8. When the data of Table 5.2 were used, a total of five iterations were required to reduce the $\sum(d_i)^2$ to its minimum value (Table 5.9), but only minor adjustments were made in the values of r_A and r_B after the second iteration. A computer program is available for carrying out this method of calculation (6).

The reactivity coefficients calculated from the data in Table 5.2 by the methods described are shown in Table 5.10. The agreement is quite satisfactory.

The least squares procedure provides a method for establishing the joint confidence limits within which the correct values of the reactivity ratios can be asserted to lie with probability $1 - \alpha$, where α is some number in the interval $0 < \alpha < 1$.

The approximate $100(1 - \alpha)$ percent joint confidence limits are the set of r'_1 and r'_2 which satisfy the equation

$$S_c = \left[\sum (d_i)^2 \right]_{\min} + 2S^2 F_\alpha(2, h) \tag{5.30}$$

where S^2 is an estimate of the experimental error, σ^2, having h degrees of freedom, $F_\alpha(2, h)$ is the critical value of F taken from tables of the F distribution, and

$$S_c = \sum (m_{Bi} - G[M_{Bi}; r'_A, r'_B])^2. \tag{5.31}$$

Table 5.8 Illustration of the Calculation Scheme

			First Iteration			Second Iteration			Third Iteration		
M_1	M_2	m_2	$10^2(d)$	$\partial G^1/\partial r_1$	$10^2(\partial G^1/\partial r_2)$	$10^2(d)$	$\partial G^2/\partial r_1$	$10^2(\partial G^2/\partial r_2)$	$10^2(d)$	$\partial G^2/\partial r_1$	$10^2(\partial G^2/\partial r_3)$
0.925	0.075	0.441	−1.984 3872	−1.514 2927	1.164 6881	−2.073 1045	−1.452 0092	1.112 8040	−2.033 2361	−1.445 3189	1.109 4550
	0.154	0.699	5.461 0395	−0.864 6152	1.581 0642	4.540 9746	−0.832 0270	1.461 2437	4.510 7696	−0.829 3356	1.454 5748
0.763	0.237	0.753	0.330 0674	−0.476 7343	1.535 6736	−0.794 9410	−0.453 9674	1.375 9614	−0.848 6722	−0.452 3962	1.367 1520
	0.326	0.828	0.756 3530	−0.259 9714	1.331 1106	−0.310 1401	−0.244 6504	1.163 1438	−0.366 2139	−0.243 6752	1.153 8826
0.579	0.421	0.864	−0.731 8432	−0.138 4032	1.080 6690	−1.638 1489	−0.128 8666	0.925 7098	−1.687 8083	−0.128 2829	0.917 1741
0.479	0.521	0.900	−0.929 0528	−0.070 4557	0.831 5158	−1.645 2383	−0.065 0106	0.701 1529	−1.685 3377	−0.064 6845	0.693 9842
0.256	0.744	0.968	0.449 3382	−0.011 7627	0.367 2981	0.115 5395	−0.010 6984	0.309 8731	0.096 4452	−0.010 6362	0.306 2355
0.133	0.867	0.983	−0.031 0805	−0.002 4858	0.179 2849	−0.191 5019	−0.002 2481	0.146 3176	−0.200 7307	−0.002 2343	0.144 5159
$\Sigma(\partial G^1/\partial r_2)^2$				3.359 7667			3.087 4926			3.961 5429	
$\Sigma(\partial G^1/\partial r_2)^2$				$1.001\,9418 \times 10^{-3}$			$8.085\,7241 \times 10^{-4}$			$7.984\,7157 \times 10^{-4}$	
$\Sigma(\partial G^1/\partial r_1)(\partial G^1/\partial r_1)$				$-4.421\,8750 \times 10^{-2}$			$-3.909\,3209 \times 10^{-2}$			$-3.875\,6420 \times 10^{-3}$	
$\Sigma(\partial G^1/\partial r_1)(d)$				$-1.909\,2007 \times 10^{-2}$			$-1.403\,8255 \times 10^{-4}$			$-4.138\,8724 \times 10^{-3}$	
$\Sigma(\partial G^1/\partial r_2)(d)$				$6.436\,8377 \times 10^{-4}$			$2.117\,3134 \times 10^{-6}$			$5.550\,2101 \times 10^{-7}$	
$\Sigma(d)^2$				$3.604\,3364 \times 10^{-3}$			$3.108\,6694 \times 10^{-3}$			$3.107\,4103 \times 10^{-8}$	
b_1				$6.615\,0633 \times 10^{-2}$			$7.376\,9994 \times 10^{-4}$			$-1.224\,1012 \times 10^{-4}$	
b_2				$9.343\,7913 \times 10^{-1}$			$6.185\,2830 \times 10^{-2}$			$1.009\,4684 \times 10^{-4}$	
S_1				$3.604\,3364 \times 10^{-3}$			$3.108\,6694 \times 10^{-3}$			$3.107\,4103 \times 10^{-3}$	
S_2				$3.251\,2085 \times 10^{-3}$			$3.107\,7535 \times 10^{-3}$			$3.107\,4099 \times 10^{-3}$	
S_3				$3.113\,1094 \times 10^{-3}$			$3.107\,4134 \times 10^{-3}$			$3.107\,4101 \times 10^{-3}$	
V				1.071 1179			1.045 2874			0.574 8503	
S_4				$3.108\,6694 \times 10^{-3}$			$3.107\,4103 \times 10^{-3}$			$3.107\,4101 \times 10^{-3}$	
r_1^{t+1}				0.090 0855			0.090 8566			0.090 8495	
r_2^{t+1}				10.000 830			10.065 484			10.065 542	

Table 5.9   Results Obtained on Each Iteration on Data of Table 5.2

j	r_1^j	r_2^j	$\Sigma(d_i)^2$
1	0.083	9.0	$3.604\,3364 \times 10^{-3}$
2	0.90\,0855	10.000\,830	$3.108\,6694 \times 10^{-3}$
3	0.090\,8566	10.065\,484	$3.107\,4103 \times 10^{-3}$
4	0.090\,8495	10.065\,542	$3.107\,4101 \times 10^{-3}$
5	0.090\,8459	10.065\,593	$3.107\,4097 \times 10^{-3}$

In general,

$$S^2 = \frac{\Sigma(d_i)^2}{n-2} \tag{5.32}$$

is a good approximation of σ^2 and has $n-2$ degrees of freedom. This allows

$$F_2(2, h) = F_2(2, n-2).$$

The solution of eq. 5.31 for r_A' and r_B' is tedious. A simpler process follows.

Let $\hat{r}_A$ and $\hat{r}_B$ be the least squares estimates of the reactivity ratio, and define

$$\hat{t}_A = \ln \hat{r}_A,$$

$$\hat{t}_B' = \ln \hat{r}_B.$$

Then the joint confidence limit envelope at the $100(1-\alpha)$ percent level is

$$(t_A' + \hat{t}_A)^2 a_{11} + 2(t_A' - \hat{t}_A)(t_B' - \hat{t}_B)a_{12} + (t_B' - \hat{t}_B)^2 a_{22} = 2s^2 F_\alpha(2, n-2) \tag{5.33}$$

Table 5.10   Evaluation of Copolymerization Data by Methods Outlined for Data of Table 5.2

	r_A	r_B	$\Delta_1 \times 10^2$	$\Delta_2 \times 10^2$
Linear Methods				
FR	0.08266	9.337	1.318	0.665
YBR	0.07439	9.061	1.382	0.698
K & T	0.07488	9.071	1.376	0.695
Nonlinear Method	0.09084	10.065		

where

$$a_{11} = (\hat{r}_A)^2 \sum \left(\frac{\partial G_i}{\partial r_A}\right)^2,$$

$$a_{12} = (\hat{r}_A \hat{r}_B) \sum \left(\frac{\partial G_i}{\partial r_A}\right)\left(\frac{\partial G_i}{\partial r_B}\right),$$

$$a_{22} = (\hat{r}_B)^2 \sum \left(\frac{\partial G_i}{\partial r_B}\right)^2,$$

where $\hat{r}_A$ and $\hat{r}_B$ are the least squares estimates of the reactivity ratios.

By shifting the origin of the coordinate system to the point $(\hat{t}_A, \hat{t}_B)$ and then rotating the coordinate axis through an angle θ given by

$$\tan 2\theta = \frac{2a_{12}}{a_{22} - a_{11}}, \tag{5.34}$$

we can reduce the above equation to

$$\frac{x^2}{A^2} + \frac{y^2}{B^2} = 1,$$

where

$$t_1' - \hat{t}_1 = x \cos \theta - y \sin \theta,$$

$$t_2' - \hat{t}_2 = x \sin \theta + y \cos \theta,$$

and

$$A^2 = \frac{2}{\alpha + \gamma - \delta},$$

$$B^2 = \frac{2}{\alpha + \gamma + \delta}.$$

Here

$$\alpha = \frac{a_{11}}{2S^2 F},$$

$$\beta = \frac{a_{12}}{S^2 F},$$

$$\gamma = \frac{a_{22}}{2S^2 F},$$

$$\delta^2 = \beta^2 + (\delta - \alpha)^2.$$

The computation of the confidence limit envelope of $\hat{r}_A$ and $\hat{r}_B$ in (r_A, r_B) space proceeds as follows:

1. Choose α and compute A and B.

2.   Compute the locus of the elliptical envelope in (x, y) space using

$$x = A \cos \phi,$$

$$y = B \sin \phi,$$

where ϕ is a dummy variable to be incremented in the range of 0 to 2π.

3. Calculate the quantities

$$t'_A - \hat{t}_A = x \cos \theta - y \sin \theta,$$

$$t'_B - \hat{t}_B = x \sin \theta + y \cos \theta.$$

4.   Calculate t'_1 and t'_2, the coordinates of the elliptical envelope in (t_1, t_2) space.
5.   Transform to

$$\hat{r}_A = \exp(t'_A),$$

$$\hat{r}_B = \exp(t'_B)$$

to obtain the joint confidence limit envelope in (r_A, r_B) space.

The confidence limits for the data of Table 5.2 are shown in Figure 5.14.

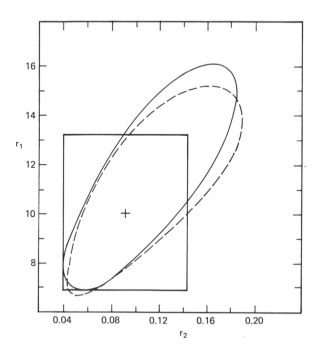

Fig. 5.14   Confidence limits for the vinyl chloride (M_1) and methyl acrylate (M_2) system: (+) least squares estimate of r_1 and r_2; (solid line) 95% confidence limits; (dashed line) approximately 95% confidence limits. The data of Table 5.2 were used. (Reprinted by permission of Ref. 9.)

Direct curve-fitting techniques using modern computers are practical for the copolymerization equations, and several authors have reported suitable programs for this purpose (10, 11) although the Tidwell–Mortimer Method (9) requires fewer iterations and makes more efficient use of computer time (12).

5.3.4 Methods for Integrated Copolymerization Equation

The integrated form of the copolymer equation to estimate reactivity ratios should be used when the relative monomer concentrations change appreciably during the course of the experiment. The differential equation governing the relative rates of consumption of two monomers M_A and M_B under steady-state conditions is

$$\frac{d[M_A]}{d[M_B]} = \frac{[M_A]}{[M_B]} \frac{r_A[M_A] + [M_B]}{r_B[M_B] + [M_A]},$$

where the concentrations are expressed in molar concentrations. Given the initial conditions that $[M_A] = [M_A]_0$ when $[M_B] = [M_B]_0$, the solution of the differential equation determines a curve of the type shown in Figure 5.15. Since all monomer concerned appears in the copolymer, if the reaction is terminated at the point $[M_A]_T$, $[M_B]_T$, the molar ratio of $[M_A]$ to $[M_B]$ in the copolymer, denoted by M_A/M_B is

$$\frac{M_A}{M_B} \approx \frac{[M_A]_0 - [M_A]_T}{[M_B]_0 - [M_B]_T} = \frac{\Delta[M_A]}{\Delta[M_B]}$$

$$\approx \frac{[\tilde{M}_A]}{[\tilde{M}_B]} \frac{r_A[\tilde{M}_A] + [\tilde{M}_B]}{r_B[\tilde{M}_B] + [\tilde{M}_A]},$$

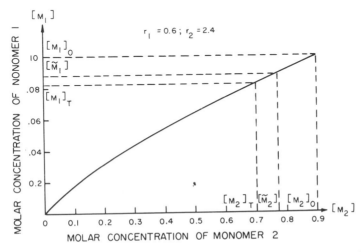

Figure 5.15 Molar concentrations in feed during copolymerization. (Reprinted by permission of Ref. 10.)

where $\tilde{M}_A$ are mean values lying somewhere in the interval ($[M_i]_0$, $[M_i]_T$). The use of mean values of the initial and final monomer feeds into the differential equation coincide closely with those from integral curves when conversions are limited to about 20% or less.

For high conversion studies the integral form must be used, and suitable computer programs have been written to estimate the reactivity coefficients (13, 14).

5.3.5 Penultimate Reactivity Ratios from Composition

Computer programs have been written and are available from the authors (15). One program permits the calculation of the penultimate reactivity ratios using input data of the initial mole fractions of each monomer, the percent conversion to polymer, the mole fractions of each monomer in the polymer, and the molecular weights of each monomer. The program starts with an initial estimate of the penultimate reactivity ratios, and it converges on the best nonlinear least squares fit. Since four parameters are involved, it is possible to converge on more than one minimum set of reactivity ratios depending on the experimental data and the initial estimates of the initial reactivity ratios. The method of minimizing this problem is to assume initially a terminal model and to calculate r_A and r_B. Then a grid of r'_A and r'_B is used in such a manner as to minimize the lowest sum of the squares. When experimental data corresponding to a terminal model are fed into this program, four reactivity ratios will be obtained, but $r_A = r'_A$ and $r_B = r'_B$.

5.3.6 Charge Transfer Reactivity Ratios from Composition Copolymerization Model

A computer program has been written (15) for calculating the six reactivity coefficients and the equilibrium constant for the charge transfer copolymerization model. The program utilizes initial estimates of the parameters and the composition of the feed and polymer. The programs allow the selection of the particular form of the charge transfer equation. In addition, the equilibrium constant can be varied. After using these programs, Pittman (15) stated that no adequate composition conversion data currently exists in the literature which can be used to test for a combined terminal charge transfer polymerization model.

5.4 USE OF PRELIMINARY DATA TO DESIGN COPOLYMERIZATION EXPERIMENTS

The choice of the proper monomer feed ratios to generate experimental conditions that yield the greatest potential for the determination of the reactivity ratios can result in a considerable saving in time and effort.

Therefore some preliminary consideration should be given to methods of selecting the proper experimental monomer ratios.

The criterion to be used in selecting the optimum is that of finding those initial feed ratios that minimize the area of the confidence region, based on the least squares estimates.

We initially attempt to select only two different initial feed ratios to be investigated as an indication of the experimental range to be covered. Unfortunately, we are dealing with a nonlinear design problem, and the values of the unknown parameters themselves appear in the expressions for the design coordinates. We are therefore in the position of having to know the answer to estimate it efficiently.

The criterion that leads to the minimization of the approximate joint confidence limits requires finding the values M_{B1} and M_{B2} of monomer B that maximize the determinant D, where

$$D = \begin{bmatrix} \dfrac{\partial G(M_{B1})}{\partial r_A} & \dfrac{\partial G(M_{B2})}{\partial r_B} \\[3ex] \dfrac{\partial G(M_{B2})}{\partial r_A} & \dfrac{\partial G(M_{B2})}{\partial r_B} \end{bmatrix},$$

and the value of G is calculated by using initial estimates of r_A and r_B. Tidwell and Mortimer (9) have computed conditions that maximize D for selected values of r_A and r_B, and these are given in Table 5.11.

Initial estimates of r_A and r_B can be obtained in a variety of ways including making preliminary runs. For example, at low concentrations of M_B, the composition of the copolymer depends almost entirely on r_A, so

$$r_A \sim \frac{M_A}{M_B},$$

and similarly for r_B. The Q and e scheme can be utilized or just "ballpark guesses" can be used to select initial values and the table used to select optimum runs for initial experimentation. This ensures that the maximum information is returned per experiment. In this fashion one can select feed ratios to provide a feedback of information for determining optimum designs and determinations of the reactivity ratios to a small confidence region. In any event this sequential approach leads to a minimum of experimental effort being expended to obtain satisfactory estimates, since at each stage current information is put to the most efficient use.

Once the sequence of experiments has led to a estimate of the reactivity ratios, careful replication of these experiments leads to an appreciation of the impact of the polymer and monomer analysis on the reactivity ratios.

Computer programs have also been written to establish the optimal monomer feed ratios for determination of penultimate reactivity ratios (15). A similar program has been written for the terminal charge transfer model (15).

Table 5.11 Experimental Conditions (mole-% M_2) Appropriate for the Precise Estimation of the Reactivity Ratios Given Preliminary Estimates (9)

						r_2						
R_1	0.05	0.10	0.15	0.20	0.30	0.40	0.50	0.60	0.70	0.80	0.90	1.00
0.05	2.441											
	97.559											
0.10	4.767	4.773										
	97.558	95.228										
0.15	6.988	6.998	7.007									
	97.557	95.223	92.993									
0.20	9.110	9.126	9.139	9.150								
	97.555	95.219	92.986	90.849								
0.30	13.082	13.109	13.129	13.140	13.138							
	97.553	95.212	92.974	90.840	86.863							
0.40	16.726	16.764	16.785	16.787	16.746	16.655						
	97.551	95.206	92.968	90.838	86.893	83.346						
0.50	20.081	20.125	20.140	20.123	20.017	19.839	19.614					
	97.549	95.202	92.967	90.845	86.945	83.473	80.386					
0.60	23.179	23.227	23.224	23.178	22.983	22.702	22.369	22.012				
	97.547	95.200	92.969	90.862	87.015	83.627	80.637	77.988				
0.70	26.048	26.093	26.064	25.979	25.682	25.283	24.837	24.377	23.919			
	97.546	95.198	92.976	90.886	87.101	83.797	80.904	78.352	76.081			
0.80	28.714	28.746	28.687	28.553	28.140	27.620	27.063	26.503	25.957	25.435		
	97.544	95.199	92.987	90.916	87.196	83.977	81.177	78.713	76.526	74.565		
0.90	31.196	31.211	31.111	30.923	30.384	29.744	29.079	28.426	27.802	27.208	26.651	
	97.543	95.200	93.000	90.952	87.298	84.163	81.447	79.066	76.955	75.061	73.349	
1.00	33.508	33.503	33.356	33.108	32.444	31.684	30.918	30.177	29.479	28.824	28.212	27.639
	97.542	95.203	93.017	90.993	87.408	84.352	81.714	79.407	77.363	75.531	73.874	72.361
1.50	43.071	42.882	42.444	41.872	40.596	39.331	38.158	37.085	36.110	35.220	34.405	
	97.540	95.234	93.132	91.237	87.976	85.260	82.942	80.921	79.129	77.521	76.062	
2.00	50.192	49.748	48.992	48.122	46.363	44.748	43.309	42.027	40.879	39.845		
	97.543	95.283	93.281	91.512	88.522	86.060	83.964	82.137	80.515	79.055		
2.50	55.680	54.955	53.913	52.804	50.699	48.846	47.233	45.814	44.556			
	97.549	95.347	93.442	91.784	89.015	86.747	84.816	83.131	81.632			
3.00	60.026	59.028	57.747	56.458	54.106	52.093	50.364	48.855				
	97.558	95.418	93.604	92.043	89.454	87.337	85.536	83.961				
3.50	63.542	62.300	60.824	59.400	56.874	54.752	52.943					
	97.571	95.492	93.761	92.285	89.844	87.850	86.153					
4.00	66.441	64.976	63.354	61.830	59.181	56.984						
	97.585	95.568	93.912	92.508	90.193	88.303						
4.50	68.869	67.214	65.476	63.878	61.143	58.895						
	97.601	95.643	94.054	92.713	90.505	88.703						
5.00	70.931	69.114	67.286	65.635	62.839							
	97.618	95.716	94.188	92.903	90.788							
7.50	77.825	75.521	73.480	71.724								
	97.714	96.049	94.750	93.665								
10.00	81.737	79.251	77.180									
	97.811	96.321	95.172									
12.50	84.269	81.737	79.695									
	97.902	96.543	95.502									
15.00	86.054	83.535										
	97.985	96.729										
17.50	87.388	84.906										
	98.059	96.886										
20.00	88.425	85.997										
	98.126	97.021										

REFERENCES

1. M. Berger and I. Kuntz, *J. Polym. Sci.*, **2A**, 1687 (1964).
2. M. Fineman and S. Ross, *J. Polym. Sci.*, **5**, 259 (1950).
3. E. C. Chapin, G. Ham, and R. Fordyce, *J. Am. Chem. Soc.*, **70**, 538 (1948).
4. T. Kelen and F. Tudos, *J. Macromol. Sci.—Chem.*, **A9-1** (1975).
5. K. Ito and Y. Yamashita, *J. Polym. Sci.*, **B3**, 625 (1964).
6. P. Tidwell and G. Mortimer, *J. Macro. Sci.—Rev.*, **C4**, 281 (1970).
7. A. L. Yezrielev, E. L. Brokhina, and Y. S. Roskin, *Vysokomol. Soedin.*, **A11**, 1670 (1969).
8. D. Margerison, D. R. Bain, K. Lindley, N. R. Morgan, and L. Taylor, *Polymer*, **16**, 278 (1975).
9. P. W. Tidwell and G. A. Mortimer, *J. Polym. Sci.*, **3A**, 369 (1965).
10. H. K. Johnston and A. Rubin, *Macromolecules*, **4**, 661 (1971).
11. D. Brown, W. Brendlein, and G. Mott, *Eur. Polym. J.*, **9**, 1007 (1973).
12. A. Rubin in *Computers in Polymer Science*, Marcel Dekker, New York, 1977, Chapter 4.
13. D. R. Montogomery and C. E. Fry, *J. Polym. Sci.*, **C25**, 59 (1968).
14. H. J. Harwood, D. W. Johnston, and H. Petrowski, *J. Polym. Sci.*, **C25**, 23 (1968).
15. C. U. Pittman, Jr., and T. Roumsefell in *Computers in Polymer Science*, Marcel Dekker, New York, 1977, Chapter 5.
16. M. Litt and F. W. Bauer, *J. Polym. Sci.*, **16C**, 1554 (1967).
17. K. Ito and Y. Yamashita, *J. Polym. Sci.*, **3A**, 2165 (1965).
18. J. Suggate, *Makromol. Chem.*, **179**, 1219 (1978).
19. J. Schaefer, R. J. Katnik, and R. J. Kern, *J. Am. Chem. Soc.*, **90**, 2479 (1968).

6

Chemical and Physical Methods of Determining Polymer Microstructure

6.1 INTRODUCTION

In the preceding chapters we have attempted to present a basis of understanding the microstructure of polymer chains from the perspective of the polymerization process. We have tried to indicate the statistical nature of the polymerization mechanism, the nature of the parameters that are determining, and the methods of the data treatment for obtaining these parameters.

Now we take a different point of view, that of the analysis. What is the actual structure of an experimental polymer? How do we determine the details of the structure? What methods do we use? Where do we start? How do we proceed?

It is not our intention to reproduce the "standard methods" that are available, but rather to offer some insight into the principles of the techniques as applied to polymers. Excellent treatises are available for all the chemical and physical methods, and their application to organic molecules, so duplication would be wasteful, but it may be worthwhile to place emphasis on the special aspects associated with polymer characterization. One needs to know the special demands placed on the technique by polymers, on the one hand, and on the other, one needs to know the special role that can be played by specific techniques. Where possible, suitable references are cited for use by the diligent reader, but no sense of completeness is anticipated. No one can hope to be complete in the sense of applications or examples, since nearly the entirety of the literature of polymer science would be required. First, we present the preliminaries required to generate a proper sample, then the general aspects of the methods themselves are given as they apply to polymers. The techniques of vibrational spectroscopy and nuclear magnetic resonance are described in separate chapters because of their relative importance.

6.2 PREPARATION AND SELECTION OF SAMPLE

Unless a suitable and representative sample is available for a structural study, that study constitutes an exercise in futility. The first requirement is that the sample be pure. For polymers this requirement is often very demanding because, almost by definition, polymers are mixtures themselves. For polymers a sample with a single structure does not exist, as the preceding chapters have indicated. Therefore the various constitutents of the polymer should be isolated in as pure a form as possible.

Usually, one first extracts a polymer with a nonsolvent, then attempts to dissolve and reprecipitate it from solution to effect purification. Successive fractionation or separation methods may be necessary. Assay of the main constituents by analysis of elementary composition is recommended. Measurements of the simplest physical and optical properties are desirable before becoming involved in an in-depth program of microstructure determination.

The second requirement of the sample is that it be representative of the preparation. I suppose the ultimate frustration is to carry out careful and sophisticated analysis on a sample that cannot be reproduced. No general or systematic approach to this problem allows one to select a representative sample, but one should at least feel through the use of simple visual and physical tests, that the selected portion of the sample is representative. Of course, duplication is expected and multiple samples recommended to the extent that time and money permit.

Methods employed for the preparation of the sample prior to the actual analysis include extraction, solution, and separation techniques. These methods should be carefully selected to prevent physical or chemical changes in the sample. Alternative solvents and procedures should be tested. Some of the most frequent impurities are residual monomers. Chromatographic techniques may be required to eliminate these monomers and other impurities.

6.3 EXTRACTION

The first step in any analysis is the removal of impurities from the sample. For polymers that exhibit low solubility due to their large molecular size, extraction with a solvent is recommended. However, to be of value the extraction procedure must be quantitative; that is, it must be complete. There is a dilemma here which was cited by Wake (1):

> *Until a completely extracted sample is available, identification of an unknown sample is impossible but, paradoxically, until the sample is identified, one does not know the correct solvent for extraction.*

The requirements of a solvent for a polymer extraction can be identified. The polymer should be swollen slightly but not dissolved. This requirement allows the solvent to penetrate the polymer and eliminates physical trapping of the impurities. Additionally, the solvent should boil well below the boiling point of any extracted material. This requirement evolves from the frequent need for analysis of the extract and quantitative recovery is most helpful. The solvent should be inert to any ingredient, as chemical reaction leads to undesirable products. Finally, the solvent should be nontoxic, cheap, and reusable where possible. As a result of these requirements, ether, acetone, benzene, and hexane are frequently used for preliminary extractions. The completeness of the extraction should be verified by a series of extractions, evaporations, and weighings of extracts.

An important aspect of extraction procedures is patience—that is, one cannot rush the process for polymer systems. Many extractions are completed in 30 min, but tests up to 8 hr should be carried out for a safety margin. Many standard procedures, particularly for rubber, require up to 72 hr.

The extraction procedure should minimize the sample size and maximize the solvent volume to ensure complete removal in the shortest time. Needless to say, the sample surface should be maximized by powdering. Details of extraction procedures are available (2).

6.4 DISSOLUTION TECHNIQUES FOR POLYMERS

Except in unusual circumstances, a solution of the polymer is ultimately required for one purpose or another, so early attention to dissolution of the polymer is desirable.

Most polymers will dissolve, but ease of solution decreases with increasing molecular size. In fact, polymer fractionation is often on the basis of decreasing solubility with increasing molecular weight (4). However, solubility depends not only on the number of units in the chain, but also on the branching, cross-linking, tacticity, and crystallinity of the polymer (4). Each polymer exhibits its own characteristic pattern with respect to solvents and nonsolvents.

The principles of polymer dissolution can be simply stated. If the free energy of dilution of a polymer in a solvent is negative over the whole range of composition, the polymer is completely miscible. When at some composition the free energy is not negative, two phases separate: a truly liquid phase consisting of a dilute solution of the polymer in the solvent, and a gel consisting mostly of polymer with its solvent. When the enthalpy of dilution (ΔH) is negative, miscibility occurs over the entire range, but for most polymers ΔH is positive, and the free energy is only negative because the entropy change is very large.

As the late Professor S. Maron explained, dissolving a polymer is like trying to get a snake into a bathtub. If it doesn't want to get in, you are not going to get it in the bathtub. But fortunately, in our case, the snake is happier because it has more degrees of freedom in the bathtub, so it only offers token resistance. This resistance to dissolution can often be overcome by an increase in temperature or exposure to the solvent for a longer time.

The choice of a solvent for dissolution is generally based on the alchemist's principle of *similis similibus solvetur*. A rather complete guide to the solubility characteristics of polymers in organic solvents is available (3). This information should also be coupled, where possible, with a knowledge of the chemical resistance and thermal characteristics of the polymer.

6.5   SEPARATION PROCEDURES (4)

Separation procedures for polymers serve two purposes: first, to separate nonpolymer impurities from the polymer itself, and second, to separate the low molecular weight fragments produced by chemical reactions. Most separations involve one form or other of chromatography. Chromatography is a separation method based on differences in partition of substances distributed between two immiscible phases: one a static phase, usually of very high surface area, and one a moving or mobile phase. Expressed in simplest terms, chromatography involves obtaining a sample, pretreating it in some fashion so that it is capable of being analyzed on the chosen column, injecting it onto the column, waiting for the separation to take place, and detecting and displaying the results of the separation in some manner that lends itself to quantitative results. The actual separation is generally performed on one of four types of columns: liquid/solid, liquid/liquid, ion-exchange, or gel-permeation. The sample is introduced into the column in one of three common ways: (1) pyrolysis, (2) flash vaporization, and (3) on-column injection. In any case the vaporization or pyrolysis must occur in a very short period of time relative to the chromatographic time scale. With polymers we must generally start with some sort of dissolution, and special devices have been designed for this purpose, including high-speed blades to quickly disintegrate and disperse the sample in the solvent. Thin layer chromatography (TLC) has become extremely popular for analysis, as it is simple, rapid, and effective (5). The technique involves the application of microgram quantities of sample in the form of dilute solution (1 to 5%) to a thin layer of stationary phase bound to a supporting plate, usually glass. The stationary phase is often a silica gel or alumina. The plate is placed in a developing tank. The mobile phase moves through the layer by capillary action. In doing so it carries with it sample components, which travel at different rates depending on the distribution coefficients between the mobile and stationary phases. Detection, if not already visible, is accomplished by viewing under ultraviolet light or spraying with a developing chemical. Unfortunately, TLC yields very little polymer. Gel permeation is becoming more useful. The principal problem with fractionating polymers is the limitation in the amount of sample in each fraction recovered. Recent advances in polymer analytical techniques have improved this situation. Preparative polymer fractionation has improved the feasibility of collecting narrow molecular weight polymer fractions for further study.

However, a caution must be noted here. It is an explicit concern that, once a sample has been received, it be subject to the minimum of contamination. In this the chromatographic methods fail in principle. The polymer is treated with large excesses of foreign compounds, that is, solvent, support phase, packing, and so forth. The amounts of these compounds used relative to a trace amount of a component being analyzed may be staggering: it can be as high as 10^8. Interference by this extraneous material or one of its constituents—previously adsorbed and retained samples, impurities, pyrolysis or chemical transformation products—thus becomes likely. Thus the separation process itself can introduce "chemical" noise that cannot be distinguished from the sample.

6.6 IDENTIFICATION PROCEDURES FOR POLYMERS (2, 7)

The initial study of any polymer sample involves seeking its identification; that is, do we have a new material or a familiar one? If the material is familiar, we can turn to the extensive polymer literature for guidance on procedures and techniques for subsequent analysis of the microstructure. If the polymer falls into a general class of systems such as polyamides, polyvinyls, and polyesters, we again have the advantage of being able to avail ourselves of the valuable work done in the past and present. If the polymer is entirely new, we want confirmation of this fact by comparison with all other known polymers in order to guide not only our own structural work but patent people as well. The identification of polymers is a sophisticated field, as you can well imagine from the variety (nearly 1500 types) of resins, plastics, paints, coatings, and so forth, that are available. Additionally, the samples can be found in all forms, sizes, and shapes. We cannot even attempt to present a tried and true procedure for any polymer sample. However, excellent monographs (7) are available on the subject, with the book by Haslam and Willis (2) coming immediately to mind. Please consult this and other books for details. For specific industries, such as rubber (1), textile (8), coatings (9), silicone (10), and resin (11), procedures have been described which will yield the identification of the majority of systems. However, with the introduction of new polymers every year, these procedures become rapidly outdated. Fortunately, one can keep informed by examining the annual updates found in journals such as *Analytical Chemistry*, which is particularly complete if altogether too brief to be helpful, without return to the original references. Most industrial laboratories are blessed with researchers who specialize in identification work of this kind, and their particular talents should be used wherever possible.

6.7 CHEMICAL ANALYSIS OF GROSS COMPOSITION

6.7.1 Elemental Analysis

Conventional chemical analysis has always been used for the determination of the gross composition of polymers and copolymers. In spite of numerous difficulties, elemental analysis is the usual basis for comparison or calibration of other techniques. But because of the low solubility and chemical resistance, even elemental analysis procedures are difficult for polymers (6). Combustion techniques may give low results because of incomplete combustion, and correction factors are often required. Chlorine analysis of copolymers has suffered from these problems, and only 94% of the expected halogen has been recovered in many cases. With silicones precautions must be taken in analysis for silicon to assure that silicon carbides are not formed.

In recent years commercial semiautomatic and automatic carbon and hydrogen analyzers have become available. These devices are useful for polymers, but to obtain accurate and reproducible results, an additional delay timer to prolong the combustion period is required to ensure complete oxidation of the sample. Combustion methods exist for simultaneous combustion for several elements in a single sample. These procedures all suffer from the drawback that conditions are never optimized for a single element; the method therefore must necessarily compromise to permit multielement determinations. This generally reduces the accuracy and scope of the analysis.

For nitrogen the Dumas technique is recommended with care that complete oxidation has occurred.

The oxygen content of polymers is usually calculated by difference if reliable analyses for the other elements present are available. Chemical analysis for oxygen in any organic material is difficult. Neutron-activation oxygen analysis is a possibility and this technique is rapidly developing.

Although the presence of trace elements contribute little, if anything, to a knowledge of the polymer structure, their presence can often account for unusual behavior, such as color formation. On the other hand, a complete analysis to assure the absence of toxic or other dangerous elements is comforting and is probably a necessary precaution under present OSHA surveillance. Atomic spectroscopy is particularly suitable for this aspect of the analysis, as every element gives rise to an electronic emission spectrum that is unique. The choice of the spectroscopic method, be it optical emission, flame, atomic absorption, or atomic fluorescence, depends on the nature of the system and usually represents a compromise to achieve useful information with minimal effort.

6.7.2   Functional Group Analysis

Since most polymers are organic in nature, the functional group analysis procedures used are the same as for other organic molecules (12). Limitations of solubility must always be kept in mind. The chemical methods involve the reaction of a specific group with a known reagent, followed by a determination of either the reaction product or the consumption of the reagent. For the purposes of determining the reaction products, reactions releasing acids, bases, oxidizing agents, or reducing agents are particularly desirable, since quantitative measurements of these substances are simple. Color-producing reactions are also of particular value, as the resulting spectrophotometric determinations are fast and accurate. A summary of the more important analytical methods for functional groups in polymers has been given, as well as a listing of the useful color-forming reagents (13).

Spectroscopic techniques such as IR, Raman, NMR, UV, and visible are extremely valuable in this area of functional group analysis if proper consideration is given to the usual effects of sequencing of units on the spectrophotometric parameters. For example, when UV spectroscopy is used for the analysis of styrene copolymers, attention must be paid to the effect of styrene sequence length on the frequency and intensity of the absorption modes. Similarly, with IR spectroscopy proper calibration of the technique is required. These matters are discussed more fully later.

6.8   CHEMICAL METHODS OF DETERMINING POLYMER MICROSTRUCTURES (14)

Chemical methods can be used to determine the microstructure of polymer chains. All these methods depend on degradation, cyclization, or preferential reactivity of sequences. With our modern sophistication in instrumentation, we tend to look down our noses at these chemical techniques, but we must remind ourselves that these chemical techniques were responsible for much of our early insight and are the basis of calibration of many of our modern methods. Today chemical reactions lead to many of the products that are subsequently fractionated and analyzed by our modern physical methods. Therefore a knowledge of these chemical techniques is essential to the modern polymer chemist. We describe the methods using copolymer notation, but the extension to isomerism, tacticity, and branching is obvious.

6.8.1   Selective Degradation for Microstructure Determination

In homopolymers and copolymers structural abnormalities frequently react differently compared to the repeat unit structure, and preferential selective degradation of the main polymer backbone occurs at these points.

If chain scission occurs in a polymer molecule in the absence of volatization with s scissions, the final average degree of polymerization is

$$\overline{DP}_t = \frac{\overline{DP}_0}{s+1},$$

or

$$s = \frac{\overline{DP}_0}{\overline{DP}_t} - 1.$$

The fraction of bonds broken in the system, α, is given by

$$\alpha = \frac{s}{\overline{DP}_0} = \frac{1}{\overline{DP}_t} - \frac{1}{\overline{DP}_0}.$$

If chain scission occurs at random, then α as a function of time is given by

$$\alpha = kt.$$

Thus, for purely random scission, a plot of α versus time should be linear and pass through origin. If the chains contain weak links, then

$$\alpha = \beta + kt,$$

where β is the fraction of "weak" links. In this fashion the presence of reactive bonds can be found. For example, in the copolymers of styrene-methyl methacrylate, the weak links are associated with isolated styrene units. With styrene-acrylonitrile copolymer, the paired acrylonitrile units degrade approximately 40 times faster than the paired styrene units.

Selective Degradation of Intramonomer or Intermonomer Bonds Only. The most obvious chemical method for obtaining information about the number of intermonomer (AB) or intramonomer (AA) units is to discover a reaction that selectively scissions these bonds. Thus a reaction breaking intermonomer bonds (AB) leads to

$$\sim A^{\downarrow}B^{\downarrow}AAAAA^{\downarrow}B^{\downarrow}A \sim \; \rightarrow \; \sim A + AAAAA + 2E + A\sim,$$

and the amount of B formed is related to the fraction of A—B bonds, $P(AB)$, in the total.

A reaction breaking intramonomer (AA or BB) bonds leads to

$$\sim ABA^{\downarrow}A^{\downarrow}A^{\downarrow}ABA \rightarrow \; \sim ABA + 2A + ABA\sim,$$

and the amount of A formed is proportional to the fraction of AA bonds, $P(AA)$.

It follows that other microstructures can be calculated from the determination of $P(AB)$ and composition:

$$P(AA) = P(A) - \tfrac{1}{2}P(AB),$$

$$P(BB) = P(B) - \tfrac{1}{2}P(AB),$$

$$P(AAA) = \frac{[P(A) - \tfrac{1}{2}P(AB)]^2}{[P(A)]^2},$$

$$P(AAB) = \frac{P(AB)[P(A) - \tfrac{1}{2}P(AB)]}{[P(A)]^2},$$

$$P(BAB) = \frac{P(AB)^2}{4[P(A)]^2}.$$

These expressions, in conjunction with analogous expressions for β-centered triads, are useful for characterizing the microstructure of the various polymerization models.

Selective oxidation and hydrolysis reactions are examples of such scission reactions. A simple example is taken from one of the earliest studies on copolymer structure (15), the ozonolysis of a 50:50 butadiene (B)-methylacrylate copolymer:

The BB linkages are measured by the recovery of succinic acid and succinaldehyde. The BMB triads could be measured by the amount of trimethyl-2-methylbutene-1,2,4-tricarboxylate recovered.

Selective hydrolysis has been used to study the sequence distribution in copolymers derived from 3,3-bischloromethyloxacyclobutane and β-propiolactone (16):

$$
\begin{array}{cc}
A & B \\
\end{array}
$$

$$(ClCH_2)_2\!-\!\underset{\underset{H_2C\!-\!O}{|\quad|}}{C\!-\!CH_2} + \underset{\underset{H_2C\!-\!O}{|\quad|}}{H_2C\!-\!CO}$$

$$\downarrow$$

$$\overset{\displaystyle CH_2Cl \qquad A}{\underset{\displaystyle CH_2Cl}{\sim\!\!\sim CH_2\!-\!\underset{\displaystyle |}{C}\!-\!CH_2\!-\!O\!-\!CH_2}}\;\; \overset{\displaystyle B}{\!-\!CH_2\!-\!CO(OCH_2CH_2CO)_n\!-\!OCH_2}\;\;\overset{\displaystyle B}{\sim\!\!\sim}$$

$$\downarrow (n+1)H_2O$$

$$\overset{\displaystyle CH_2Cl}{\underset{\displaystyle CH_2Cl}{\sim\!\!\sim CH_2\!-\!\underset{\displaystyle |}{C}\!-\!CH_2\!-\!O\!-\!CH_2\!-\!CH_2\!-\!COOH}} + nHOCH_2\!-\!CH_2COOH + HOCH_2$$

$$\beta\text{-hydroxypropionic acid}$$

The $P_2(BB)$, determined by the amount of β-hydroxypropionic acid liberated, indicated that these polymers have a blocklike structure.

Preferred Intermonomer and One Intramonomer Bond.   It is also possible to have selective degradation in which only B—A (not A—B) and BB linkages in the copolymer are broken in a given reaction, resulting in products of the type $A_n B$ (where $n = 0, 1, 2, \ldots$). This is the case where ester-ester copolymers (18, 19) are hydrolyzed and when synthetic poly-peptides undergo enzymatic degradation:

$$\sim\!A\!-\!B\overset{\downarrow}{-}A\!-\!B\overset{\downarrow}{-}A\!-\!A\!-\!A\!-\!B\overset{\downarrow}{-}A\!-\!B\overset{\downarrow}{-}AA\!-\!B^{\downarrow}B^{\downarrow}B$$

$$\downarrow$$

$$AB + AB + AAAB + AB + AAB + 2B$$

The products obtained for this type of selective degradation, as well as their probabilities, are shown in Table 6.1. A computer program has been written to calculate the yields of bond $A_n B$ fragments to be obtained from the selective degradation of a copolymer prepared in any specified degree of conversion (17).

Table 6.1   Products Obtained from Selective Degradation of Copolymers

Product	Structure in Polymer	Probability Expression for B units in Standard Environment
B	~B—$\overset{*}{\text{B}}$~	$P_2(BB)$
AB	~B—A—B*~	$P(BA)P(AB)$
AAB	~B—A—A—$\overset{*}{\text{B}}$~	$P(BA)P(AA)P(AB)$
AAAB	~B—A—A—A—$\overset{*}{\text{B}}$~	$P(BA)P(AA)^2P(AB)$

Complete Depropagation at Internal or External Monomer Bonds. Some copolymers degrade by depropagation or unzipping mechanisms; for these cases detailed information concerning the copolymer microstructure is available. Consider a homopolymer of the following structure:

$$\begin{array}{ccccccc}
& \text{H} & & \text{H} & & \text{H} & & \text{H} \\
& | & & | & & | & & | \\
-\text{CH}_2-\text{C}-&\text{CH}_2-\text{C}-&\text{CH}_2-\text{C}-&\text{CH}_2-\text{C}- \\
& | & & | & & | & & | \\
& \text{X} & & \text{X} & & \text{X} & & \text{X}
\end{array}$$

If the site of initiation is at internal or external bonds, the monomers $\text{CH}_2=\text{CHX}$ are recovered, and ordinarily one cannot distinguish how the decomposition occurred. Consider the case of an alternating copolymer of the following type:

$$\begin{array}{ccccccc}
& \text{H} & & \text{H} & & \text{H} & & \text{H} \\
& | & & | & & | & & | \\
-\text{CH}_2-\text{C}-&\text{CH}_2-\text{C}-&\text{CH}_2-\text{C}-&\text{CH}_2-\text{C}- \\
& | & & | & & | & & | \\
& \text{X} & & \text{Y} & & \text{X} & & \text{Y}
\end{array}$$

Since breaking at either internal or external bonds produces the same species, the arrangement does not yield any new structure, as in the homopolymer. However, for an alternating copolymer of the type

$$\begin{array}{cccccccc}
\text{H} & \text{H} & \text{H} & \text{H} & \text{H} & \text{H} & \text{H} & \text{H} \\
| & | & | & | & | & | & | & | \\
-\text{C}-&\text{C}-&\text{C}-&\text{C}-&\text{C}-&\text{C}-&\text{C}-&\text{C}- \\
| & | & | & | & | & | & | & | \\
\text{H} & \text{X} & \text{Y} & \text{Y} & \text{H} & \text{X} & \text{Y} & \text{Y}
\end{array}$$

decomposition at internal bonds produces "hybrid" monomers, that is, species composed of fragments from each of the original units (i.e., $CHY=CH_2$, $CHX=CHY$). Provided both methods of decomposition were equally likely, we would obtain 25% each of the species $CH_2=CHY$, $CHY=CHY$, $CHY=CH_2$, and $CHX=CHY$.

In general, in any copolymer there exist sequences of A units, $CH_2=CHX$ bounded by B, $CHY=CHY$, as follows:

$$
\begin{array}{cccccc}
H & H & H & H & H & H \\
| & | & | & | & | & | \\
-C-C-C-C-C-C- \\
| & | & | & | & | & | \\
Y & Y & H & X & Y & Y
\end{array}
$$

Thus the hybrid units are formed only at boundaries of the sequences of i units and are proportional to $P_2(AB)$, whereas the remainder produces monomer units. A parameter θ is defined as follows:

$$\theta = 2\frac{\text{internal monomer bonds broken}}{\text{internal monomer bonds in polymer}}.$$

Note that when $\theta = 1$, all bonds are broken with equal probability (random);
when $\theta = 0$, only external bonds are broken, so the yield of A $= P(A)$;
when $\theta = 2$, only internal bonds broken, and one A unit per each sequence of A units between B units is lost.

The probability $P_n(A_n)$ is given by

$$P_n(A_n) = P(AB)P(AA)^{i-1}P(BA),$$

and

$$M_A = M_B \sum_{n=1} nP_n(A_n).$$

When pyrolysis is complete,

$$Y(A) = M_B \sum_n \left(n - \frac{\theta}{2}\right)P_n(A_n),$$

so

$$Y(A) = M_A\left[1 - \frac{\theta}{2}\left(\frac{1}{r_A x + 1}\right)\right],$$

$$Y(B) = M_B\left[1 - \frac{\theta}{2}\left(\frac{x}{r_B + x}\right)\right].$$

Thus the yields of monomer A and B can be used as a measure of the polymer microstructure.

Table 6.2 Thermal Decomposition Reactions for A—B Copolymer

Type	Reaction	Products
I	$\cdot$CHA—CH$_2$—CHA—CH$_2\sim$	$\rightarrow$ CHA=CH$_2$ + $\cdot$CHA—CH$_2\sim$
II	$\cdot$CHA—CH$_2$—CHB—CH	$\rightarrow$ CHA=CH$_2$ + $\cdot$CHB—CH$_2\sim$
III	$\cdot$CH$_2$— CHA— CH$_2$—CHA	$\rightarrow$ CH$_2$=CHA + $\cdot$CH$_2$—CHA$\sim$
IV	$\cdot$CH$_2$—CHA—CH$_2$—CHB	$\rightarrow$ CH$_2$=CHA + $\cdot$CH$_2$—CHB$\sim$

Consider the pyrolysis of an A—B copolymer, A-type monomer units may split off from the chain ends as shown in Table 6.2 (20). Similar reactions may be written for the formation of B monomer units. There is considerable difference between the dissociation energies for reaction I (AA) and reaction II, since the resulting chain radicals differ in stability. Therefore, in the unzipping reaction of the vinyl-type copolymer, the penultimate unit has an important influence on the probability that a C—C bond will close and liberate a monomer unit. The yield of A monomer units obtained in a pyrolysis experiment, $Y(A)$ as the molar fraction of all units initially present, may be written (20) as

$$Y(A) = [M_A - \tfrac{1}{2}P_2(AB)P_2(AA)] + \tfrac{1}{2}P_2(AB)^2,$$

where M_A is the molar fraction of A units on the copolymer. Note that for a homopolymer $P(AA)$ is the fractional yield from the homopolymer of $A_j Y_0(A)$, since $P_2(AB) = 0$, and θ_A is unity. Hence

$$P_2(AA) = Y_0(A).$$

The difference between reaction I and II can be expressed in terms of a boundary effect parameter β_A, defined as

$$\beta_A = \frac{P(AB)}{P(AA)},$$

where $P(AB)$ and $P(AA)$ are, respectively, the probabilities that A type monomer will result from decomposition of the radicals at I and II. Using this definition, we obtain

$$Y(A) = P_2(AA)[M_A - \tfrac{1}{2}P(AB)(1 - \beta_A)].$$

If the boundary effect parameter has been previously determined from degradation studies, then

$$P_2(AB) = \frac{2[M_A - Y(A)/Y_0(A)]}{(1 - \beta_A)}.$$

Selective Degradation to Specific Sequences. In actual practice few, if any, copolymers are known that produce only monomeric species and, in fact, some systems exhibit highly selective chain scissions yielding particular

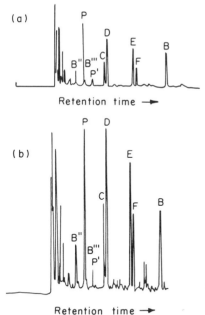

Figure 6.1 Chromatogram of C_8–C_{13} hydrocarbons from pyrolysis of propene-1-butene copolymer: (*a*) pyrolysis under N_2; (*b*) pyrolysis and hydrogenation under H_2. (Reprinted by permission of Ref. 21.)

fragments. For example, propylene-butene copolymers undergo a unique and useful decomposition under flash photolysis (21). The dominant mechanism is random chain breaking, which initiates a radical chain reaction with intramolecular transfer on the fifth carbon atom, followed by a chain breaking process in C_6–C_7; this gives a trimer olefin molecule. The distribution of the trimer is related to the sequence of monomer units in the copolymer chain. When these copolymers are pyrolized, they lead to the same light products (C_1–C_7) as the homopolymers. However, new pyrolysis products are observed in the range C_9–C_{12}. A typical pyrogram is illustrated in Figure 6.1, where four new peaks labeled C, D, E, and F are observed. Mass spectra show that these peaks are two C_{10} (C and D) and two C_{11} (E and F) olefins which correspond to trimers including either two propene (p) units (C_{10} olefins) or two butene (b) units (C_{11}). Table 6.3 shows the assignments of the peaks to the trimer sequence. Noting the intensity $I(X)$ of the peak X in a pyrogram therefore yields the following:

$$\frac{I(p)}{P_3(ppp)} = \frac{I(C)}{P_3(ppb)} = \frac{I(D)}{P_3(pbp)} = \frac{I(E)}{P_3(bpb)} = \frac{I(F)}{P_3(bbp)} = \frac{I(B)}{P_3(bbb)}.$$

Ultimately one can calculate the probabilities of all of the triads from measurements of the relative error of the peaks in the chromatograms.

In rare cases chain scission can occur in such a manner that only AB units are broken, and the sequences BA_nB for the values of n can be directly determined. Such is the case in ethylene oxide-maleic anhydride

Table 6.3   Trimers from the Copolymer Sequences

Sequence[a]	Olefins	Paraffins	Attribution to Peak
| ppb block | ```
c = c—c—c—c—c
    |   |   |
    c   c   c
            |
            c
``` | ```
c—c—c—c—c—c—c—c
    |   |
    c   c
``` | C |
| | ```
c = c—c—c—c—c
    |   |   |
    c   c   c
    |
    c
``` | ```
c—c—c—c—c—c—c—c
      |   |
      c   c
``` | C |
| pbp alternate | ```
c = c—c—c—c—c
    |   |   |
    c   c   c
        |
        c
``` | ```
c—c—c—c—c—c—c
    |   |
    c   c
        |
        c
``` | D |
| bpb alternate | ```
c = c—c—c—c—c
    |   |   |
    c   c   c
    |       |
    c       c
``` | ```
c—c—c—c—c—c—c—c—c
      |   |
      c   c
``` | E |
| bbp block | ```
c = c—c—c—c—c
    |   |   |
    c   c   c
    |   |
    c   c
``` | ```
c—c—c—c—c—c—c—c
      |   |
      c   c
          |
          c
``` | F |
| | ```
c = c—c—c—c—c
    |   |   |
    c   c   c
        |   |
        c   c
``` | ```
c—c—c—c—c—c—c—c
    |   |
    c   c
        |
        c
``` | F |

[a]   p = propene unit; b = butene unit.

(EO-MA) copolymers (22). The chains are hydrolyzed and the glycol ethers are separated by gas-liquid partition chromatographic (GLPC) analysis. The chromatograms in Figure 6.2 are of the trimethylsilyl ether derivatives of the glycols for a molar feed ratio of 6:1 of a high conversion EO-MA copolymer. The numbers along the horizontal axis indicate the number of ethylene oxide units in the glycol. A calibration curve for each glycol was obtained and appropriate molar response factors determined so that the relative molar run concentrations could be determined. These results are shown in Table 6.4. It is to be observed that for these EO-MA copolymers there is no repeating pattern of the ethylene oxide sequence concentration,

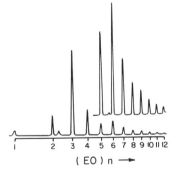

Figure 6.2 Chromatograms of the trimethylsilyl ether derivatives of the glycols obtained from hydrolysis of a high conversion EO-MA copolymer made from a charge ratio of 6:1 using SnCl$_4$ as catalyst. The numbers along the horizontal axis indicate the number of ethylene oxide units in the glycol. The insert was taken at higher gain. (Reprinted by permission of Ref. 22.)

hence the sequence probabilities are nonrandom and irregular. No finite Markovian description can predict this type of behavior. The author suggested a mechanism involving the ability of the catalyst to coordinate different combinations of monomers prior to their addition to the chain and then to add these monomers in two different ways (22).

Table 6.4 The Relative Molar Run Concentrations, [B(A)$_N$B], in High Conversion, SnCl$_4$-Catalyzed Ethylene Oxide-Maleic Anhydride Copolymers as Determined by GLPC Analysis[a] (22)

N	Relative Concentration		
	I	II	III
1	0.090	0.252	0.215
2	0.118	0.098	0.065
3	0.459	0.480	0.542
4	0.108	0.089	0.080
5	0.049	0.029	0.047
6	0.076	0.030	0.029
7	0.039	0.012	0.011
8	0.021	0.005	0.006
9	0.018	0.003	0.003
10	0.011	0.001	0.001
11	0.007		
12	0.003		

[a] The initial EO:MA charge ratios are 6:1, 3:1, and 1:1, respectively, for columns I, II, and III. The mole fractions of EO in copolymer are 0.80, 0.74, and 0.73, respectively, for columns I, II, and III.

6.8.2 Cyclization Reactions

Cyclization reactions between monomer types can provide information about sequence distribution. Cyclizations including rings of lactones, imides, lactanes, α-tetralenes, and endone ring formation are possible cyclization techniques that may be used in sequence studies. The extent of cyclization that occurs randomly between reactive units in an A—B type of copolymer has been calculated by Alfrey and co-workers (23) for a structure of the general type

$$\sim A\!-\!B\!-\!A\!-\!A\!-\!B\!-\!A\!-\!A\!-\!B\!-\!A\!-\!B\!-\!A\!-\!B\!-\!.$$

The fraction F_B of B units that would react is given by

$$F_B = \left\{ \cosh\left[\frac{P(AB)^2}{4P(A)P(B)}\right] - \left(\frac{P(A)}{P(B)}\right)^{1/2} \sinh\left[\frac{P(AB)^2}{4P_1(A)P_1(B)}\right]^{1/2} \right\}^2.$$

Harwood (17) has written a computer program to calculate F_B as a function of $P_2(AB)$ for a series of copolymer compositions. The results are shown in Figure 6.3. It can be seen that F_B is very sensitive to $P_2(AB)$, indicating that cyclization reactions can, in principle, be very valuable for sequence characterization.

Copolymers of methyl methacrylate and vinyl halides undergo lactonization (24, 25) on heating, liberating methyl halides in the following

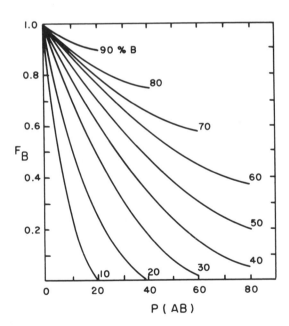

Figure 6.3 Variations of f_B with $P_2(AB)$ for copolymers of various compositions. (Reprinted by permission of Ref. 17.)

manner:

$$\underset{\substack{|\\H_3COOC}}{\overset{\substack{CH_3\\|}}{\sim CH_2-C-CH_2-CH\sim}} \quad \xrightarrow{150°C} \quad \underset{\substack{|\\CO\text{————}O}}{\overset{\substack{CH_3\\|}}{\sim CH_2-C-CH_2-CH\sim}} + CH_3Cl$$

and this intersequence cyclization process can be used to characterize the polymers. Later work (25) has utilized pyrolysis-gas chromatography to extend the accuracy of this technique by heating the polymer in vacuum at 200°C until lactonization is complete. From the weight loss the fraction of the methyl methacrylate units cyclized can be calculated. Lactone formation of vinyl alcohol (V) acrylic acid junctions has been used to characterize these copolymers (26). In calculating the amount of lactone formation, pentads were considered. Thus the acid units $\overset{*}{A}$ centered in AA$\overset{*}{A}$AA, AA$\overset{*}{A}$AV, or VA$\overset{*}{A}$AV pentads cannot form lactones. Acid units $\overset{*}{A}$ in VV$\overset{*}{A}$VV, VV$\overset{*}{A}$VA, VV$\overset{*}{A}$AA or VV$\overset{*}{A}$AV are almost certain to lactonize, since they have at least one neighboring alcohol group. Acid units centered in AA$\overset{*}{A}$VA and VA$\overset{*}{A}$VA pentads have a 50:50 chance of reacting. Finally, all acid units centered in AV$\overset{*}{A}$VA type pentads were assumed to be converted to lactone. The experimental results showed excellent agreement with the theory (26).

6.8.3 Sequence Analysis by Cooperative Reactions

When neighboring groups inhibit or enhance the reactivity of monomer units in copolymers, these cooperative effects can be used as a tool for sequential analysis. Autocatalysis or autoinhibition has been noted in the hydrolysis or pyrolysis of a number of systems. The utility of this effect is limited because the cooperative effects are very sensitive to chain configuration and activity. An excellent example of these cooperative effects is the alkaline hydrolysis behavior of methacrylic acid-methacrylamide copolymers (27). The experimental results can be interpreted on the basis that isolated amide groups are unreactive:

$$\sim acid-amide-\underset{\downarrow}{amide}-amide-acid$$

$$\sim acid-acid-\underset{\downarrow}{amide}-amide-acid$$

$$\sim acid-acid-\underset{\downarrow}{amide}-acid-acid\sim$$

no further reaction

The proportion of unaccessible amide groups thus equals the ratio of amide runs to amide units in the copolymer, or $P_2(AB)/P_1(A)$.

The photobromination of styrene (s)- fumaronitrile (f) copolymers is preferential for the styrene unit in the middle of sss triads for copolymers containing less than 60 mole-% styrene (28). Such a correlation might be expected if styrene units adjacent to a brominated styrene do not react with bromine.

Recently, a theoretical treatment of these cooperative reactions in copolymers has appeared (29). The probability that an arbitrary unit reacts in the time interval dt is assumed to be equal to kdt if its two adjacent units are unreacted, equal to $a_i kdt$ if only one of the units has reacted and is of type i, and equal to $c_{ij} kdt$ if it is between two reacted species i, j. The reaction probability of an arbitrary unit in the middle of a triplet depends on the four possible states of the two neighboring units, that is, whether none, one, or both units have reacted. Although the general solution is quite complex, in the special case with one completely protecting unit ($a_A = 0$) and the other unprotecting ($a_A = 1$), the extent of reaction, $y_A(t)$, is given by

$$y_A(t) = \tfrac{1}{2} P_1(A) P(AA)^{-1} (1 - e^{-2z})$$

and

$$Y_B(t) = P(A)^{-1} P(AA)^{-3} [P(AA) - P(A)] z^2 +$$
$$2 P_1(A) P_1(B)^{-1} P_2(AA)^{-3} [1 - P(AA)][P(AA) - P_1(A)](1 - e^{-z}) +$$
$$\tfrac{1}{2} P(A)^2 P(B)^{-1} P(AA)^{-3} [1 - P(AA)]^2 (1 - e^{-2z}),$$

where

$$z = P(AA)(1 - e^{-t}).$$

This result has been applied to guoran, which consists of 56% mannase units that are branched with galactose units and 44% unbranched units (29). An oxidized branched mannase unit completely protects its neighbors ($a_t = 0$), and oxidized unbranched mannase provides partial protection ($a_B > 0$).

The ratio of unoxidized mannase to unoxidized unbranched mannase after a long term reaction is 1.75. The total oxidation was 0.58 for this period. Numerical analysis indicated that the extent of reaction should be 1.05, which is different from the experimental value. However, an assumption concerning the rate constants limits one's ability to make firm deductions concerning the structure of guoran (29).

6.9 PHYSICAL METHODS OF DETERMINING MICROSTRUCTURE OF POLYMERS

At one time or another every physical technique has been used for the characterization of the microstructure of polymers. Consequently, it would be foolhardy to attempt to give a general approach involving all possible techniques. Rather, a few of the more prominent and useful techniques are discussed with pertinent examples. Vibrational and NMR spectroscopy are treated in subsequent chapters.

The physical techniques can be applied to polymeric molecules in the same fashion as low molecular weight compounds when proper consideration is given to the limited solubility and heterogeneous nature of the samples. A variety of sources can be consulted for information on the role of specialized techniques for use with polymers.

6.9.1 Ultraviolet (UV) and Visible Spectroscopy

The absorption of light by polymers in the visible and UV region involves promotion of electrons from the ground state to higher energy states. The exact electronic structures of the high energy states of molecules that have absorbed ultraviolet or visible light are not well understood. An isolated functional group called a chromophore can exhibit specific absorption in the UV or visible range. These chromophores generally absorb at very nearly the same wavelength and have nearly the same molar extinction coefficient in a variety of molecules. Hence unsaturated and oxygenated groups generally show strong UV absorption. The carbonyl group of aldehydes and lactones is easily seen in the 275–295 nm range. When two chromophoric groups are conjugated, the absorption band is generally shifted 15–45 nm to longer wavelength with respect to the unconjugated chromophore. The addition of still another chromophoric group or more in conjugation results in a shift of the position of absorption by the system to longer wavelengths. The extinction coefficient generally increases in a stepwise manner depending on the number of additional conjugated chromophoric groups. It is this effect that makes UV spectroscopy effective in the study of sequences. This technique has been applied to copolymers of 1-halogenated butadiene-styrene monomers (31). Elimination of HX from such copolymers provides unsaturated structures with a sequence length that is related to the number of conjugated double bonds in the sequence. Each different sequence gives rise to its own characteristic UV absorption. The spectrum of a copolymer of 1-chloro-1,3-butadiene with styrene after HCl elimination is shown in Figure 6.4 for a feed ratio of 0.153. The UV spectra of model polymers containing from 3 to 12 conjugated double bonds show that the principal feature of any polyene is a set of three bands of about equal intensity, separated by from 12 to 25 nm. Thus the three bands at 298, 312, and 327 nm are in the correct location for four conjugated double bonds representing a dyad derived from 1-chloro-1,3-butadiene residues. Similarly, the absorption bands at 347, 367, and 388 nm correspond to six double bonds (triad). Only two of the three expected bands at 412 and 437 nm are observed for the tetrad and only one at 473 nm for the 10-polyene content.

The specific absorbances of all bands increase with increasing number of double bonds in sequence. The bands within each group exhibit the

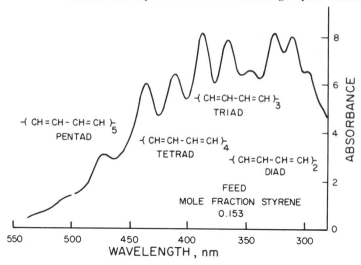

Figure 6.4 UV spectrum of 1-halogenated butadiene styrene copolymers from 0.153 feed after HCl elimination. (Reprinted by permission of Ref. 31.)

same relative intensities regardless of the copolymer composition, indicating that they arise from one particular structural unit. The results of the UV analysis are shown in Table 6.5. The control band of each structural set was used to calculate the copolymer sequence distribution as shown in Table 6.6. In this manner UV spectroscopy can be used for determination of the microstructure of polymers.

Table 6.5 Ultraviolet Data of Copolymers after HCl Elimination

	Specific Absorbance[a] at λ (nm)								
f_1	298	312	327	347	367	388	412	437	473
0.947	46	46	26		5.8	2.8			
0.892	40	47	41		6.3	3.5			
0.861	64	63	55		9.1	3.9			
0.829		74	66		13	5.1			
0.734	82	71	62		19	7.7			
0.698	110	107	101		44	30	9.3	4.1	
0.649	102	106	101		50	35	11	5.3	
0.602	107	97	96		50	32	13	5.9	
0.551	151	154	150		77	53	20	7.9	
0.490	153	151	144		78	71	42	30	
0.448	151	190	192		138	139	80	77	34
0.324	155	193	195		165	170	97	91	54
0.247	166	198	202	142	183	187	104	100	67
0.221	135	168	169	100	103	102	85	73	40
0.153	149	186	191	133	165	162	98	90	55

[a] 1% solution in dimethylformamide.

Table 6.6 Distribution of 1-Chloro-1,3-butadiene (M_2) in Copolymer

f_1	% M_2 Sequences Containing n Units				Calculated Sequence Ratio[a]	Observed Ratio of Specific Absorbance	
	$n=1$	$n=2$	$n=3$	$n=4$		312:367 nm	367:412 nm
0.947	94.6	5.01	0.28	0.01	18.9	7.9	
0.892	89.0	9.78	1.07	0.12	9.10	7.4	
0.861	85.9	12.14	1.72	0.24	7.07	6.9	
0.829	82.6	14.36	2.50	0.43	5.75	5.7	
0.734	73.0	19.70	5.32	1.44	3.71	3.7	
0.698	69.4	21.24	6.50	1.99	3.26	2.4	4.7
0.649	64.4	22.91	8.15	2.90	2.81	2.1	4.5
0.602	59.7	24.05	9.69	3.90	2.48	1.9	3.8
0.551	54.6	24.79	11.25	5.11	2.20	2.0	3.9
0.490	48.5	24.98	12.86	6.62	1.94	1.9	1.9
0.448	44.3	24.68	13.74	7.65	1.79	1.4	1.7
0.324	32.0	21.75	14.79	10.07	1.46	1.2	1.7
0.247	24.3	18.41	13.93	10.54	1.32	1.1	1.8
0.221	21.8.	17.03	13.32	10.42	1.28	1.6	1.2
0.153	15.0	12.78	10.85	9.22	1.18	1.1	1.7

[a] The calculated sequence ratio is the ratio of M_2 sequences containing n units to those containing $n+1$ units.

6.9.2   Mass Spectroscopy (32)

Every organic molecule is characterized by its molecular mass and by specific decomposition behavior when bombarded with electrons of 10–100 eV energy. The mass distributions of the decomposition products obtained by plotting the masses and abundancies of the charged fragments are unique to every molecule. Since a basic problem in the mass spectrometry of polymers is the nonvolatility of the molecules, the samples must be degraded either chemically or physically, so for a polymer the volatile fragments are investigated. The primary limitation of mass spectrometry to polymers in the past has been the continued fragmentation of the molecular species induced by the electron impact ionization. This fragmentation was superimposed on the pattern of the primary species of pyrolysis, yielding a highly complex pattern. These problems have been reduced by applying gas chromatographic techniques to the pyrolyzate prior to the mass spectrometer. Field ionization techniques reduce the amount of secondary fragmentation and considerably simplified spectra result. More recently, pyro-field techniques in mass spectrometry have been developed especially for polymers and the results are quite promising (33). With field ionization mass spectrometry, molecules are ionized in very strong electric fields of some 10^{10} V/m. Field ionization of a molecule proceeds very rapidly and this, as well as the acceleration of the ion in the field, alleviates further fragmentation. The strongest peak in the field-ion mass spectrum is the parent peak of the molecule.

The application of mass spectroscopy to copolymers depends on the mechanism of degradation. When the interlinks are weaker than the intrasequence links, the polymer should break up preferentially at these intersequence links and homofragments should be observed. When the bond strengths of the copolymer chain are approximately equal, the appearance of homo- and heterosequence fragments should be in proportion

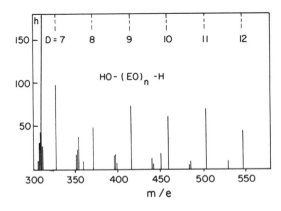

Figure 6.5 Mass spectrum of poly(ethylene glycol). (Reprinted by permission of Ref. 34.)

to their frequency in the chain. When the intersequence bonds are stronger than the bonds in the homosequence, the fragments should consist of small homosequences together with larger heterosequences. For example, when polysulfones of the type $-M_n-SO_2-$, where M can be an α-olefin, diene, styrene, and others, the C—S bond is considerably weaker than the C—C, so the polysulfones decompose into SO_2 and the other component. When a polystyrylsulfone with 12.1% styrene is examined, it is found to contain M_2-H_2 (208) and M_2 (206) as the heaviest and strongest masses, indicating that the 2:1 polystyrylsulfone has an alternating structure (32).

Some polymers exhibit unzipping to the monomer with pyrolysis, leading to the presence of a large amount of monomer in the pyrolyzate. For these copolymers the monomer peaks are directly proportional to the number of monomer units in the copolymer. For copolymers of α-methylstyrene a linear relationship is found, while for methyl methacrylate some deviation from linearity arises from the change in the decomposition behavior caused by isolated MMA units.

In general, however, the decomposition processes are quite complicated and the mass spectrum is not easy to interpret. Exceptions occur for copolymers of ethylene oxide (EO) and propylene oxide (PO), whose modes of fragmentation under electron impact are particularly suited to analysis of the copolymer in terms of EO units, the number of PO units, the order or randomness of the blocks, and the average molecular weight (34). Block copolymers are prepared by adding EO or PO in the first batch to form a homopolymer (in the presence of sodium hydroxide under 6 atm nitrogen at 150°C):

$$n\text{CH}_2\!\!-\!\!\text{CH}_2 + \text{C}_4\text{H}_9\text{OH} \rightarrow \text{C}_4\text{H}_9\text{O}(\text{CH}_2\text{CH}_2\text{O})_n\text{H}.$$
$$\diagdown \diagup$$
$$\text{O}$$

The second block is formed by adding a second batch of the other alkoxide to the alcohol end group:

$$\text{C}_4\text{H}_9\text{O}\!-\!(\text{CH}_2\text{CH}_2\text{O})_n\!-\!\text{H} + m\text{C}_3\text{H}_6\text{O}$$
$$\rightarrow \text{C}_4\text{H}_9\text{O}\!-\!(\text{CH}_2\text{CH}_2\text{O})_n\!-\!(\text{C}_3\text{H}_6\text{O})_m\!-\!\text{H}.$$

This process can be continued to prepare block copolymers of any kind. Random copolymers can also be prepared of these two alkoxides. In the high resolution mass spectrometer, all mass measurements up to m/e 900 could be made accurate to ± 10 ppm, permitting computation of a unique combination of atoms and the atomic composition $\text{C}_x\text{H}_y\text{O}_z$. Since this corresponds to some combination of $(\text{EO})_n-(\text{PO})_m$, then n and m are unique integers obtained by solving the equations $x = 2n + 3m$, $y = 2x$, and $z = n + m$. For a PO homopolymer with $\bar{M}_n \sim 600$ the mass spectra showed a repetitive fragmentation pattern repeating every 44 mass units (corresponding to the monomer unit $\text{C}_2\text{H}_4\text{O}$) up to a m/e of 635 corresponding to $\text{HO}\!-\!(\text{EO})_{14}\text{H}_2^+$ (Figure 6.5). Similar results were obtained for a homopolymer of PO with the repeating unit being 58 (C_3H_6O) (Figure 6.6). For the block copolymer ($\bar{M}_n = 537$) with the first block EO followed by an equimolar block of propylene oxide, Figure 6.7 shows the distribution of EO units in the first block. Figure 6.8 shows the PO distribution of the second block, which ranges between one and eight PO units with the maximum probability of either four or five units. The random copolymers can be analyzed in a similar fashion. Thus the mass spectra can be used to classify the sample as either random or blocked and to determine the lengths of the blocks.

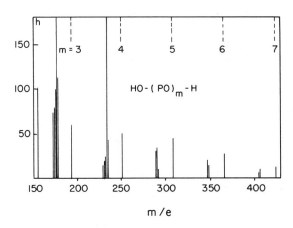

Figure 6.6 Mass spectrum of poly(propylene glycol). (Reprinted by permission of Ref. 34.)

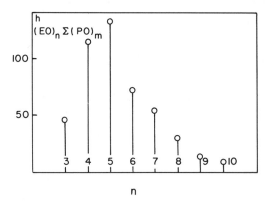

Figure 6.7 Distribution of EO units from the block copolymer $CD_3O(EO)_n(PO)_mH$. (Reprinted by permission of Ref. 34.)

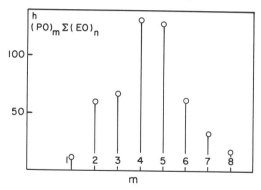

Figure 6.8 Distribution of PO units from the block copolymer $CD_3O(EO)_n(PO)_mH$. (Reprinted by permission of Ref. 34.)

6.9.3 Electron Spectroscopy for Chemical Applications (ESCA) (35)

ESCA is a relatively new technique that is exhibiting phenomenal growth and has great potential for the future, particularly for intractable materials and specific surface effects. ESCA involves the measurement of binding energies of electrons ejected by interactions of a molecule with a mono-energetic beam of soft x-rays. An ESCA spectrum consists of a plot of the number of electrons of given kinetic energy arriving at the detector as a function of the kinetic energy. In general, the ESCA spectrum consists of well resolved peaks superimposed on a background arising from inelastically scattered electrons. Since ESCA can, in principle, study the core and valence levels of any element, it can provide the chemical composition of polymers and information on the microstructure. The kinetic energy of the electrons is specific for the chemical nature of the element. In Figure 6.9 the C_{1s} and O_{1s} spectra of PET are shown. The O_{1s} spectrum shows two kinds of ester oxygen and the C_{1s} spectrum distinguishes the ester, aliphatic, and aromatic carbon levels. Table 6.7 shows the binding energies of the C_{1s} levels for homopolymers of ethylene and fluoroethylene. The shifts in the C_{1s} binding energies have been investigated for primary and secondary substituent effects and the results are of predictive value (35).

The C_{1s} and F_{1s} spectra of a species of ethylene-tetrafluoroethylene copolymers are shown in Figure 6.10. The copolymer composition calculated from ESCA results agreed well with the elemental analysis. Further, the C_{1s} spectrum was deconvoluted into seven component peaks that were assigned to the various pentad sequences. The major peaks arise from the heterocomponent (AB), while lines at 292 eV and 285.0 were $(A)_5$ and $(B)_5$, respectively. Other peaks arose from BBABB and BBAAB units and finally from a AABAA unit. Thus it appears that ESCA has potential for the investigation of the microstructure of polymers, particularly surfaces and coatings.

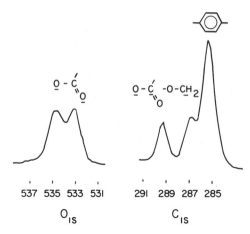

Figure 6.9 ESCA of the C_{1s} and O_{1s} levels for poly(ethylene terephthalate) film. (Reprinted by permission of Ref. 35.)

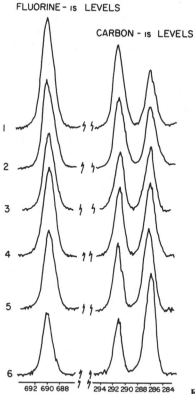

Figure 6.10 ESCA for the F_{1s} and C_{1s} levels for a series of copolymers of ethylene-tetrafluoroethylene. (Reprinted by permission of Ref. 35.)

146

Table 6.7   Binding Energies of the Homopolymers of Ethylene and the Fluoroethylenes

		C_{1s}	$\Delta(C_{1s})$	F_{1s}	$\Delta(F_{1s})$
I $[CH_2—CH_2]n$		285.0	(0)	—	—
II $[CHF—CH_2]_n$	—CFH—	288.0	3.0	689.3	(0)
	—CH$_2$—	285.9	0.9	—	—
III $[CFH—CFH]_n$		288.4	3.4	689.3	0.0
IV $[CF_2—CH_2]_n$	—CF$_2$—	290.8	5.8	689.6	0.3
	—CH$_2$—	286.3	1.3	—	—
V $[CF_2—CFH]_n$	—CF$_2$—	291.6	6.6	690.1	0.8
	—CHF—	289.3	4.3	690.1	0.8
VI $[CF_2—CF_2]_n$		292.2	7.2	690.2	0.9

REFERENCES

1. W. C. Wake, *Analysis of Rubber and Rubber-like Polymers*, Maclaren and Sons, London 1958.

2. J. Haslam and H. A. Willis, *Identification and Analysis of Plastics*, D. Van Nostrand Co., New York, 1965.

3. J. Brandrup and E. H. Imergut, Eds., *Polymer Handbook*, Interscience Publishers, New York, 1966.

4. M. S. Cantow, Ed., *Polymer Fractionation*, Academic Press, New York, 1967.

5. R. Amos, *Talanta*, **20**, 12 (1973).

6. F. E. Critchfield and D. P. Johnson, *Anal. Chem.*, **33**, 1834 (1961).

7. M. Kline, Ed., *Analytical Chemistry of Polymers*, Part III: Identification Procedures and Chemical Analysis, Interscience Publishers, New York, 1962.

8. C. E. M. Jones, Ed., *Identification of Textile Materials*, 4th ed., The Textile Institute, Manchester, England, 1958.

9. O. D. Shreve, "Synthetic Organic Coatings" in *Organic Analysis*, Vol. III, Interscience Publishers, New York, 1956.

10. A. Lee Smith, *Analysis of Silicones*, Interscience Publishers, New York, 1974.

11. C. P. A. Kappelmeier, Ed., *Chemical Analysis of Resin-Based Coating Materials*, Interscience Publishers, New York, 1959.

12. S. Siggia, *Quantitative Organic Analysis via Functional Groups*, 3rd ed., John Wiley & Sons, New York, 1963.

13. N. Bikales, *Characterization of Polymers*, Encyclopedia Reprints, Wiley-Interscience, New York, 1971, p. 91.

14. H. J. Harwood, *Angew. Chem.*, **4**, 394 (1965).

15. R. Hill, J. R. Lewis, and J. Simonsen, *Trans. Faraday Soc.*, **35**, 1073 (1939).

16. K. Tada, T. Saegusa, and J. Furukawa, *Makromol. Chem.*, **71**, 71 (1964).

17. H. J. Harwood, N. W. Johnston, and H. Piotrowski, *J. Polym. Sci.*, **C25**, 23 (1968).

18. K. Tada, T. Saegura, and J. Furukawa, *Makromol. Chem.*, **71**, 71 (1964).

19. K. Ito and Y. Yamashita, *J. Polym. Sci.*, **A3**, 2165 (1965).

20. Y. Shibasaki, *J. Polym. Sci.*, **A5**, 21 (1967).

21. J. C. Verdier and A. Guyot, *Makr. Chemie*, **175**, 1543 (1974).

22. J. Schaefer, D. A. Bude, and R. J. Katnik, *Macromolecules*, **2**, 289 (1969).

23. T. Alfrey, C. Lewis, and B. Mogel, *J. Am. Chem. Soc.*, **71**, 3793 (1949).

24. N. L. Zutty and F. J. Welch, *J. Polym. Sci.*, **1**, 228 (1963).

25. M. Tanaka, F. Nishimura, and T. Shono, *Anal. Chim. Acta*, **74**, 119 (1975).

26. I. Sakeuada and K. Kawashima, *Chem. High Polym.* (*Tokyo*), **8**, 142 (1951).

27. S. H. Pinner, *J. Polym. Sci.*, **10**, 379 (1953).

28. L. Rodriguez, *Makromol. Chem.*, **12**, 110 (1954).

29. J. J. Gonzales and P. C. Hammer, *Polym. Lett.*, **14**, 645 (1976).

30. J. L. Acosta and R. Sastre, *Rev. Plast. Mod.*, **26**, 67 (1973).

31. A. Winston and P. Wichackeewa, *Macromolecules*, **6**, 200 (1973).

32. D. O. Hummel, H. D. Schuddemage, and K. Rubenacker in *Polymer Spectroscopy*, D. O. Hummel, E. D., Verlag Chemie, Munich, 1974, p. 355.

33. D. O. Hummel, H. Dussel, K. Rubenacker, and T. Schweren, *Makromol. Chem.* **145**, 259 (1971).

34. A. K. Lee and R. D. Sedgwick, *J. Polym. Sci., Polym. Chem. Ed.*, **16**, 685 (1978).

35. D. T. Clark and W. J. Feast, *J. Macromol. Sci.*, **C12**, 191 (1975).

7

Vibrational Spectroscopy of Polymers

7.1 INTRODUCTION

Vibrational spectroscopy, particularly IR spectroscopy, is one of the most important methods for the identification and characterization of polymeric materials. IR and Raman spectroscopy have been particularly valuable because of their flexibility, simplicity, and sensitivity in structural investigations. Sample handling techniques have been developed for both IR and Raman such that virtually any polymer sample can be studied. Instrumentation is sufficiently advanced that quality spectra are obtained easily and rapidly. Computerization of the instruments has allowed further increases in the sensitivity of the techniques through improved data acquisition and processing. As a result all laboratories concerned with polymers recognize the importance of these vibrational spectroscopic tools.

Vibrational spectroscopy has been widely used to identify polymers, to quantitatively analyze chemical composition, and to specify steric and geometric isomerism, configuration, conformation, branching, end groups, and crystallinity. In addition the identification and analysis of additives, residual monomers, fillers, and plasticizers have been accomplished. Chemical reactions including polymerization, vulcanization, degradation, and weathering have been studied using vibrational techniques. The role of physical variables such as temperature and pressure have been studied, as well as the differences induced by change of state. Finally, the degree of crystallinity and orientation of semicrystalline polymers can be measured.

7.2 BASIC THEORY

7.2.1 Basic IR Spectroscopy (1)

When continuous light (i.e., light that consists of all possible wavelengths) is passed through a prism, the light is dispersed into its component wavelengths. When a sample is placed between this source and a detector, the light transmitted by the sample is no longer continuous; that is, some of the light frequencies are absorbed by the sample. The missing frequencies can be detected by allowing the light to fall on the wide band detector. This procedure, called *absorption* spectroscopy, is diagrammed in Figure 7.1. The recorded changes in the percent of the light transmitted by the sample as a function of wavelength or frequency are called a *spectrum*. When light is considered as a wave, the relationship between wavelength λ and frequency $\tilde{\nu}$ is

$$\lambda \tilde{\nu} = c,$$

where c is the velocity of light (3.0×10^{10} cm/sec). For purposes of spectroscopy it is easier to consider light energy as a packet of photons traveling with a high velocity. The relationship between the energy of a photon and the frequency of light $\tilde{\nu}$ (cycles/sec) is

$$E = h\tilde{\nu},$$

where h is called Planck's constant and has a value of 6.63×10^{-27} erg-sec. In this notation wavenumber ν is a wave characteristic that is proportional

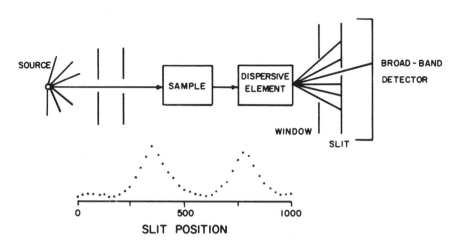

Figure 7.1 Diagram of absorption spectroscopy experiment.

to energy and is defined as the number of waves per centimeter;

$$\nu(cm^{-1}) = \frac{\tilde{\nu}}{c}.$$

For wavelength expressed in microns instead of centimeters

$$\nu(cm^{-1}) = \frac{1}{\lambda}(microns) \times 10^4.$$

The term *infrared spectroscopy* is associated with the portion of the electromagnetic spectrum with wavelength from 2 to 25 microns (5000–40 cm^{-1}). An absorption spectrum arises from the absorption of specific frequencies of light by a molecule while the remainder of the frequencies are passed unchanged. A quantum of energy (a photon) is absorbed when a molecule is excited from a lower to a higher molecular energy level. Absorption may occur when the frequency of light is equal to the energy level differences of the molecule:

$$h\nu = \Delta E = E_2 - E_1.$$

Not all frequencies are absorbed by the molecule because the energy levels of the molecules are quantized and not continuous. Thus the absorption spectrum is an energy portrait of the molecule. Since each different molecule has its own peculiar set of energy levels, it gives its own characteristic absorption spectrum.

   While the visible region of the spectrum measures the energies arising from motions of electrons in molecules, the IR spectrum measures energies corresponding to motions of the nucleus of the atoms making up the molecule. Because molecules are not rigid, their flexibility results in vibrational motion. For purposes of calculating the vibrational energies, the molecules are considered as point masses held together by bonds acting like springs undergoing harmonic motion. For a diatomic molecule (AB) consisting of masses M_A and M_B connected by a spring with stiffness k, the frequency of vibration is a bond stretching mode like

$$\begin{array}{cc} M_A & M_B \\ \leftarrow O \;\text{---}\!\!\!\sim\!\!\!\sim\!\!\!\sim\!\!\!\sim\!\!\!\sim\!\!\!\text{---}\; O \rightarrow \end{array}$$

where the frequency

$$\nu_{AB} = \frac{1}{2\pi c}\sqrt{\frac{k}{\mu}}$$

and

$$\mu = \frac{M_A M_B}{M_A + M_B}.$$

The frequency of a C—H bond calculated on this basis is 3100 cm^{-1}, or 3.2 microns. For a diatomic molecule this frequency corresponds to the stretching of the bond and is the only vibrational degree of freedom. The

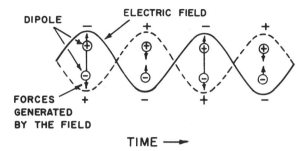

Figure 7.2 Diagram of the interaction of an oscillating electric field (light) with a vibrating dipole.

translation and rotation of the molecule as a whole causes no change in the internal energy of the molecule, yielding a null or zero frequency. When light of the proper frequency is radiated on the sample, absorption can occur only if there is some method of interaction between the vibrational energy levels and the light. The interaction required is an oscillating electric dipole induced by the vibration which can interact with the oscillating frequency field of the light. A visual picture of the interaction process is shown in Figure 7.2.

An IR absorption band may be observed for each vibrational degree of freedom for a molecule, called normal modes, if a change in the dipole moment occurs for the vibration, if the frequency is sufficiently resolved from other frequencies, and if the intensity is strong enough to be detected. For large molecules the number of degrees of vibrational freedom is $3 N^{-6}$, where N is the number of atoms in the molecule. For large molecules with low symmetry many IR bands are observed. The IR spectrum of a molecule is a unique physical property and has been termed its fingerprint on this basis.

From this discussion of how a particular vibration absorbs radiation, it is apparent that the intensity depends on two vector quantities: the vector direction of the dipole change for the vibration, and the vector direction of the electric field component of the radiation. For gases, liquids, and most solids the molecules are randomly oriented and the radiation is unpolarized, so no dichroic effects are important.

With solids some preferred orientation may be obtained. In these instances, the intensity of the absorption is a function of two vector quantities: the transition moment vector, and the electric vector of the light (usually linearly polarized for this purpose). We can linearly polarize the light in the vertical direction and make an absorbance measurement. This is given the notation of $A_{\parallel}$ while the corresponding measurement in the perpendicular or horizontal direction is denoted by $A_{\perp}$. The dichroic ratio is defined as

$$\rho = \frac{A_{\parallel}}{A_{\perp}}.$$

For single crystals the dichroic ratio can be used to define bond direction. The relative orientation of the polymer chains in oriented samples can be measured by using the dichoric method.

7.2.2 Basic Raman Spectroscopy (3)

Light incident on a molecule can be scattered and not absorbed. Sir C. V. Raman of India observed in April of 1928 that, when an intense beam of light strikes a sample, the scattered light contains not only the incident frequency but very weak satellite lines at higher and lower frequencies. He surmised that these weak lines arise from an exchange of energy through inelastic scattering between the sample and the light beam. He also soon discovered that these weak satellite lines are universal but have different frequencies for different molecules. The magnitude of the frequency shifts between the satellite lines and the exciting frequency correspond to energy levels arising from vibrational motion of the molecule.

The Raman effect arises from inelastic photon scattering. The exchange of energy between the molecule and the incident photon gives the scattered photon a new energy or frequency that is different from the initial one and is therefore measurable.

Let us consider using a laser monochromatic frequency of 488.0 nm or 20,492 cm^{-1} as a source. When the incident light of 20,492 cm^{-1}, ω_0, falls on the molecule, the molecule is raised to a virtual state. The only requirement of this virtual state is that it does not correspond to an electronic energy level of the molecule. From this virtual state the molecule can either emit light of energy $h\omega_0$ and return to its ground state or it can emit light energy of $h(\omega_0 - \omega_1)$ and drop to its first excited vibrational level. These emissions produce two lines: one with 0 cm^{-1} shift and the other shifted in frequency by ω_1, the vibrational energy level of the molecule. If the molecule is initially in its excited state, the molecule can emit the photon $h\omega_0$ or return to the ground state, emitting a photon of energy $h(\omega_0 + \omega_1)$. In Figure 7.3 we see how the Raman scattering interacts

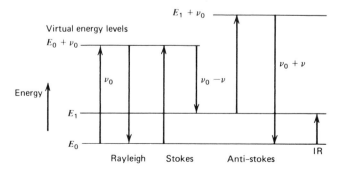

Figure 7.3 Energy level diagram showing transitions induced in the Raman experiment.

with the vibrational energy levels. The elastic scattering, called Rayleigh lines (ω_0), has the same energy as the incident photons, provides no information of interest in this context, and must be separated from the weaker Raman lines. The lines occurring as a result of accepting a quantum of energy from the photon appear at higher frequency ($\omega_0 + \omega_1$) and are called *anti-Stokes lines*, whereas the lines resulting from the sample sacrificing energy to the photon appear at lower frequencies ($\omega_0 - \omega_1$) and are called *Stokes lines*. Since, from the Boltzmann distribution, a larger number of molecules always exist in the lower energy or ground state, the Stokes lines have higher intensities and are usually the lines measured experimentally. The anti-Stokes lines are weaker for this reason and merely duplicate the frequency information contained in the Stokes lines. Since the efficiency of the elastic scattering is much larger than the inelastic, the Rayleigh line is much more intense than the Stokes lines, as illustrated in Figure 7.4. These frequencies and energy levels are as characteristic of the molecule as a fingerprint is characteristic of a person. For some molecules many of the observable Raman shifted frequencies are identical to the frequencies observed by IR spectroscopy for the same molecule. The recognition of the identity in frequencies of the Raman effect and IR spectra suggested that, like IR spectroscopy, Raman spectroscopy could be used to provide fundamental information on the architecture of molecules.

In addition to the intensity and frequency of the Raman lines, one can also measure the polarization character of the lines. In fact, it was the unique polarization properties of this "new radiation" that led C. Raman to believe he was observing a new phenomenon. Normal emission lines are

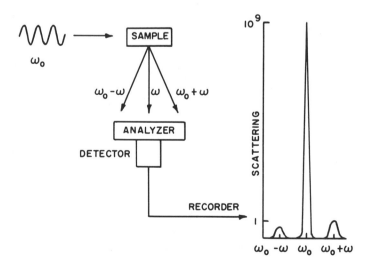

Figure 7.4 Diagram of the Raman experiment.

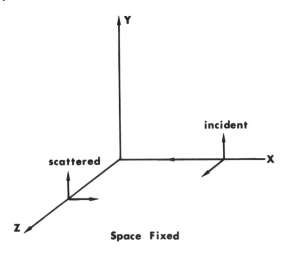

Figure 7.5 Geometry of the Raman experiment.

not polarized. In the usual Raman experiment the observations are made perpendicular to the direction of the incident beam (Figure 7.5). The *depolarization ratio* is defined as the intensity ratio of the two polarized components that are parallel and perpendicular to the direction of the incident light. The polarization of the Raman lines of CCl_4 are shown in Figure 7.6. Since the laser beam is polarized and highly directional, polarization measurements are easily made with accuracy. Theoretically, the depolarization ratio can have values ranging from zero to $\frac{3}{4}$ depending on the nature and symmetry of the vibration. Nonsymmetric vibrations

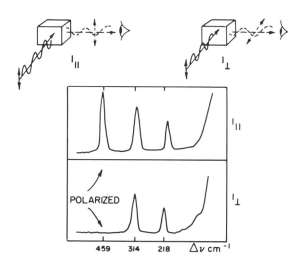

Figure 7.6 Raman depolarization measurement on carbon tetrachloride.

give depolarizations of $\frac{3}{4}$. Symmetric vibrations have depolarizations ranging from 0 to $\frac{3}{4}$ depending on the polarizability changes of the respective chemical bonds making up the molecule. Accurate values of the depolarization ratio are important for determining the assignment of a Raman line to a symmetric or asymmetric vibration. Potentially, polarization measurements can yield information about the nature of chemical bonds.

For solids preferred orientation is possible, with the result that more detail can be obtained. In a single crystal each molecular species is oriented with respect to every other one. Therefore the molecular polarizability (α_{i_j}) ellipsoids are also oriented along definite directions in the crystal. Since the electric vector of the incident laser beam is highly directional or polarized, the directionality in the crystal can be utilized to excite and obtain Raman data from each element of the polarizability ellipsoid. In this manner six different spectra can be generated because of the α_{xx}, α_{yy}, α_{zz}, α_{xy}, α_{xz}, and α_{yz} components. In the IR region only the three different directions of the dipole moment can be excited (μ_x, μ_y, and μ_z) to yield polarization information.

To better understand how the polarization data of single crystals are obtained, consider Figure 7.7. For the purpose of our experiment the z-axis can be chosen such that it is collinear with the optical axis of the crystal. We can also choose our laser polarization to be along the z-axis. If we place a polarization analyzer in the path of the collected light, a specific polarization can be selected to be measured. In Figure 7.7, with the analyzer in the z-direction, light with $\vec{E}$ in the z-direction is passed. With the laser polarization along z and collection along z, a spectrum from the α_{zz} component of the tensor is obtained. By rotating the analyzer 90°,

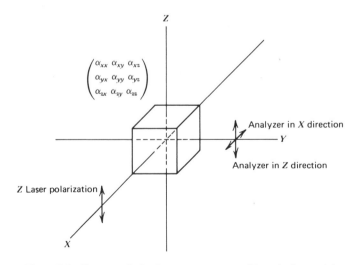

Figure 7.7 Raman polarization measurement with a single crystal.

thereby collecting x polarized light, while still exciting along z, α_{zx} is obtained. To obtain spectra from the remaining tensor components, both the laser polarization and the crystal can be rotated. For example, if the laser polarization is rotated to be in the y-direction, α_{yz} and α_{yx} are obtained. By rotating the crystal its defined axes are now rotated, and α_{xx} and α_{yy} can also be obtained. The activities of lines in the different spectra are used to assign the symmetry of the different crystal vibrations and to obtain the crystalline symmetry. In addition to the α used in generating the spectrum, information on how the sample was excited and the light collected is usually given. For the examples sketched in Figure 7.7 the notation is $X(zz)Y$ and $X(zx)Y$, where X is the direction of propagation of the laser beam, first z is the laser polarization, z and x are the analyzer positions, (zz) and $(zx = xz)$ identify the component of the polarizability tensor, and Y is the direction of collection.

7.2.3 Basis of Raman and IR Spectroscopy

Common molecules like chlorobenzene have many vibrational degrees of freedom and the Raman spectrum is rich in lines. Likewise, the IR spectrum of a molecule like chlorobenzene contains many IR absorption bands. Careful comparisons of the spectra demonstrate that, although many frequencies are common to both IR and Raman spectra, many frequencies are found in the Raman that are not present in the IR and vice versa. A particularly striking example is given for *trans*-1,4-polychloroprene, shown in Figure 7.8 (4). These differences between the Raman and IR spectra are a result of the difference in the physical basis of scattering and absorption and are determined by the structure of the molecule.

    In order for a sample to either scatter or absorb an incident beam of photons, there must be some mode of interaction with the oscillating electric field of the electromagnetic radiation. Since we are interested in the vibrational energy levels, the vibratory motion must produce a change in the molecule which allows the light to be absorbed or scattered. It is possible for the vibrational motion of the molecule to be such that the light is not absorbed in spite of the coincidence of the vibrational energy and the frequency of the IR light. As we have seen, for IR absorption the vibratory motion must produce a change in the dipole moment (a separation of charge). On this basis the stretching of the bonds of a heteronuclear diatomic molecule result in absorption of IR light, while the same motion in homonuclear diatomic molecules does not give rise to IR absorption since no permanent dipole exists. On the other hand, the Raman scattering occurs for those vibrational motions, producing a polarization or distortion of the electron charge of the chemical bond. Thus the stretching motion of a homonuclear diatomic molecule is active in the Raman effect.

    These physical observations can be given a classical treatment. In the

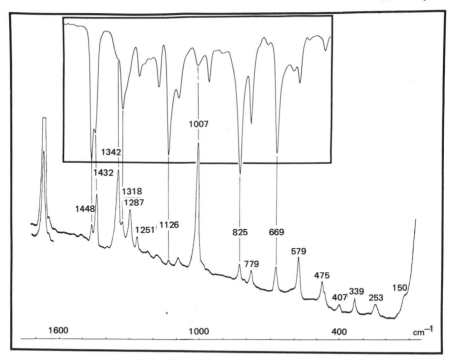

Figure 7.8 Raman spectrum of *trans*-1,4-polychloroprene (polymerized at −20°C) Insert: IR spectrum of the same sample. (Reprinted by permission of Ref. 4.)

IR region we can express the dipole moment change in a space fixed coordinate system with M_x, M_y, and M_z components of the dipole moment M (vibrational extreme) and M_x^0, M_y^0 components of the dipole moment M^0 (equilibrium position) by the following expansions:

$$M_x = M_x^0 + \sum_k \left[\left(\frac{\partial M_x}{\partial x_k} \right)_0 x_k + \left(\frac{\partial M_x}{\partial y_k} \right)_0 y_k + \left(\frac{\partial M_x}{\partial z_k} \right)_0 z_k \right] + \cdots,$$

$$M_y = M_y^0 + \sum_k \left[\left(\frac{\partial M_y}{\partial x_k} \right)_0 x_k + \left(\frac{\partial M_y}{\partial y_k} \right)_0 y_k + \left(\frac{\partial M_y}{\partial z_k} \right)_0 z_k \right] + \cdots,$$

$$M_z = M_z^0 + \sum_k \left[\left(\frac{\partial M_z}{\partial x_k} \right)_0 x_k + \left(\frac{\partial M_z}{\partial y_k} \right)_0 y_k + \left(\frac{\partial M_z}{\partial z_k} \right)_0 z_k \right] + \cdots.$$

If expressed in terms of normal coordinates Q_i, we find that the dipole moment of the molecule changes with frequency ν_i of a normal mode if

$$\left(\frac{\partial M_x}{\partial Q_i} \right)_0, \left(\frac{\partial M_y}{\partial Q_i} \right)_0, \left(\frac{\partial M_z}{\partial Q_i} \right)_0$$

are nonzero. Furthermore, the intensity of the IR absorption is proportional to the square of the above nonzero derivatives. Therefore if we plot dipole moment values against internuclear distance, the slope of the curve at the equilibrium position gives the intensity. Thus the amplitude of the vibration is not primarily responsible for the intensity. Those transitions with large changes in amplitude are often rather weak. Intensity depends principally on the magnitude of the dipole moment change and the form of the vibration.

The Raman effect may be expressed in classical terms as

$$\mu = \alpha E,$$

where μ is the induced dipole moment, E is the electric field, and α is the polarizability of the molecule.

Here

$$\mu_x = \alpha_{xx}E_x + \alpha_{xy}E_y + \alpha_{xz}E_z,$$

$$\mu_y = \alpha_{yx}E_x + \alpha_{yy}E_y + \alpha_{yz}E_z,$$

$$\mu_z = \alpha_{zx}E_x + \alpha_{zy}E_y + \alpha_{zz}E_z.$$

In general all six components of the polarizability tensor change when the nuclei are displaced from equilibrium positions. For small displacements we may again expand to give

$$\alpha_{xx} = \alpha_{xx}^0 + \sum_k \left[\left(\frac{\partial \alpha_{xx}}{\partial x_k} \right)_0 x_k + \left(\frac{\partial \alpha_{xx}}{\partial y_k} \right)_0 y_k + \left(\frac{\partial \alpha_{xx}}{\partial z_k} \right)_0 z_k \right] + \cdots,$$

as we did for the dipole moment components and also for the other components of α.

Utilizing similar expressions for α_{xy} and α_{xz} plus conversion to normal coordinates gives

$$\alpha_{xx} = \alpha_{xx}^0 + \sum_i \left(\frac{\partial \alpha_{xx}}{\partial Q_i} \right) Q_i + \cdots$$

Then substituting the following:

$$Q_i = Q_i^0 \cos(2\pi \nu_i t),$$

$$E_x = E_x \cos 2\pi \nu t,$$

$$E_y = E_y^0 \cos 2\pi \nu t,$$

$$E_z = E_z^0 \cos 2\pi \nu t,$$

we obtain

$$\mu_x = (\alpha_{xx}^0 E_x^0 + \alpha_{xx}^0 E_y^0 + \alpha_{xz}^0 E_z^0) \cos 2\pi \nu t +$$

$$+ \sum_i \left[\left(\frac{\partial \alpha_{xx}}{\partial Q_i} \right)_0 E_x^0 + \left(\frac{\partial \alpha_{xy}}{\partial Q_i} \right)_0 E_y^0 + \left(\frac{\partial \alpha_{xz}}{\partial Q_i} \right)_0 E_z^0 \right] Q_i^0 \times$$

$$\times \tfrac{1}{2} [\cos 2\pi(\nu + \nu_i)t + \cos 2\pi(\nu - \nu_i)t]$$

also for μ_u and μ_z.

Thus a fundamental vibration appears in the Raman effect if one or more of the six components of the change of polarizability, $(\partial \alpha_{xx}/\partial Q_i)_0, \ldots$, and so on, differ from zero.

The interaction of the laser light beam and the sample is based on the vibrational motion producing a bond polarization or an induced dipole moment rather than a change in the permanent dipole moment. On this basis IR spectroscopy shows strong absorption bands for polar chemical groupings such as C=O, OH, CH, whereas strong Raman lines appear for nonpolar and homonuclear groupings such as C=C, S—S, C—S. Oxygen (O_2) and nitrogen (N_2) have strong Raman lines but no IR bands for the frequencies corresponding to the motion of stretching of the bond. The dimerization of mercury ion in aqueous solution was established based on the observation of a Raman frequency for the Hg^+—Hg^+ bond; no Raman frequency appears for isolated Hg^+ ions.

The differences in the sensitivity of the IR and Raman effect to the type of bonds and the motion give rise to vibrational patterns of the same molecule which appear strikingly different if observed by IR or Raman. One portion of the molecule may produce strong Raman scattering in one frequency region and another portion may give strong IR absorption in another frequency region. This is illustrated for *trans*-1,4-polychloroprene in Figure 7.8.

Not only is the polarity of the chemical bonds a determining factor in the type of spectra observed, but the structure, particularly the symmetry, of the molecule also has a role in the type of spectra observed. Using the tools of group theory, one can easily predict, based on the symmetry of the molecule, which vibrational motions are active in the Raman or IR region, or both. When the molecule is sufficiently symmetric to have a center of symmetry, no vibrational modes of the molecule are common to both IR and Raman; that is, the rule of mutual exclusion holds. A center of symmetry is a point in the molecule where, if an arrow is drawn from this point to any atom, a reflection of this arrow back through the point results in the arrow pointing to an atom indistinguishable from the original one. Ethylene contains a center of symmetry at the center point between the two carbons. Benzene also has a center of symmetry. Both an isolated, perfect, all trans polymethylene chain and a crystalline system have a center of symmetry. Interestingly, the selection rules predict that the Raman effect always contains the same or more frequencies than the IR. Experimentally, the Raman effect is richer in observable fundamental frequencies than is the IR.

7.2.4   Method of Structure Determination with Raman and IR Spectroscopy (6)

Let us consider a simple example such as an AB_4 molecule. There are seven different structures possible for any given AB_4 molecule. Each of

Table 7.1 Selection Rules for AB$_4$ Type Molecules

AB$_4$	Point Group	Total	Number of Vibrations			Coincidence	Examples
			Raman polarized	depolarized	IR		
Tetrahedral	T_d	4	4	1	2	2	CCl$_4$
Planar Square	D_{4h}	7	3	1	3	0	XF$_4$
Square pyramid	C_{4v}	7	7	2	4	4	
All angles 90 and one bond different	C_{3v}	6	6	3	6	6	POCl$_3$
Two types of bonds and angles unequal	C_{2v}	9	9	4	8	8	SF$_4$
Four bonds equal, angles unequal	D_{2d}	7	7	2	4	4	
All bonds and angles different	C_1	9	9	9	9	9	

these model structures exhibits different vibrational patterns in the IR and Raman regions. Fortunately, it is possible from the symmetry of each model structure to predict the number of fundamentals expected in the IR and Raman, as well as the polarization properties of the Raman lines and the number of coincidences, that is, the number of frequencies found in both IR and Raman. For AB_4 the theoretical predictions are given in Table 7.1.

Utilizing this information as a guide, we can usually determine which of the model structures is most likely for a molecule. For example, carbon tetrachloride shows four Raman lines (one polarized) and two IR fundamentals, with two coincidences. An examination of Table 7.1 indicates that this pattern results when the AB_4 molecule is tetrahedral, so carbon tetrachloride must have the tetrahedral structure. Xenon tetrafluoride has no frequencies that are common to both IR and Raman, and must therefore be a planar square molecule.

Let us consider the problem of determining whether the 1,2-disubstituted ethylenes have the substituents cis or trans to the double bond. The predictions for the cis molecule are 12 Raman lines (5 polarized) and 10 coincidences between the IR and Raman frequencies. However, the predictions for the trans molecule are 6 Raman lines (5 polarized) but no coincidences between the Raman and IR spectra. A casual glance at the IR and Raman spectral frequencies allows one to determine the structure of the 1,2-disubstituted ethylenes.

Other examples could be cited, but the procedure should be obvious. One should not have the idea that the determination of structure is always as easy as in the given examples. Problems arise from the appearance of more lines than predicted due to breakdown of selection rules or fewer than expected because of the limited sensitivity and frequency region of the spectrometers. The polarization measurements are not always decisive. Accidental degeneracies of frequencies cause coincidences to occur that are not predicted. Frequently, measurements may not be sufficiently accurate to decide whether an IR band and a Raman line are coincident. These problems complicate the decision-making process. Other techniques and basic chemical knowledge must be used whenever possible. A complete and highly readable description of the basis of predicting the spectra of different molecules is available (6).

7.2.5 Selection of Vibrational Spectroscopic Method (7)

The motivation for using vibrational spectroscopy is the identification and determination of the structure of molecules. As has been demonstrated, the combination of IR and Raman spectra is an effective tool for the determination of the structure of molecules. The Raman effect provides more information about the nonpolar portions of the molecules and represents the richer half (in theory) of the vibrational spectra, while the IR effect yields information about the polar portions of the molecule. Due to the complementary nature of the two types of spectroscopy, they should both be used whenever possible, since Raman spectroscopy enhances the effectiveness of IR for solving chemical structure problems and vice versa.

There may be compelling *experimental* reasons for using IR or Raman spectroscopy for certain problems. Since the Raman effect is a light *scattering* rather than a light *transmission* process (as is IR), the transparency, size, and shape of the sample are relatively unimportant. Thus one can run large samples or extremely small samples with comparative ease in the Raman, whereas with IR one must often resort to a variety of specially developed techniques to make the sample sufficiently transparent to the IR beam to obtain quality spectra. For example, a powdered sample is dispersed in potassium bromide and pressed into a light transmitting disk. A powdered sample in Raman spectroscopy is simply compacted (since the signal is proportional to the density or number of scattering centers) and run by reflection of the beam from the surface of the powder. For the study of liquids or solutions in the IR region, one must use cells with windows of materials that transmit throughout the IR frequency region of interest, so materials such as NaCl, KBr, and CsI are used. These materials are expensive, fragile, hygroscopic, and easily scratched. For Raman studies the liquids are contained in a glass capillary or melting point tube, since glass is a weak Raman scatterer. The Raman cells are cheap, disposable, and require no special storage or handling procedures. Solutions are handled in a similar manner experimentally. Raman studies of aqueous solutions can be made, as water is nearly ideal as a solvent for Raman studies since water is a weak Raman scatterer but a strong IR absorber. For solution studies the utilization of the double beam technique in IR to subtract the spectrum of the solvent from the solution is a distinct advantage over the single beam Raman technique. Eventually, computerization of data from the Raman system will allow the subtraction of different spectra, but currently this is not a widespread practice among Raman spectroscopists.

If the sample size is very small, as with a coating, a fiber, or a finish, Raman spectroscopy has a considerable advantage over IR, since it is only necessary to fill the focused beam of size 10 microns in diameter to obtain a Raman spectrum. Often, with biological samples obtained by extraction the supply of sample is limited. However, IR techniques of microsampling allow examination of nanogram quantities. In the case of a finish such as a coating or paint on a substrate like a metal or wood, it is necessary with IR spectroscopy to remove the coating or to extract a sample film capable of transmitting the IR beam. With metal substrates attenuated total reflection (ATR) cannot be used as the substrate is conducting. With Raman spectroscopy the spectra can easily be obtained by front surface reflection. Sometimes the polymer sample cannot be made into a thin film required for taking an IR spectrum because the polymer is intractable. Often the polymer cannot be compression molded into a film because it is infusible, and films sometimes cannot be cast from a solvent because the polymer is insoluble. For polymers of this type, like the thermosetting resins, cross-linked rubbers, some polyamides, and polyimides, Raman spectroscopy can be used on the available piece of sample without further modification.

In many cases it is desirable to examine spectroscopically a sample without pretreating it in any way. Since the thermal history is an important factor is determining the properties of a plastic, it is obviously undesirable to melt or dissolve the sample to obtain a spectrum. With Raman spectroscopy this pretreatment is often unnecessary. It is possible to examine directly injection molded pieces, pipe and tubing, blown film, cast sheet, or monofilaments. If the sample should be studied in its natural environment, *in vivo*, for example, Raman spectroscopy can be used.

Filled polymers, like composites, present difficulty for IR spectroscopists, since the fillers such as glass, clay, and silica are strong IR absorbers that block the IR spectrum of the polymer. These particulate fillers (glass, clay, and silica) are poor Raman scatterers, so the Raman spectrum of the polymer is obtainable without removal of the filler. However, it is unlikely that the Raman spectrum could be obtained with carbon black filled composites, such as rubber, because the temperature rise caused by the absorption of the laser beam by the carbon would

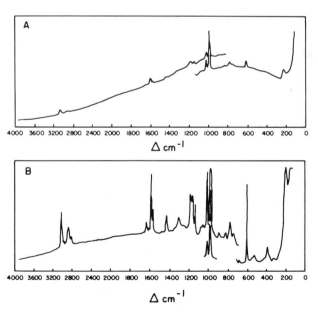

Figure 7.9 Raman spectrum of 8% *cis*-butadiene-modified polystyrene: (*a*) taken immediately after placing sample in instrument; (*b*) Raman spectrum of the same polymer after several hours of drench-quench treatment. (Reprinted by permission of Ref. 8.)

destroy the sample. However, ATR IR measurements can be made on carbon filled samples.

One very large advantage of IR spectroscopy is that the spectrum of nearly any sample can eventually be obtained by one special technique or another, but with Raman spectroscopy this is not the case.

The samples that cannot be examined by Raman spectroscopy are those samples that absorb the laser beam. If the absorption by the sample is strong, exposure of the sample to the laser beam results in instantaneous destruction of the sample through uptake of heat or photolysis. If the laser absorption process is weak, as when it arises from an impurity, the energy can be reemitted as fluorescence. Unfortunately, the fluorescence occurs in the same frequency range as the Raman effect and can be many times stronger. Fortunately, in many cases continued exposure to the beam may result in photolysis of the impurity or optical bleaching, with a resulting decrease in the fluorescence. An example of this process is shown in Figure 7.9 (8).

7.3   SAMPLING TECHNIQUES FOR POLYMERS

7.3.1   IR Spectroscopy

Regardless of the sophistication and expense of a spectrometer, the quality of the spectrum depends on proper care in sample preparation. For IR spectroscopy a uniform, light-transmitting sample is required except for special samples when reflection techniques are used. The range of thickness varies between 1 μm and 1 mm, and depends on the specific polymer being examined. If the samples are too thick, the spectrum totally absorbs in many frequency regions and much useful information is lost. On the other hand, if the samples are too thin, many of the important features of the spectrum are lost in the background. Although the samples need to be uniform across the cross-section of the beam (approximately 15 × 5 mm with dispersion and 15 × 9 mm with FT IR) to prevent nonlinear effects caused by a wedge-shaped sample, the samples cannot be uniform to the extent that interference fringes appear. For most polymers a thin self-supporting film is ideal and procedures are available for preparing such samples. Other forms of samples can also be used. However, except for those samples fabricated as commercial films, the solvent casting or compression molding does transform the sample and, not only is the thermal and processing history of sample lost, but physical and chemical changes can be induced that can be misleading. For solvent casting techniques care must be taken to remove residual solvents. A complete description of techniques for film preparation exists (9). For a polymer of unknown composition a series of solvents have been prescribed to facilitate determination (10). For example, ethylene chloride dissolves most thermoplastics, especially vinyls and acrylics, whereas toluene is useful for olefin polymers. Methyl ethyl ketone is helpful for butadiene copolymers. Resins rich in OH, COOH, or NR_3 groups may be susceptible to water.

Since most commercial polymers have a variety of additives, some preliminary extraction or purification procedures are recommended. No general procedure is available due to the extreme variety of commerical samples. For textile samples a recommended (11) solvent scheme is shown in Table 7.2.

For insoluble polymers other methods must be used. In Table 7.3 the method of treatment is prescribed based on the initial form of the material (12). If the polymers melt without decomposition, degradation, or oxida-

Table 7.2 Appropriate Solvents for Various Textile Additives (11)

Solubility in Various Solvents	Soluble Additives
Soluble in carbon tetrachloride	Oils
	Fats
	Waxes
	Phosphate, phthalate, sebacate
	Adipate, etc., plasticizers
	Poly(vinyl acetate)
	Ethyl cellulose
Insoluble in carbon tetrachloride; soluble in denatured alcohol	Glycols
	Glycerol
	Sulfated oils
	Rosin salts
	Cellulose nitrate
Insoluble in carbon tetrachloride and denatured alcohol; soluble in water	Starch
	Gums
	Casein
	Gelatin
	Animal glue
Insoluble in carbon tetrachloride, denatured alcohol, and water; soluble in dioxane	Cellulose acetate
	Cellulose nitrate
	Poly(vinyl acetate chloride)
	Poly(vinyl chloride)
Insoluble in carbon tetrachloride, denatured alcohol, water, and dioxane; soluble in or decomposes to give soluble products in 2% HCl	Urea-, melamine-, and casein-formaldehyde resins
	Starch
	Glue
	Casein
Insoluble in organic solvents released from coated fabric by solvents that dissolve film-forming components	Clays, ferric oxides, zinc oxide, titanium dioxide, Prussian Blue, etc.

Table 7.3   Occurrences and Treatment of Insoluble Polymers[a] **(12)**

Form of Material	Method of Treatment
1. Films/Sheets (up to 1 mm)	No treatment necessary if thickness correct and no opacity due to fillers
Change thickness by:	Heat-pressing
	Cold-rolling
	Scraping surface for dispersion
If fillers present	Removal by chemical attack
	Attenuated total reflection
2. Thick sheets, blocks, and fabricated products	Microtome sectioning
	Heat-pressing
	Abrasion to powder (surface and interior)
Foams	Disintegration and ball-milling
3. Coatings on metal, wood, ceramics, plastics, or fibres	Attenuated total reflection
	Abrasion to powder (surface only)
	Reflection from metal (transmission spectrum results)
4. Powders, dry or slurry coagulated granules	Attenuated total reflection
	Heat-pressing
Chips or lumps	Ball-milling
	Solvent swelling to gel
	Heat degradation to soluble
	Rolling in jeweller's rolls
	Pressure sintering to a pellet in metal die
5. Lattices or gels	Cast-on plate dry and sinter
6. Fibers	Squashing between plates
	Attenuated total reflection

[a]   Attenuated total reflection methods may be used in most instances; only the most appropriate are listed above.

tion, heat processing or compression molding can be used. Shims of desired thickness or teflon-coated aluminum foil can be used to prepare suitable films. Rapid quenching usually leads to less translucent samples and is recommended. Powders can be dispersed in KBr and pressed into pellets to give useful spectra. Powders can also be mulled with liquid dispersion materials such as paraffin oil (Nujol) and hexachlorobutadiene. The mull is squeezed between KBr or NaCl plates, and the spectra are then obtained. If it is desired to obtain spectra of samples without changing the state by dissolving or melting, sometimes the sample can be microtomed to a suitable thickness. However, considerable expertise is required to obtain suitable samples by microtoming.

Many polymeric samples that cannot be produced in suitable form for transmission measurements can be studied by attenuated total reflection (13). Fibers, rubbers, fabrics, and paints are often examined using ATR techniques.

A prism of high refractive index is pressed in good optical contact with the sample being examined. Depending on the difference in refractive index, a critical angle of incidence can be found where total internal reflection occurs in the sample. The ATR spectrum is slightly different from the transmission, as the intensity of the absorption bands increases with increasing penetration. One advantage of ATR is that only a few microns of the surface are penetrated by the beam, so that surface species are enhanced relative to the bulk phase. As a result ATR has been used to study reactions of polymer surfaces, surface coatings, sizes, adhesives, and so forth.

With the development of FT IR sampling requirements are less restrictive due to the high energy throughput of the instrument; for example, suitable spectra can be obtained with only 1% transmission through the sample. Additionally, very thin samples (approaching monolayers) can be examined due to the high signal-to-noise ratio available and ordinate scale expansion capabilities. However, the benefits of FT IR can only be achieved with proper sample preparation.

The discussion of sampling techniques has been restricted to qualitative analysis. When quantitative measurements are required, the sampling requirements are more restrictive. Wedge-shaped samples lead to deviations from the Beer-Lambert law. The samples must be randomly oriented, as preferential orientation in any of the three directions leads to the measurement of absorptivities that are not proportional to concentration.

7.3.2 Raman Sampling (7)

Sampling for Raman spectroscopy is extremely simple relative to IR spectroscopy. First, Raman is a scattering rather than an absorbance technique, so the sample can be opaque. Second, since the excitation source is in the visible region, Raman cells can be made of glass, which is not hygroscopic, fragile, or expensive compared to IR windows of KBr and NaCl.

The geometries of the samples relative to the laser beam are shown in Figure 7.10 (7). With liquids cylindrical glass cells with optically flat bottoms are used. Mirrors can be used to generate a more scattered signal by multipassing the beam through the liquid. With careful alignment and a homogeneous liquid or solution, multipassing can generate a threefold increase in signal.

With bulk solids including fibers focusing the beam on the surface of

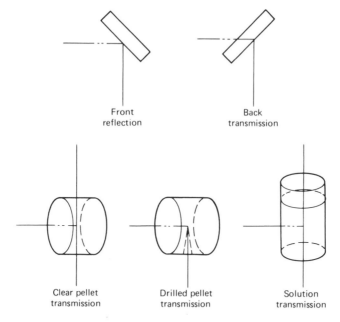

Front
reflection

Back
transmission

Clear pellet
transmission

Drilled pellet
transmission

Solution
transmission

Figure 7.10 Raman sampling techniques. (Reprinted by permission of Ref. 7.)

the sample, which is aligned at an angle of approximately 60°, is usually the
optimum method. For samples generating low levels of scattered intensity,
a small hole drilled at the point where the laser focus occurs is sometimes
helpful. For powders packing into a capillary tube of small diameter
(usually melting point tubes) is sufficient.

What one gains in simplicity of actual sampling, one often loses
through the generation of luminence by the laser which manifests itself as a
broad continuum in the Raman spectrum of many samples. In many cases
the fluorescence emission causes serious interference since, even if one
observes the Raman spectrum superimposed on the fluorescence background, the noise contribution of the fluorescence degrades the signal-to-
noise ratio of the Raman spectrum.

There are several methods of dealing with fluorescent samples—but
no general method for all samples. If the fluorescence occurs, the various
methods can be tried successively until a Raman spectrum of high signal-
to-noise ratio is obtained. In general, for polymers the fluorescence arises
from impurities in the sample, as strong absorption for polymers in the
visible is the exception rather than the rule. Therefore most of the methods
deal with eliminating the obnoxious impurity. This appears to be a simple
process except that we almost never know the impurity that generates the
fluorescence emission. Second, the level to which the impurity must be
reduced is very low, as the efficiency of the absorbance—emission is very
high relative to the Raman emission. The simplest and most often used

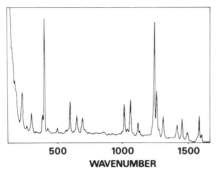

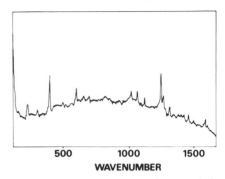

Figure 7.11 Raman spectrum of tetramethyl-thurium disulfide (purified). (Reprinted by permission of Ref. 21.)

Figure 7.12 Raman spectrum of tetramethyl-thurium disulfide (as received). (Reprinted by permission of Ref. 21.)

technique of lowering the fluorescence background is the so-called *drench quenching* technique. This technique simply involves soaking the sample in the laser beam until the transient luminescence background decays to a reasonable level. The results are illustrated in Figure 7.10 (8). One supposes that the impurity is photolyzed and destroyed by the laser beam. In the process heat is generated and the polymer sample may be affected. An obvious "bleaching" of the portion exposed to the beam sometimes occurs.

If the drench quenching technique does not succeed within a reasonable time of exposure to the laser, other purification techniques may be

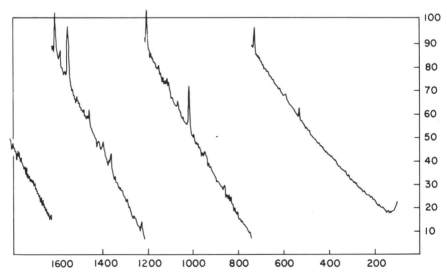

Figure 7.13 Raman spectra of indene with different excitation frequencies: (*a*) indene, practical, 488.0 nm excitation; (*b*) indene, practical, 514.5 nm excitation; (*c*) indene, practical, 647.1 nm excitation. (Reprinted by permission of Ref. 8.)

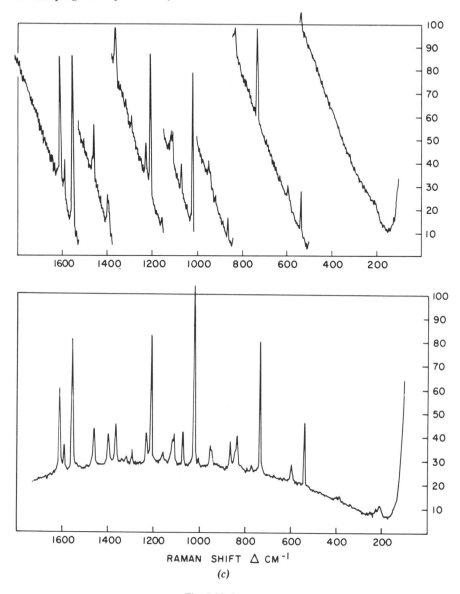

RAMAN SHIFT Δ CM^{-1}

(c)

Fig. 7.13 Contd.

tried. Techniques such as solvent extraction, recrystallization, distillation, sublimation and filtration may be useful. Filtration of the sample solution through activated charcoal has also been helpful. An example of the effect of recrystallization of tetramethyl thurium disulfide is shown in Figures 7.11 and 7.12. Of course, these purification techniques complicate con-siderably the process of obtaining a Raman spectrum.

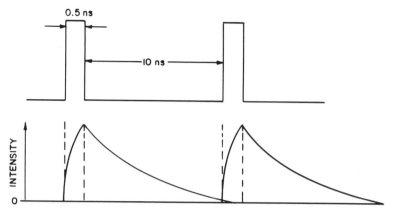

Figure 7.14 Theoretical model for Raman-fluorescence time discrimination. (Reprinted by permission of Ref. 8.)

Another useful technique when suitable lasers are available is to use an exciting frequency tuned away from the fluorescence absorption. If one can obtain a fluorescence excitation curve, one can pick an exciting line that yields the minimum fluorescence. With the current availability of tunable lasers, this method may become more useful. Figure 7.13 shows the effect of changing frequency on the spectrum of indene (8). It is clear that 647.1 nm excitation yields the only acceptable spectrum.

A more recent but electronically demanding method is to sample the Raman signal before the fluorescence emission rises to its maximum value. Since the Raman signal is nearly instantaneous and the fluorescence lifetimes are in the nanosecond range, if one can pulse fast enough one should be able to differentiate between the fast and slow signals. The idea is demonstrated in Figure 7.14 (8). If one uses a mode-locked laser to generate pulses of about 0.5 nsec with pulse intervals of about 10 nsec, by gating the detector for the signal only for the duration of the laser pulse, in theory one can observe all the Raman signal and none or very little of the fluorescence signal. This technique has been demonstrated to improve the Raman signal by a factor of three.

When fluorescence is absent, Raman spectroscopy is simple. In its presence Raman spectroscopy can often be frustrating.

7.4 INSTRUMENTATION FOR VIBRATIONAL SPECTROSCOPY

7.4.1 IR Dispersion Instruments

In 1945 commercial IR spectrometers became available, and since that time many improvements have been made. Except for the FT IR spectromete⁻

they all have a similar design. An excellent discussion of the spectrometer components by Potts (1) should be consulted by the inquisitive. The spectrometer consists of a source, a dispersive element, a slit system, a detector, and a recorder.

Sources of IR radiation consist of an object heated to a high temperature of 1300–1500°K. The element can be a wire (Nichrome), a silicon carbide (Globar), or a rare earth oxide (Nernst Glower). The radiation is imaged on an entrance slit with a collimating mirror.

The dispersing element of the spectrometer is designed to disperse the continuous blackbody radiation into its component frequencies in a fashion similar to the atmospheric dispersion of sunlight into a rainbow. In the early instruments a prism was used, but current instruments use plane diffraction gratings. A grating disperses the radiation because the angle of diffraction is a function of the wavelength of incident radiation. To confine the dispersed frequencies to a predetermined frequency domain or resolution, entrance and exit slits are required. The slits are bilateral, and both jaws move away from the center of the slits. The light, after transversing the sample and being dispersed by the monochromator, falls on a detector. For IR spectroscopy two types of detectors are used: thermal and photoconductive. A thermocouple is most often used and generates a potential difference that increases with increasing temperature. However, a direct current signal is not desirable, so the light beam is interrupted by a chopper generating an ac current that is easily amplified. In addition, extraneous noise frequencies are not amplified. When this is coupled with an appropriate recorder, one has a single beam IR spectrometer. However, the energy of the source is not constant throughout the frequency range. In addition, the atmosphere has molecules like H_2O and CO_2 which absorb the radiation. Finally, the recorded signal includes all the instrumental absorptions and distortions, so a blank spectrum must be recorded and carefully compared with the sample spectrum. As a result of these problems, double beam spectrometers were developed. In fact, optical null techniques were used to allow one to record the energy *difference* between the two beams without any relation to the energy content in either beam. The optical diagram for a double beam spectrometer is shown in Figure 7.15. A chopper is inserted in the incoming beam to produce two identical beams and the chopper spins. In the reference beam a light attenuator or comb is inserted. As a result of the chopping process the signal at the detector consists of a square wave with one portion arising from the sample beam and the other from the reference beam. When both beams are equal, no difference in cycles exists. When the sample absorbs some of the incident radiation, the sample beam generates a difference in the square wave signal, and this difference is used to drive the attenuator into the reference beam to make the square wave detector signal equal to zero. The same signal passed through a servomotor drives the recorder pen so that the position on the recording chart is a direct measure of how much of the

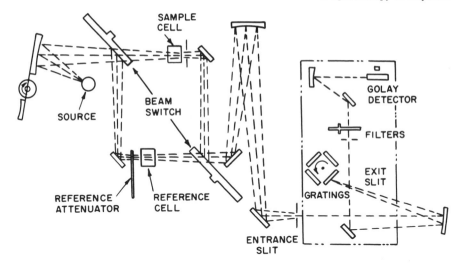

Figure 7.15   Commercial optical null types spectrometer.

attenuator is required to null the sample absorption. In this manner instrumental effects like changes in source energy are eliminated, as well as atmospheric absorptions.

With this system the signal strength at the detector depends on the square of the slit opening (1), while the noise depends on the amplifier gain necessary to produce sufficient voltage to drive the servo system for recording. Narrower slits lead to better frequency resolution but less signal for the same source intensity, while wide slits lead to poor resolution and loss of spectral definition. To quote Potts (1), "What is gained in resolution by narrowing the slits is lost in increased recorded noise level, and, conversely, what is gained in lower recorded noise level by widening the slits is lost in resolution." An alternative approach is to increase the signal at the detector by increasing the measurement time. Theory indicates that the signal-to-noise ratio is proportional to the square root of measurement of scanning time. However, resolution is proportional to the fourth root of the scan time, so "to obtain a twofold increase in resolution at an equivalent signal-to-noise ratio, the scan time must be increased by the obviously impractical factor of 16" (1).

For studies of polymers whose spectral band widths are quite large, resolution is not an important factor. On the other hand, due to the broad, weak bands in the IR spectra of polymers, a high signal-to-noise ratio is required for detection of spectroscopic differences and quantitative measurements.

7.4.2   FT IR Spectrometer (14)

IR spectroscopy is widely used in polymer chemistry today. However, if it were not for the energy limitations of a conventional IR spectrometer, even

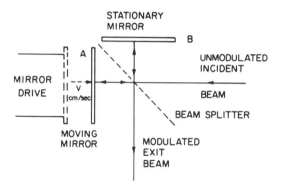

Figure 7.16 Diagram of a Michelson interferometer.

wider use would be expected. For polymer analysis, where the bands are generally broad and weak, this energy limitation is particularly severe.

The energy limitation can be minimized by using interferometers of the Michelson type rather than the conventional prism and grating instruments. This technique has often been called *interferometric spectroscopy*, but it is more properly called *Fourier transform spectroscopy*, as there are many different interferometric techniques in spectroscopy. The manner of obtaining the spectral information is quite different from a prism or grating instrument. The distinguishing feature is that the spectrum is obtained by taking the Fourier transform of an interferogram.

The Michelson interferometer is shown schematically in Figure 7.16. It consists of two perpendicular plane mirrors, one of which, B, is stationary, while the other, A, moves at a constant velocity in the direction shown. Between the two is a semireflecting film, or beam splitter, at which the incident beam is divided and later recombined after a path difference has been introduced between the two beams. Assume, for simplicity, a monochromatic input to the interferometer. When the interferometer is adjusted so that the optical path length of each arm is identical, the two light beams will be in phase when they return to the beam splitter and they will constructively interfere. The detector will view a maximum signal. If the mirror is moved back one-quarter of the wavelength, the two beams will be 180° out of phase, they will destructively interfere, and the detector will see no signal. If the mirror is continuously moved, the signal will oscillate from strong to weak for each quarter wavelength movement of the mirror. The signal generated at the detector will be simply a cosine wave whose strength depends on the intensity of the monochromator light. When we use a broad line spectral input, each frequency may be treated independently, and the output will be the summation of all cosine oscillations caused by all of the optical frequencies in the source. At zero path length all waves are in phase and, as the mirror is moved away from the zero position, they rapidly scan out to a steady average value. The resulting signal put out by the detector is the interferogram shown in Figure 7.17.

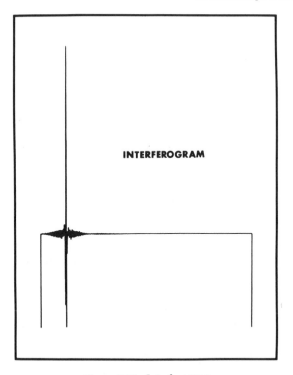

Figure 7.17   Interferogram.

This result can be represented mathematically as

$$I(x) = \int_{-\infty}^{+\infty} B(\nu) \cos(2\pi x\nu)d\nu,$$

where $I(x)$ is the interferogram where x is the mirror displacement, and $B(\nu)$ is the intensity of the source as a function of frequency (ν), or the spectrum.

This equation is one-half of a cosine Fourier transform pair, the other half being

$$B(\nu) = \int_{-\infty}^{+\infty} I(x) \cos(2\pi x\nu)dx.$$

By means of this pair of equations we have a relationship between the interferogram $I(x)$ and the spectrum $B(\nu)$. We simply have the computer take the Fourier transform as indicated and we have the spectrum. This is the basic process of FT IR spectroscopy.

There are several advantages of an FT IR over dispersive spectrometers, including speed, sensitivity, and data processing. The major disadvantage is the cost of FT IR, which ranges to $100,000. The fundamental advantage of FT IR over the monochromator is that all frequen-

cies of the spectrum are gathered simultaneously. A complete interferogram can be measured in the time (~1 sec) it takes to measure one
frequency element with a slit spectrometer. Stated another way, for the
same measurement time the signal-to-noise ratio of a spectrum measured
interferometrically is improved by a factor of $N^{1/2}$, where N is the number
of frequency elements. The second fundamental advantage of FT IR is the
optical throughput. The radiation reaching the detector of a dispersive
spectrometer must pass through two narrow slit systems, whereas, except
the size of the mirrors, no similar limitation exists. This represents an
improvement of as high as 80 times the throughput for a dispersive
instrument.

The final and perhaps most important advantage for working with
polymers is that the spectrum is in digital form that may be processed
easily with a minicomputer (15). In this way spectra can be coadded,
subtracted, or ratioed or stored for future use. This allows a comparison of
every frequency element of the spectra, and extremely small differences in
the spectra are easily detected.

The processing and storage of the digital data allow a remarkable
advantage over systems where the data are recorded in analog form.

7.4.3 Raman Spectroscopy

The experimental apparatus for Raman spectroscopy is quite simple, and
is outlined in block form in Figure 7.18. Required are (1) a powerful light
source (a laser is currently used), (2) an illuminating chamber for the

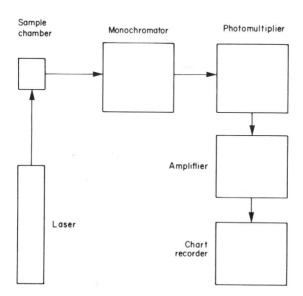

Figure 7.18 Diagram of a Raman instrument. (Reprinted by permission of Ref. 7.)

sample, (3) a high performance light dispersion system to resolve the more intense elastically scattered light from the weak inelastically scattered Raman signal, (4) a light detection and amplification system capable of detecting weak light levels, and (5) a recorder.

Since the Raman effect is very inefficient and only one photon in a million is inelastically scattered, powerful light sources such as lasers are needed. The most important asset of a laser as a light source for Raman spectroscopy is its coherency. Since the light waves of the coherent light of a laser are all in phase, the additive effect produces a focused light intensity that is 10^4 times as great as for normal incoherent light of the same power. The coherent nature of light also yields a collimated beam that does not spread out with distance in the manner of a beam from a flashlight. A flashlight beam only travels a limited distance before its luminosity is lost, since the beam looses its intensity through spreading. A laser beam can travel large distances without the beam size increasing substantially. In fact, a laser beam shone from the earth to the moon as a part of our space research spread to a diameter of only one-half mile and the reflected beam was detected in Houston, Texas. The laser beam is monochromatic in frequency, with a very narrow bandwidth, which is also advantageous as a Raman source. Additionally, the laser beam is linearly polarized, which yields an additional gain factor of two, since one-half of normal light is lost in passing through a polarizer. Since a continuous source is desirable for purposes of spectroscopy, gas lasers such as helium-neon, argon ion and krypton ion lasers are used. The helium-neon lasers are more stable, have a longer life, and are cheaper than the argon and krypton ion lasers but have much lower power output. Since the Raman signal increases with the power of the incident beam, high power is desirable. Pulsed lasers can be used and offer some advantages at the detection end of the system, but pulsed lasers are not normally used except for special purposes such as remote and resonance Raman spectroscopy.

The illuminating chamber is simple and is built to facilitate the focusing of the beam on the sample whether it is a gas, liquid, or solid. The beam can be multipassed through a liquid or gas to enhance the Raman signal. The scattered light is condensed using a lens system and is focused on the slits of the light dispersion system.

Special monochromator systems are used to reduce the scattered light level sufficiently that the Raman signal can be detected. Since the Raman signal is 10^{-6}–10^{-9} as strong as the Rayleigh line, high dispersion is required. Two and three grating monochromators in tandem have been used to reduce the scattered light to an acceptable level. Various configurations of the monochromater are possible, but the requirement of frequency matching in the double or triple monochromators presents a challenging coupling problem for frequency scanning systems. A mismatch of the frequency viewed by the monochromators is disastrous, since everything one had hoped to gain using tandem systems is lost.

For detection of the small number of scattered photons, high performance photomultiplier tubes are used. Modern tubes have low internal noise and high gain. The amplification methods have evolved from lock-in amplification, direct current amplification, to pulse counting. Currently, most instruments are equipped with a pulse counting system. The individual photoelectron pulses from the phototube are detected and individually processed through an adjustable grating system, then subsequently counted over a predetermined time period. The output is recorded on a strip chart recorder as a function of wavelength or frequency.

A number of instrument companies supply Raman instruments as a complete integrated package, along with a broad range of sampling devices that allow the spectra to be obtained easily and under special conditions such as at high and low temperatures and under high pressure.

7.5 IDENTIFICATION TECHNIQUES
USING VIBRATIONAL SPECTROSCOPY

The first question that is asked as a rule is the nature of the polymer. The use of vibrational spectroscopy in the identification of materials is based on the fact that each molecule has its peculiar energy levels unique to itself and, since the vibrational spectroscopic techniques carefully map these energy levels, each full frequency range spectrum is a fingerprint of the particular molecule.

At the high frequencies the stretching and bending of chemical bonds occur, and as the frequency decreases the motion involves a larger number of the atoms in the molecule until at the lowest frequencies all of the atoms are involved. Thus each molecule has a unique spectrum in both the IR and Raman regions.

A molecular identification is accomplished when the unknown spectrum is matched with the spectrum from a reference collection. This matching can be carried out in several ways. The most precise method is a point-by-point comparison throughout the entire frequency range. Only with digitized spectral data have such point-by-point comparisons been possible, but with increasing use of computers on FT IR as well as dispersive instrumentation, this method will be used more often. When digital difference spectra are calculated between an unknown sample and a stored reference spectrum and no difference bands are detected, the spectra are identical to the limits of sensitivity of the spectra. Even in courts of law this spectral comparison would constitute an unequivocal identification and evidence of identity of samples. With the high storage capacity of computers, many polymer spectra can be stored, searched, and compared. The potentials of this technique are numerous, including industrial espionage of competitors' products. Unambiguous identification of

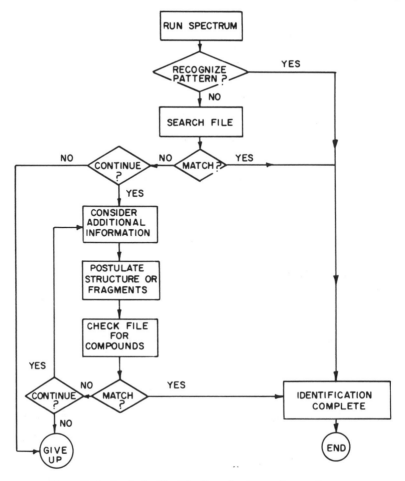

Figure 7.19 Logic for identification of polymers by spectroscopy.

one's own product can be achieved by labeling with an identifying component. The future holds the possibility of purchasing magnetic tapes or disks containing digitized spectra of a series of polymers, plasticizers, additives, and so forth, in the same manner as is presently available for catalogs of spectra on cards or microfilm.

In the ordinary case the polymer spectroscopist does not have available digitized reference spectra and must resort to his own ingenuity; the logical process that he would normally follow is shown in Figure 7.19. From the first examination of the spectrum, he determines whether he has a familiar spectrum. If not, he searches his reference file. If still unsuccessful, he considers other information and attempts to put the pieces of the puzzle together until he has successfully found a spectral match and identification. An excellent guide to this process for polymers and their additives is available (10).

Another aid in identification of commercial plastics and resins is available in the use of computerized search systems. Over 130,000 spectra are on computer tapes from the reference collection. The computer searches the data banks and prints out the spectrum number and reference catalog of the 20 closest matches to the unknown. The total time per search is usually less than 15 min.

When one is working with other than pure compounds, personal encoding of the spectral data is required, and one can establish one's own personal reference file relative to the particular field of interest whether that be fibers, elastomers, or plastics. A representative sample of the entire process, from encoding the spectra to the computer printout, is available in the literature (24).

The future step will be the utilization of a "learning machine" computer program for analysis of IR spectra. In the data correlation step the computer accepts digitized spectral data on systems with known structure and generates identification vectors. After such generation these identification vectors can be applied to data from unknown materials to compile a list of structural features.

7.6 DETERMINATION OF THE CHEMICAL FUNCTIONALITY OF POLYMERS USING VIBRATIONAL SPECTROSCOPY

The comparative analysis of the spectra of a great number of molecules has brought the realization that certain chemical groups give rise to modes that fall within a certain determined, at times quite restricted, range of frequencies, independent of the rest of the molecule. These characteristic frequencies are called *group* frequencies and can be used to detect the presence of these groups in a molecule or polymer (16, 17).

The vibrational energy levels can be calculated from first principles, and as a result some of the factors influencing the spectra have been discovered. These factors include bond stiffness and the masses of the atoms as well as the geometry and interaction between the chemical bonds. The motions, such as the stretching of diatomic pairs or bending of the triatomic groups, are relatively unaffected by the nature of the rest of the molecule and therefore have nearly the same vibrational energy regardless of the attached molecular skeleton. These modes are termed *internal* modes and give rise to the characteristic or group frequencies. For example, a mode appears between 3300 and 2700 cm^{-1} for a C—H bond regardless of the nature of the rest of the molecule. The particular type of C—H bond, that is, aryl or alkane, results in a specific absorption frequency within this range and is different for an aryl or alkane C—H bond.

As a result of these characteristic frequency ranges, vibrational spectroscopy can be used for qualitative organic analysis simply by recognizing the presence of these frequencies. Conversely, the absence of a known characteristic frequency from the spectrum of a molecule can be considered definitive evidence for the absence of that chemical grouping. In fact, small differences in frequency within the range of the characteristic group vibration and observation of supplementary bands allow distinctions to be made between similar types of chemical bonds. For example, a C—H band has different, but nearly overlapping frequencies, for methyl, methylene, or methane groups.

The chemical group can exhibit, in addition to the characteristic frequencies or internal vibrations, external vibrations. These are vibrational motions in which the entire molecule is involved. These external vibrational energy levels are extremely sensitive to the geometry of the molecule, and the frequencies can occur over a broad range. It is these external frequencies, which are characteristic of the entire molecule, that give rise to the unique or "fingerprint" portion of the spectra. In a few cases extensive empirical studies have allowed correlations of these external vibrations with general molecular types of structures. The sensitivity of these motions to changes in the geometry of a molecule makes them extremely useful in comparing structural differences such as geometric isomerism.

In review, measurement of the IR frequencies of a molecule can serve three useful purposes. First, the complete spectrum is unique for each molecule and serves to identify the molecule. The characteristic group frequencies can be used to define the organic functional groups within the molecule and, finally, in some cases the external modes can be used to define the geometric variations present in the molecule. For polymer analysis we avail ourselves of all three of these valuable aspects.

The general procedure for analyzing IR spectra is to examine the high frequency region (4000–2000 cm^{-1}) to determine the simple chemical groups present, such as C—H, O—H, N—H. Then the structural units containing these bonds such as CH_3, CH_2, and so forth, are determined by combination of the above information with structural correlations in the lower frequency region. Finally, spectral information is sought for information about the neighbors of the various structural units, where possible. The sequences of units are combined to form the overall molecule. Finally, the postulated structure is drawn. The file of reference spectra is searched to determine whether the complete spectrum of such a structure has been previously studied. If so, the original spectrum is retrieved and compared with the current spectrum for final confirmation of the molecular structure, utilizing the remaining external modes in the low frequency range.

These factors can be illustrated using simple hydrocarbons. The different types of C—H bonds show absorption within well defined areas in the C—H stretching region (3300–2700 cm^{-1}). The approximate positions of the C—H bands for different types of C—H groups are shown in Table 7.4 (18). The C—H stretching modes for unsaturated and aromatic

Table 7.4. Characteristic Group Frequencies of Hydrocarbons[a]

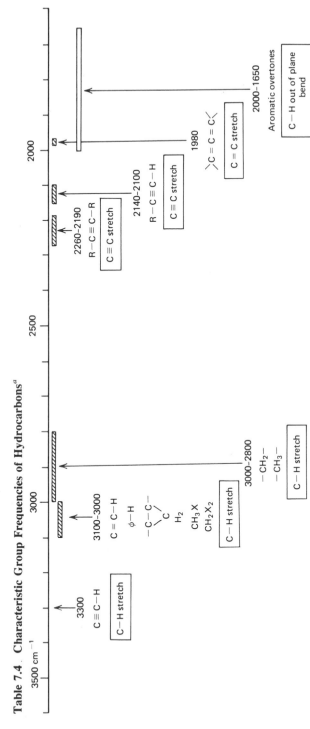

[a] From *Atlas of Spectral Data and Physical Constants for Organic Compounds*, Vol. 1, 2nd ed., J. G. Grasselli and W. M. Ritchey, Eds., CRC Press, Cleveland, Ohio, 1975, p. 351.

molecules occur above $3000 \, \text{cm}^{-1}$, while the aliphatic and aldehyde C—H entities absorb below $3000 \, \text{cm}^{-1}$. The single band from the methine C—H arises from the bond "stretching." This absorption occurs near $2900 \, \text{cm}^{-1}$, but is usually lost among other aliphatic absorption stretching modes. The two bands for the methylene arise from the fact that two C—H bonds are present. When chemically equivalent groups occur, they interact to produce different vibrational energy levels. When the two C—H stretching modes vibrate in phase (termed symmetric) such as

$$\begin{array}{c} \text{H} \\ \diagup \\ \diagdown \text{C} \\ \diagup \\ \diagdown \\ \text{H} \end{array}$$

the band occurs at $2850 \, \text{cm}^{-1}$. The out-of-phase motions (or asymmetric) mode, drawn as

$$\begin{array}{c} \text{H} \\ \diagup \\ \diagdown \text{C} \\ \diagup \\ \diagdown \\ \text{H} \end{array}$$

occurs at $2930 \, \text{cm}^{-1}$. In the same vain for the methyl group, one expects three C—H stretching modes, but only two are observed. The reason is that the two asymmetric modes

$$\begin{array}{cc} \text{—C}\!\equiv\!\text{H} & \text{—C}\!\equiv\!\text{H} \end{array}$$

have the same energy (termed degenerate) and so absorb at the same frequency, $2960 \, \text{cm}^{-1}$. As the number of methyl, methylene, or methine groups in a molecule increases, the relative intensities of the stretching modes increase proportionately.

After ascertaining that these bands exist, hence that the corresponding methine, methylene, and methyl groups are present, one should examine other regions of the spectrum to confirm these results and seek other useful structural information.

The deformation mode for the methine group absorbs weakly in hydrocarbons at $1350–1315 \, \text{cm}^{-1}$. As a result of band overlap of the stretching mode and the weakness of the deformation mode, the C—H bond is not particularly easy to characterize by IR spectroscopy.

The methylene group has a series of molecular vibrations. These modes include the stretching and bending modes, with the addition of three

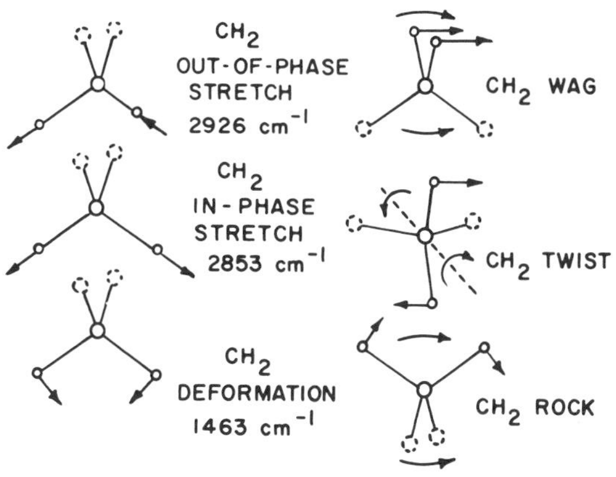

Figure 7.20 The vibrations of the CH₂ group.

rotations of the group on the carbon chain termed *twisting, wagging,* and *rocking.* The vibrations are illustrated in Figure 7.20. The CH_2 deformation mode is a good internal group frequency occurring near $1463 \, cm^{-1}$ in alkanes, independent of the number or sequence of methylenes. This mode shifts to $1440 \, cm^{-1}$ when the CH_2 group is adjacent to unsaturated carbons, and is shifted further to $1425 \, cm^{-1}$ when it is adjacent to a carbonyl. The CH_2 twisting, wagging, and rocking modes are extremely sensitive to the detailed environment of the unit. Detailed experimental and theoretical studies have established that these modes are sensitive to the number of methylene sequences, the geometric structure (trans or gauche), and crystalline packing of the sample. When the methylene is isolated, $X—CH_2—Y$, the twisting mode occurs at 1175, the wagging at 1170, and the rocking mode at $770–785 \, cm^{-1}$, depending on the nature of the substitutent.

Detailed correlations for the organic functional groups have been made, tabulated, and discussed for IR (16) and Raman spectroscopy (17). In general, for polar groupings IR is preferred, and for nonpolar, Raman spectroscopy is best. The Raman effect offers high sensitivity for nonpolar groupings, including the single and multiple bonds of carbon, sulfur, and nitrogen. Cyclic and aromatic structures are particularly strong scatterers.

Raman spectroscopy has been useful in the study of isomeric and branched hydrocarbons where the carbon–carbon stretching modes in the $800–1150 \, cm^{-1}$ region are strong. The primary, secondary, tertiary, and cyclic compounds can be differentiated.

For polymers containing unsaturated groups, Raman spectroscopy is useful in differentiating between internal and external double bonds, cis and trans isomerism, and conjugated unsaturation. One of the earliest applications of Raman spectroscopy was the study of keto-enol tautomerism. For butadiene and isoprene rubbers the type of unsaturation can be determined from the intense Raman scattering of the C=C stretching modes. The trans and cis 1,4-polybutadiene structures scatter at 1664 and 1650 cm^{-1}, respectively, whereas the 1,2-vinyl structure scatters at 1639 cm^{-1} (19). For polyisoprene the cis and trans 1,4-structures scatter at 1662 cm^{-1}, the 3,4-structure at 1641 cm^{-1}, and the 1,2-vinyl at 1639 cm^{-1} (20). Quantitative measurements of the relative amounts of unsaturation and the change during cross-linking can be made with Raman spectroscopy (21). The Raman lines for the vinyl unsaturation of addition polymers are sufficiently strong to suggest their use for end group analysis.

Polymers containing sulfur can be characterized by Raman spectroscopy, since the S—S and C—S linkages are strong scatterers. The line at 506 cm^{-1} in proteins can be assigned to the S—S band of the cysteine residue. Mercaptans, polysulfides, and cyclic sulfur compounds can be studied (21). The Raman spectrum of polyethylene sulfide shows that the 756 and 724 cm^{-1} lines, corresponding to the C—S stretching modes, are nearly 10 times as strong as the hydrocarbon portion of the spectrum (25). The C—S—C bending modes at 337 and 317 cm^{-1} are also strong scatterers. These results suggest that Raman spectroscopy may be useful for the characterization of sulfur vulcanization, and progress has been reported. An example of the Raman spectroscopic differences due to the change in functionality of a polymer as a result of the vulcanization reaction is shown in Figure 7.21 (21).

The use of difference spectroscopy by comparing digitized spectra has greatly enhanced the ability to detect changes in chemical functionality of polymers. In the difference spectrum absorbance bands above the base line reflect an increase in a particular chemical species, while those bands below the base line reflect a decrease in an absorbing species due to the chemical process. The IR difference spectrum for a polyethylene film irradiated in N$_2$ for 50 hr is shown in Figures 7.22 and 7.23 (23). The increase in absorption bands at 1716 and 965 cm^{-1} caused by the irradiation process is more evident in the difference spectrum. Moreover, in the difference spectrum there is a decrease in absorbance of the vinyl end groups, R—CH=CH$_2$, at 1642, 991, and 909 cm^{-1}. There are other more subtle changes evident in the difference spectrum. There is a noticeable decrease in the 1742 cm^{-1} band which is indicative of aldehydic carbonyl groups. This change in the 1742 cm^{-1} band was not evident from a comparison of the irradiated and unirradiated spectra because of the interference caused by overlapping of the 1716 cm^{-1} band. After subtraction, however, the absorbance change in the 1742 cm^{-1} band becomes evident. In the 1000–1400 cm^{-1} region spectral changes have also occurred that are not observable without the difference spectrum. Qualitatively, the difference spectrum for air irradiation is very similar to the irradiation in nitrogen with the exception that the 1716 cm^{-1} ketonic carbonyl mode is stronger for the sample irradiated in air (23). A band at 1410 cm^{-1} appears and increases in intensity as the irradiation conditions become more severe. This band is assigned to a methylene deformation influence by an adjacent carbonyl

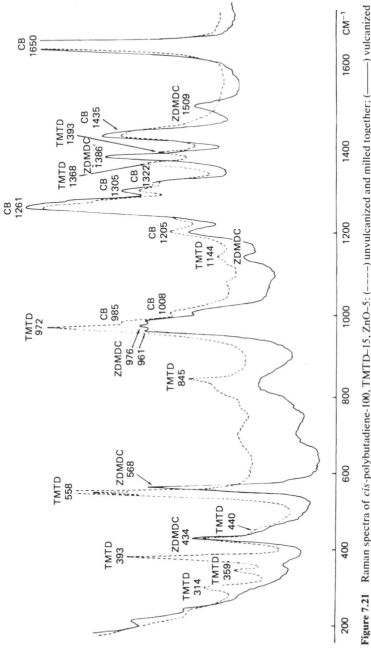

Figure 7.21 Raman spectra of *cis*-polybutadiene-100, TMTD-15, ZnO-5: (– – –) unvulcanized and milled together; (———) vulcanized for 1 hr at 150°C and extracted for 48 hr with MEK. (Reprinted by permission of Ref. 21.)

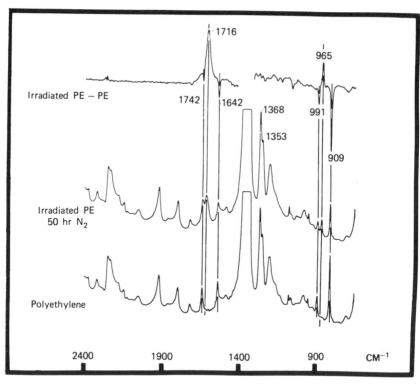

Figure 7.22 IR absorbance spectra for polyethylene film irradiated in nitrogen: (bottom) unirradiated polyethylene; (middle) polyethylene irradiated for 50 hr in nitrogen; (top) difference spectrum (irradiated polyethylene–polyethylene). (Reprinted by permission of Ref. 23.)

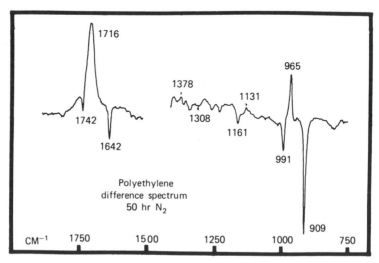

Figure 7.23 Difference spectrum for polyethylene film irradiated for 50 hr in nitrogen. (Reprinted by permission of Ref. 23.)

group

$$R-CH_2-\overset{\overset{\displaystyle O}{\|}}{C}-CH_2-R'$$

An increase in methyl content in the polyethylene films also occurs, as reflected by an increase of the symmetric methyl deformation mode at 1378 cm^{-1}. Bands at 1131 and 1068 cm^{-1} increase upon irradiation and are indicative of C—O stretching vibrations. The amorphous phase in polyethylene—specifically, as reflected in the 1368, 1353, 1308, 1261, and 800 cm^{-1} vibrational bands—is affected by irradiation in preference to the crystalline phase.

7.7   EFFECT OF CHAIN AND SEQUENCE LENGTH IN VIBRATIONAL SPECTROSCOPY (25, 26)

When chemical units are repeated on a chain in a regular fashion, all of the units have the same energy, so they are potentially capable of resonating or coupling their vibrational motions. This intramolecular vibrational coupling can lead to development of a series of resolvable vibrational modes characteristic of the length of the coupled units.

For paraffins these observations have been confirmed, as shown in Figure 7.24, where the frequency ranges for the different CH$_2$ modes are shown as a function of the chain length. The methyl modes have a constant

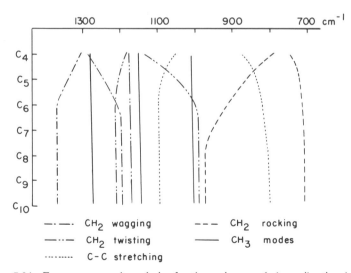

Figure 7.24   Frequency zone boundaries for the various methylene vibrational types.

frequency, since the coupling to the end methyl groups is constant. This plot shows only the boundaries of the vibrational modes with no indication of the number of modes within each boundary. For each type of vibrational mode it is expected that an additional mode of the same type will be generated with the addition of each methylene unit. Unfortunately, due to resolution limitations, selection rules, and weak intensities, one seldom observes all the additional modes. In fact, as shown later, when the chain becomes infinite, the selection rules are highly restrictive and only a single optical mode for each vibrational type of the repeating unit is expected.

Thus the vibrational pattern for a regular polymer chain depends on the number of coupled oscillators or units, the normal modes of an isolated repeating unit, and the extent of coupling of the vibrational modes with other repeat units.

These salient features of the spectra of ordered polymers may be demonstrated by analyzing a uniform one-dimensional lattice of point masses (25). The N frequencies for a linear chain of N atoms acting as parallel dipoles with fixed ends (including only nearest neighbor interactions) are given by the following equations:

$$\omega_s^2 = \omega_0^2 + \omega_1^2(1 + \cos\theta),$$

$$\theta = \frac{s\pi}{N+1},$$

where ω_0 is the frequency of the uncoupled mode, ω_1 is the interaction parameter, and s is an integer from 1 to N.

The N frequencies (ω_s) are a function of θ, which physically corresponds to the phase difference of the vibrational mode of adjacent atoms or cells. When $\theta = 0$ (not possible with finite chains), all atoms have the same motion and, as θ increases, the difference in phase increases to a limiting value of 180° corresponding to the atoms on one unit reaching their maximum displacement at the same time while the equivalent atoms in the adjacent unit are reaching their minimum position. Physically, each value of s corresponds to a standing wave of different wavelength in the molecule. If the interaction parameter ω_1 is small, the normal modes fall in the neighborhood of ω_0 and only a single unresolved mode is observed. Thus the vibrational mode with small ω_1 has no dependence on chain length. This result reminds one of the concept of group frequencies, where the "internal" modes were relatively independent of their environment. This is the case, for example, with the carbon–hydrogen stretching modes, which occur at the same frequencies for all chain lengths of paraffin and in fact for the polymer.

When the interaction parameter is sufficiently large, the energy levels of the N coupled harmonic oscillators are resolvable. The vibrational modes generated as the chain length increases are shown diagrammatically for chains with N up to 8 in Figure 7.25. An additional mode is expected for each unit added to the chain.

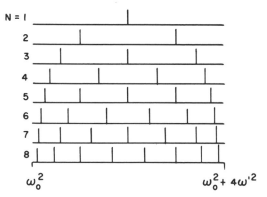

Figure 7.25   Normal vibrational frequencies for a chain of N coupled oscillators, $N = 1$, $2, \ldots, 8$. (Reprinted by permission of Ref. 25.)

Not all of these coupled modes are observed because of selection rule restrictions, but sufficient data do exist to verify the theoretical predictions. In Figure 7.26 the square of the observed frequency for some of the s modes are plotted versus $1 + \cos \theta$ for $C_{22}H_{46}$ (28). The limiting frequency extrapolates to 716 cm^{-1}, which corresponds to the proper frequency, while the interaction constant is 423 cm^{-1}, giving 1107 cm^{-1} as the other extreme value of the rocking frequency. The relative intensities of the modes as s increases are substantially different, making detection of the modes difficult.

The relative IR intensities for a coupled chain of $N = 8$ are shown in Figure 7.27 (25). Often only the limiting mode is observed, but this mode is

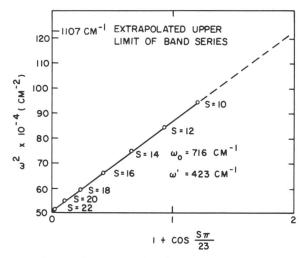

Figure 7.26   Frequencies of the modes plotted versus the phase factor. (Reprinted by permission of Ref. 27.)

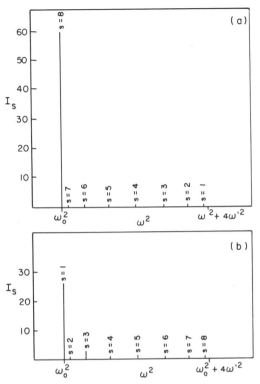

Figure 7.27 Expected IR absorption spectrum of eight coupled dipoles with free ends: (*a*) parallel dipoles; (*b*) antiparallel dipoles. (Reprinted by permission of Ref. 25.)

dependent on the ordered chain length up to a limiting value of N that is different for each mode. The lowest frequency is always the strongest band in the series, while the second strongest is calculated to be from 25% to 3% of the strongest.

This analysis and comparison with experimental data leads to useful conclusions. First, some vibrational modes show a chain length dependence, particularly for short chains. Second, some vibrational modes are independent of chain length and are useful for structure-frequency analysis in the same manner as for other organic molecules.

The final question meriting our attention is the question of what happens when the regular chain structure is disrupted by structural defects such as different geometric or steric isomer or chemical defects. In general these defects decouple the vibrational modes, causing a dramatic change in the spectrum. An excellent example is shown in Figure 7.28, where the IR spectra of *n*-butane is shown as an ordered solid and as a liquid (27). The introduction of the rotational isomer has completely decoupled the two halves of the molecule as well as added the spectrum of the gauche isomer. The increase in band width arises from changes arising from thermal effects. For longer chains the effect is not as obvious but is real and detectable. Hence one can conclude that the coupling only occurs through

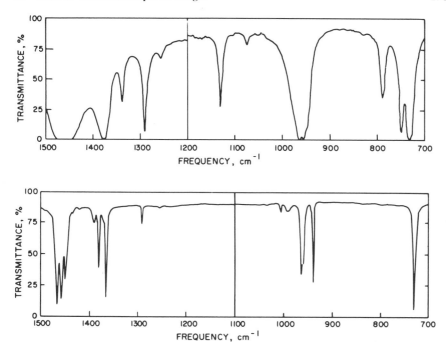

Figure 7.28   IR spectrum of n–C_4H_{10}: (a) n–C_4H_{10} at $-70°C$: (b) n–C_4H_{10} at $-196°C$. (Reprinted by permission of Ref. 28.)

repeat units having the same chemical and geometric structure. Thus structural defects in the chain produce vibrational spectra of shorter sequences. In the case of copolymers of ethylene with other monomers IR bands arising from ethylene sequences have been observed as follows:

	Frequency of rocking mode (cm^{-1})
$\overset{X}{\underset{\vert}{C}}-CH_2-\overset{X}{\underset{\vert}{C}}-$	815
$\overset{X}{\underset{\vert}{C}}-CH_2-CH_2-\overset{X}{\underset{\vert}{C}}$	752
$\overset{X}{\underset{\vert}{C}}-CH_2-CH_2-CH_2-\overset{X}{\underset{\vert}{C}}$	733
$\overset{X}{\underset{\vert}{C}}-CH_2-CH_2-CH_2-CH_2-\overset{X}{\underset{\vert}{C}}$	726
$\overset{X}{\underset{\vert}{C}}-CH_2-CH_2-CH_2-CH_2-CH_2-\overset{X}{\underset{\vert}{C}}$	722

These vibrational modes have been used to characterize the sequence distribution of ethylene-propylene copolymers (29).

Bands characteristic of various sequences in vinylidene chloride (VDC)–vinyl chloride (VC) copolymers have been assigned as follows:

	Frequency of vibrational mode (cm^{-1})
–VDC–VC–VDC	1197
–VC–VC–VDC	1235
–VC–VC–VC	1247

Vibrational modes of all the possible sequence lengths are certainly present in the spectra but, because of the small coupling frequency shifts of the larger sequences, they are lost in the shadows of the stronger bands characteristic of shorter sequences. Difference spectroscopic measurements as well as low temperature (to sharpen bands) can improve the length of sequences that are detectable.

Let us consider the spectrum of an infinite chain of dipoles (30). Physically we will never encounter such an infinite chain but, as we have just seen, above a certain chain length the coupling effects can be considered as arising from an infinite chain, as addition of more units does not measurably influence the results. In other words, when the chain length reaches a critical value, the repeat unit can be considered as part of an infinite chain. For an infinite chain of atoms of mass m spaced a distance d apart and held together by a force K between atoms, the frequency function is given by (30)

$$\nu = \frac{1}{\pi}\left[\frac{K}{2m}(1 - \cos 2\pi k d)\right]^{1/2}$$

$$= \frac{1}{\pi}\sqrt{\frac{K}{m}}(\sin \pi k d).$$

Thus the frequency is a periodic function of the wave number k, the period being equal to $1/d$. The results are plotted in Figure 7.29 (30). The

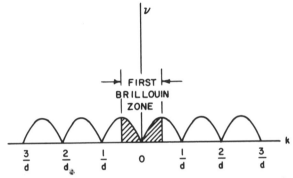

Figure 7.29 Plot of frequencies for the monatomic chain. (Reprinted by permission of Ref. 30.)

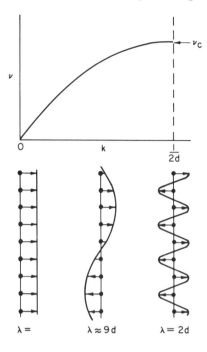

Figure 7.30 Dispersion curves for the monatomic chain, showing forms of the vibrational modes for certain values of k. (Reprinted by permission of Ref. 30.)

frequency goes through a maximum and has a period equal to $1/d$. The first period is called the *first Brillouin zone*. The maxima occur at

$$\nu_m = \frac{1}{\pi}\sqrt{\frac{K}{m}},$$

which is the limiting frequency. A plot of frequency ν versus the wave number k is known as a *dispersion relation*. The dispersion relation for a monatomic chain is shown in Figure 7.30 (30). The motion exhibited by the infinite chain can be visualized as standing waves of wavelength $\lambda = 1/k$. When $k = 0$ ($\lambda = \infty$), all of the atoms are in phase. As shown in Figure 7.30, at intermediate values of k a standing wave motion occurs, and when $k = \pi$ ($\lambda = 2d$), all of the particles have the same amplitude of oscillation but alternate ones are out of phase. For a monatomic chain no spectral activity is expected.

Let us consider a linear diatomic one-dimensional chain. For the special case in which all bonds are identical and the atoms are equally spaced but have different masses, the result is

$$\nu^2 = \frac{k}{4\pi^2}\left\{\left(\frac{1}{m_1}+\frac{1}{m_2}\right)\pm\left[\left(\frac{1}{m_1}+\frac{1}{m_2}\right)^2-\frac{4}{m_1 m_2}\sin^2\pi kd\right]^{1/2}\right\}.$$

The dispersion curve is shown in Figure 7.31 (30). Two curves result for the longitudinal vibrations. The curve that passes through the origin is called the *acoustical branch*, because the frequencies fall in the region of

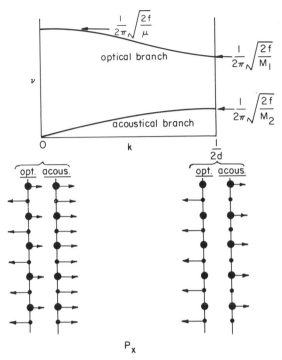

Figure 7.31   Dispersion curves for the longitudinal vibrations of the diatomic chain. (Reprinted by permission of Ref. 30.)

sonic or ultrasonic waves. The upper curve is the *optical branch* and falls in the IR spectral region. There is a frequency gap where no frequencies occur. The forms of the vibrations for each branch are shown in Figure 7.31 for limiting values $k = 0$ and $k = \frac{1}{2}d$. For the optical branch at $k = 0$, the optical branch represents a simple stretching of the bond between two given atoms in which their center of mass remains fixed. This mode is obviously optically active (IR and Raman), since the two atoms are assumed to be different. The acoustical vibration at $k = 0$ does not result in a change in dipole moment and is optically inactive. Frequencies corresponding to the intermediate values of k for the optical branch are spectrally inactive, since for every atom exhibiting a positive dipole displacement in the infinite chain there is a corresponding atom moving in the opposite direction so that the total is zero.

When the results are extended to three dimensions for the motion of a diatomic chain, there is one pair of dispersion curves (one optical and one acoustical branch) for each direction in space. For the isolated chain the two transverse directions are equivalent and each branch is doubly degenerate.

A generalization of these results to a molecular chain is easily

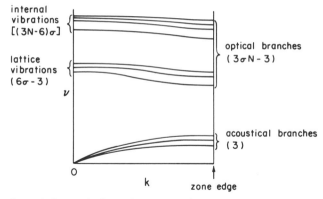

Figure 7.32 General form of dispersion curves for a molecular crystal. (Reprinted by permission of Ref. 30.)

comprehended. When the repeating unit contains m atoms, there are $3m$ separate branches. Three of these modes correspond to acoustical modes, so $3m - 3$ optical branches exist. Of course, some of these branches can be degenerate. The general result is shown in Figure 7.32 (30). Many of the modes of the isolated repeat unit are severely modified by the insertion in a chain, since they are internal frequencies. Others are coupled in the chain. Solid state physicists prefer to call these modes *lattice modes*.

Normal coordinate analysis of polymer chains has been made and complete dispersion curves obtained as shown in Figure 7.33 (31) for an

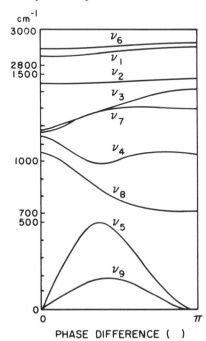

Figure 7.33 Dispersion curves of an infinite polymethylene chain, in-plane modes. (Private communication from Prof. Tasumi.)

infinite polymethylene chain. There are nine branches, since the methylene repeating unit has three atoms. The three translations and one allowable rotation correspond to the zero modes on the acoustical branches. The IR and Raman active modes for polyethylene correspond to the frequencies at the intercepts when $\theta = 0$ and π. The C—H stretching modes and angle bending modes are independent of the phase, as expected, while the twisting, wagging, rocking, and C—C stretching branches show effects due to the chain.

Interestingly, from the complete dispersion curve of a polymethylene chain, the frequency distribution of all lower oligomers can be predicted. This is accomplished by calculating the appropriate θ for each value of s for the N oligomer. Vertical lines drawn at these values of θ intersect the dispersion curve at the expected frequency of the oligomers.

The theoretical vibrational spectrum, IR and Raman, of an infinite polymer chain is simple. For polymethylene only 14 modes (5 IR, 8 Raman, 1 inactive) are expected. Additionally, no rotational fine structure is expected, so the bands should be narrow. The actual situation is somewhat different. First, all polymer chains contain end groups that often appear in the spectra. For polyethylene the vinyl end groups are easily seen. Second, the chains are highly distorted, and third, many of the polymers are semicrystalline, having crystalline and amorphous domains.

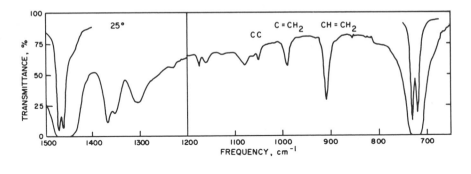

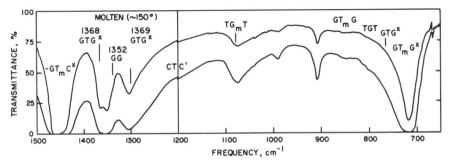

Figure 7.34 IR spectrum of polyethylene with vinyl end groups at 25°C and molten (150°C). (Reprinted by permission of Ref. 41.)

Table 7.5 Assignment of Bands in the Spectrum of Polyethylene

	Polyethylene			
Solid (25°C)		**Melt (>130°C)**		
I	II	I	II	Assignment
1472, vs	1472, vs			δ, cryst
		1463, vs	1463, vs	δ, $-GT_mG*-$, m large
1462, vs	1462, vs			δ, cryst
		1455, s, sh	1455, s, sh	δ, $-GT_mG*-$, m small
~1438, m, b, sh	~1440, m, b, sh	~1438, s, b, sh	~1438, s, b, sh	δ, $-GG-$
1413, vvw				
	1378, m		1378, m	U
1366, m	1368, m	1368, s	1367, s	W, $-GTG*-$
1352, m	1353, m	1349, s	1352, s	W, $-GG-$
	1344, w, sh		1344, m, sh	W, $-TG$
1338, w, sh				W, $-GTTG*-$
1309, m	1308, m	1303, s	1305, s	W, $-GTG*-$
~1270, w, b, sh	~1270, w, b, sh	≈1270, m, b, sh	≈1270, m, b, sh	W, $-GT_mG*-$, $m \geq 3$
1229, vw	1220, vvw			
1175, w	1176, vw			W, cryst
1160, w				
1129, vvw		1125, vvw?		
~1085, w, b, sh	~1087, vw, b, sh	~1087, w, sh, b	~1087, w, sh	$R + W$, $-TG_mT-$, $m \geq 2$
1078, w,b	1079, w, b	1076, m, b	1073, m, b	$R + W$, $-TGT-$
1062, vw	1062, vw			T, cryst
1050, w	1050, vw			T, cryst
989, w		989, w		$RCH\!\!=\!\!CH_2$
	966, vvw, b		968, vvw, b	
953, vvw				
908, m	909, vvw	907, m	910, w	$RCH\!\!=\!\!CH_2$
888, vw	888, w		888, w	β
~858, vw, b		~855, vw, b	~850, vw, b	$-TG_mT-$, $m > 2$, (?)
~808, vw, b		~805, vw, b, sh	~800, vw, b, sh	$-T_mGT_n$, m, n large, (?)
	~775, w, b, sh		~775, w, b, sh	P, $-TG$
~745, w, b, sh		~745, m, b, sh	~745, m, b, sh	P, $-GTG*-$
730, vs	731, vs			P, cryst
		718, s	720, s	P, $-GT_mG*-$, $m > 2$
719, vs	719, vs			P, cryst
620, vw, b		~625, vw, b		

^{a} Polyethylene I has vinyl end groups; polyethylene II has methyl end groups. δ, W, T, and P are methylene bending, wagging, twisting, and rocking modes; U and β are methyl symmetric bending and rocking; R is C—C stretching. $G*$ means that the bond can be either right or left gauche or that it leads to a methyl group terminating the chain.

But at a cursory examination considering only the strong bands the spectrum of polyethylene appears relatively simple compared to that of other organic molecules. The IR spectrum of polyethylene is shown in Figure 7.34 (27). The spectrum is dominated by the methylene bending ($\sim 1450\,cm^{-1}$) and rocking modes ($\sim 725\,cm^{-1}$), as expected, with the end groups appearing at 1375 (CH_3) and 909 (vinyl). The twisting and wagging regions ($1300-1400\,cm^{-1}$) appear as broad bands. If one examines the spectrum more closely, however, many interesting observations can be made, including the splitting of the bending and rocking modes due to the crystalline phase, as shown in Figure 7.34. When the polymer is melted, the bands are broad, but many band assignments to different sequences of trans and gauche isomers can be made, which upon reexamination of the solid sample reveal their presence there as well as reflecting the structure of the amorphous phase.

An assignment of these bands has been made and is shown in Table 7.5 (27). In fact, the spectrum of polyethylene is understood fairly well and a great deal of information can be obtained from it. However, we are continually learning more about it, as reflected by the half dozen or so papers per year on the vibrational spectra of polyethylene.

7.8   QUANTITATIVE SPECTROSCOPIC METHODS FOR POLYMERS

7.8.1   IR Spectroscopy (1)

Traditionally, when the spectroscopist has succeeded in identifying a substance or component of a mixture, he is immediately asked to determine the concentration or amount of material. If a material is to be analyzed, it must have one or more unique, reasonably strong modes. Quantitative analysis is accomplished by comparing the absorption or scattering of a component of the mixture or unknown with a corresponding measurement for a pure substance or a mixture with known concentration. For IR spectroscopy we measure the absorbance

$$A = \log_{10} \frac{I_0}{I},$$

where I is the intensity passing through the sample, and I_0 is the intensity impinging on the sample. It is common practice to take I as the peak of the analytical band, and I_0 is determined by drawing an appropriate base line.

An absolute photometric accuracy of about $\pm 0.2\%\ I$ is attainable with an excellent IR instrument properly aligned and with calibration to ensure a linear photometric response of the mechanical attenuator and associated servomechanism. In addition to photometric accuracy, one must also recognize effects caused by reflection losses, beam convergence, polarization, and nonuniformity of sample.

From an accurate measurement of absorbance we use the Beer–Lambert law:

$$A = k_a Cb,$$

where k_a is the specific absorptivity, C is the concentration, and b is the path length (cm).

Deviations from the linearity between absorbance A and the concentration may arise if the sample exhibits association or dissociation over the concentration range. For polymers an additional complication arises because the intensity of the analytical band can be influenced by changes in the ordering of sequence, whether this be chemical (copolymer), geometric, or stereoregularity. The spectral changes arising from these defects are the same and induce changes in the intensity. In general, these defects produce asymmetric broadening of the bands with significant changes in intensity. By analogy with the frequency dependence on chain or sequence length it has been suggested that the change in k_n on the chain length is a gradual change, becoming constant for large n. On this basis a nonlinear relationship is expected, and it is necessary to establish empirical relationships between absorbance and concentration. This simply reinforces the need for careful calibration of the system through the concentration range of interest.

There are basically four methods of calibrating the absorption coefficient k for a polymer structure. Chemical analysis such as titration and elemental analysis can be used on various systems. Physical methods such as radioactive labeling, NMR, and UV can be most useful. Radioactive labeling of monomers and subsequent counting in the copolymer are particularly recommended. Model compounds can sometimes be used. Model polymers such as natural polymers and others can be used as standards for unsaturation. Blends of homopolymers are useful for developing analysis methods for block copolymers but are insufficient for random copolymer systems. Oligomers are often useful, particularly for end group analysis and branching. Long chain hydrocarbons have often been used for these systems. In some cases the monomers can be used as model systems if they have sufficient size that coupling effects are nil.

Finally, utilizing recent computer methods of data processing, one can sometimes generate the spectra of a pure component from a mixture (32). The IR spectrum of a two-component mixture can be represented by

$$M(\nu) = f_1(\nu) + f_2(\nu),$$

where $M(\nu)$ is the spectrum of a mixture of components, and $f_i(\nu)$ is the spectrum of a pure component i. The spectrum of a mixture of the same two components in different proportions is represented as

$$M_1(\nu) = a_1 f_1(\nu) + a_2 f_2(\nu).$$

Solving the equations for $f_1(\nu)$ and $f_2(\nu)$, one obtains

$$f_1(\nu) = \frac{1}{a_1 - a_2} M_1(\nu) - \frac{a_2}{a_1 - a_2} M(\nu),$$

$$f_2(\nu) = \frac{1}{a_2 - a_1} M_1(\nu) - \frac{a_1}{a_2 - a_1} M(\nu).$$

The ratio spectrum

$$R(\nu) = \frac{M_1(\nu)}{M(\nu)} = \frac{a_1 f_1(\nu) + a_2 f_2(\nu)}{f_1(\nu) + f_2(\nu)}$$

defines the coefficients a_1 and a_2 by means of "flat areas" (1) in a spectral region where $f_1(\nu) \gg f_2(\nu)$, $R(\nu) \approx a_1$. Conversely, if $f_2(\nu) \gg f_1(\nu)$ in any region, $R(\nu) \approx a_2$. The accuracy to which $f_1(\nu)$ and $f_2(\nu)$, the pure component spectra, are calculated depends solely on the accuracy to which a_1 and a_2 are determined.

In Figure 7.35 (32) are shown two mixture spectra, $M_1(\nu)$ and $M(\nu)$, each consisting of three absorbance bands. [$M(\nu)$ is shifted above the base line for clarity.] Two of the bands are representative of component A, hence have the same relative intensity in $M_1(\nu)$ and $M(\nu)$. The third band does not maintain the same intensity relative to the component A bands, hence is characteristic

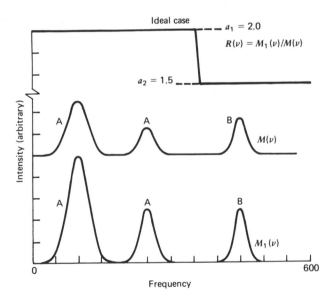

Figure 7.35   Ideal ratio of two Gaussian band spectra: (bottom) spectrum of a two-component mixture, $M_1(\nu)$; (middle) spectrum of same two components in different proportions, $M(\nu)$ (shifted above base line for clarity; (top) ratio spectrum $R(\nu) = M_1(\nu)/M(\nu)$ defining a_1 and a_2. (Reprinted by permission of Ref. 32.)

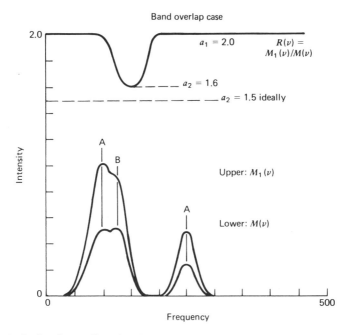

Figure 7.36 Ratio of two Gaussian band spectra containing overlapping bands: (bottom) spectra of two-component mixtures $M_1(\nu)$ and $M(\nu)$; (top) ratio spectrum showing nonideality in determination of a_2. (Reprinted by permission of Ref. 32.)

of a second component B. The ratio spectrum $R(\nu)$ reveals two ideally flat areas defining $a_1 = 2.0$ and $a_2 = 1.5$. These ideal coefficients yield $f_1(\nu)$, containing only A bands, and $f_2(\nu)$, containing the B band.

The determination of the coefficients is inaccurate if, as in Figure 7.36 (32), band overlap occurs, as one expects in polymer systems. In this case the intensities of all bands in $M_1(\nu)$ and $M(\nu)$ have been maintained as in Figure 7.36, but the component B band now overlaps with one of the component A bands. The result is that a_2 is inaccurate, and therefore the derived $f_1(\nu)$ and $f_2(\nu)$ are not spectra of pure components.

It is a simple task to determine the direction of the error in the measured coefficients. At any chosen frequency $R(\nu)$ has an *experimentally* measured value, denoted A^{exp}. If this value is an estimate of the *ideal* coefficient a_2, an expanded expression is

$$A_2^{\text{exp}} = \frac{a_1 f_1 + a_2 f_2}{f_1 + f_2},$$

where f_1 and f_2 are values of $f_1(\nu)$ and $f_2(\nu)$ at a particular frequency. It can easily be shown that

$$A_2^{\text{exp}} > a_2 \qquad \text{if } a_1 > a_2$$

and

$$A_2^{exp} < a_2 \qquad \text{if } a_1 < a_2.$$

Similarly,

$$A_1^{exp} > a_1 \qquad \text{if } a_2 > a_1$$

and

$$A_1^{exp} < a_1 \qquad \text{if } a_2 < a_1.$$

Thus one seeks the extreme values of the ratios.

Figure 7.37 shows spectra of poly(vinyl chloride) (PVC) containing both 15% and 5% by weight of a common plasticizer, dioctyl phthalate (DOP) (32). The coefficients a_1 and a_2 are obtained from the highest and lowest values in $R(\nu)$. These points are the frequencies at which DOP and PVC, respectively, most dominate the mixture spectrum. As shown in

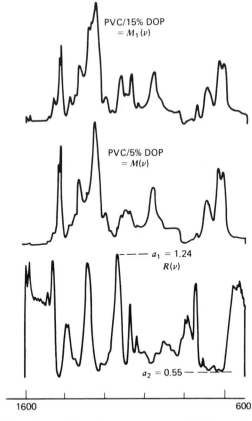

Figure 7.37   (Top) Experimental absorbance spectrum of PVC containing 15% by weight DOP; (middle) spectrum of PVC containing 5% by weight DOP; (bottom) ratio spectrum. (Reprinted by permission of Ref. 32.)

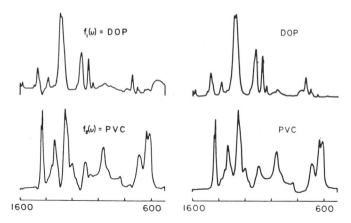

Figure 7.38 Calculated and experimental spectra of DOP and PVC: (left) DOP and PVC spectra calculated with the data of Figure 7.37; (right) experimental absorbance spectra of DOP and PVC. (Reprinted by permission of Ref. 32.)

Figure 7.38, the derived curve $f_1(\nu)$ is in fact a close approximation to the actual spectrum of pure DOP, and is accurate enough to deduce the structure of DOP. Similarly, the $f_2(\nu)$ curve compares well to the experimental spectrum of pure PVC. Slight errors in $f_1(\nu)$ and $f_2(\nu)$ are attributable to inaccuracy in a_1 and a_2 caused by band overlap and/or band shift in addition to interaction between the PVC and DOP molecules.

For solid samples two additional problems arise in quantitative IR measurements. First, the thickness, particularly in KBr or a mull, is difficult to measure. The films used for samples are often very thin, so that a small error in thickness makes for a large relative error. Second, nonrandom orientation of the polymer can lead to erroneous results, since the Beer–Lambert law assumes random orientation. Even solvent casting of polymer films can lead to nonrandom distribution of chains in the thickness direction of the film. For samples of polyethylene oxide cast from chloroform, high molecular weight samples were clearly random, while the lower molecular weights were between 60% and 80% preferentially oriented in the thickness direction. Compression molding of semicrystalline samples leads to row crystallization at the surface and preferred orientation. A method has been described to eliminate these problems, but the measurements are tedious (33).

For solid samples where the thickness is difficult to measure the thickness measurement is made by measuring an "internal thickness band," that is, a band whose intensity is proportional to the thickness of the sample. Then

$$\frac{A_s}{A_i} = \frac{k_s C_s b}{k_i C_i b} = KC,$$

where A_i is absorbance of the internal thickness band.

These internal thickness bands serve as reference bands for difference spectra. It is not always possible to find internal thickness bands, and care must be exercised in their use. Using digitized data, one can ratio a series of spectra and determine whether bands exist whose ratio remains constant over the range of analytical interest and for a range of thicknesses. It is anticipated that modes such as bond stretching and bending, which are insensitive to their environments, would be the best candidates for internal thickness bands.

7.8.2 Raman Spectroscopy

The desirability of using Raman spectroscopy to make quantitative measurements of the functional groups in polymers is recognized. It should be noted that the relationship between concentration (gm/cc) and intensity of the Raman scattering is linear, whereas changes in IR transmission are logarithmic with concentration. In Raman spectroscopy it is necessary to use the "internal standard," technique as the number of scattering sites in the laser beam cannot be determined. For solutions a known amount of a standard can be added for determination of the relative amount of the unknown material in the solution. For solids it is necessary to take the ratio of the analytical Raman line to a line considered an "internal thickness" line. In this manner quantitative results can be obtained for copolymers and other samples run in pellet form. The same problems of calibration of the scattering coefficient exist as for the specific absorptivity in IR spectroscopy. Quantitative measurements have been made on copolymers of vinyl chloride and vinylidene chloride. Linear relationships have been established between the composition, dyads, and triads and the intensity of certain Raman frequencies in the spectra of the vinyl chloride–vinylidene copolymers. The composition can be measured to an accuracy of 2% (34).

7.9 STEREOCHEMICAL CONFIGURATION OF POLYMER CHAINS (7)

Monosubstituted vinyl polymers $(CH_2, CHX-)_n$ may exist in a variety of possible stereoregular structures. They may possess either syndiotactic (substitution on the alternate carbons on opposite sides) or isotactic (substitution on the alternate carbons on the same sides) stereoregularity. It is also possible for these stereoregular structures to assume either planar or helical conformations. Alternatively, the polymer may exist in an atactic form (random placements on the alternate carbons) with a disordered conformation. Each of these structures has different selection rules for IR and Raman activity. It is possible to distinguish these models based on the spectral properties without a detailed knowledge of the molecular motions and energies. It is our intention to summarize these different selection rules, and indicate how they are used to deduce the stereochemical structure of monosubstituted vinyl polymers.

Any observed vibrational spectrum may be classified into sets of modes that can be defined as follows:

1. [R, IR]—frequencies coincident in both Raman and IR spectra.
2. [R, O]—frequencies active in Raman, but inactive in IR spectra.
3. [O, IR]—frequencies inactive in Raman, but active in IR spectra.

Further, each set of Raman and IR frequencies may be subclassified according to their individual Raman polarization and IR dichroic behaviors. The Raman lines may be polarized or depolarized, while IR bands may possess parallel or perpendicular dichroism with respect to the helix axis. The terminology to be used subsequently is based on the general expression $[a, b]$, where a indicates the polarization properties of the Raman lines, and b the dichroic behavior of the IR bands. On this basis a may be p (polarized), d (depolarized), or O (inactive), while b may be σ (perpendicular), π (parallel), or O (inactive). Therefore the expression $[p, \sigma]$ means that the frequency associated with a particular mode is coincident in both Raman and IR, being polarized in the Raman and possessing perpendicular dichroism in the IR. All observed bands must therefore fall into one of the following classifications:

$$[p, O], [d, O], [O, \pi], [O, \sigma], [p, \pi], [p, \sigma], [d, \pi], \text{ and } [d, \sigma].$$

Any spectrum can consist of any or all combinations of the above possibilities. The range of possibilities is, in principal, determined solely by the symmetry properties of the molecule under consideration. However, the picture may be distorted by the limited spectral sensitivity of the Raman and IR instrumentation, and by the presence of multiple structures causing deviations from the optical activity predicted from consideration of the symmetry of the perfect chain. Improved instrumentation can minimize the problem of sensitivity. The spectral bands associated with irregular structures in minor amounts are weak and variable from sample to sample, making their detection possible in some cases.

The various possible structures and their idealized IR and Raman spectra are shown in Figure 7.39 (7). It can clearly be seen that every structure considered possesses a unique set of spectroscopic properties. For example, the vibrational spectrum of a planar syndiotactic vinyl polymer may be classified as $[p, \sigma], [d, \pi], [d, \sigma], [d, O]$, whereas that for an isotactic 3_1 helix is $[p, \pi], [d, \sigma]$. It should therefore be possible to develop an idealized spectroscopic scheme for the unequivocable determination of the structure of any unknown monosubstituted vinyl polymer.

Polyvinyl chloride (PVC) is a monosubstituted vinyl polymer. The x-ray and IR evidence strongly supports the extended syndiotactic struc-

STRUCTURE	SYMMETRY	R IR	p π	p σ	d π	d σ	p O	d O	O π	O σ	EXAMPLE
CENTER OF SYMMETRY	D_{2h} C_{2h}						√	√	√	√	PE, PES
ATACTIC			√	√	√	√					PVF
SYNDIOTACTIC HELIX >3₁	D_n					√	√	√	√		PEO
SYNDIOTACTIC HELIX 3₁	D_3					√	√		√		
SYNDIOTACTIC HELIX 2₁	D_2				√	√	√				
SYNDIOTACTIC PLANAR	C_{2v}			√	√	√	√				PVC
ISOTACTIC HELIX >3₁	C_n		√		√		√				POLYBUTENE
ISOTACTIC HELIX 3₁	C_3		√		√						PP
ISOTACTIC PLANAR	C_s		√	√							

Figure 7.39 Selection rules for monosubstituted polyvinyl polymers.

ture, but a folded syndiotactic and threefold helical isotactic structure have also been proposed. The vibrational modes of these models obey different selection rules and have different dichroic properties that can be used to spectroscopically test each possible structure. An examination of Figure 7.39 suggests that the most obvious method of distinguishing between these three structures is through Raman depolarization measurements. If the vibrational modes that are only Raman active give polarized scattering, the folded syndiotactic model for PVC is required.

If the vibrational modes that are only Raman active give depolarized scattering, the extended syndiotactic structure is required. Thus the spectral type $[d, O]$ is unique to the extended syndiotactic model. Additionally, the spectral type $[p, \sigma]$ is also unique to the extended syndiotactic model. For the isotactic PVC model the polarized Raman lines possess parallel IR dichroism, that is, the type $[p, \pi]$.

The Raman spectrum of ordinary PVC has been obtained (35). The polarized Raman lines occur at 363, 638, 694, 1172, 1335, 1430, and 2914 cm^{-1}. Bands are found in the IR region at each of these frequencies. This latter result rejects the folded syndiotactic structure as an acceptable model for PVC, since this structure requires the polarized Raman lines to be unique. The IR bands at frequencies corresponding to the polarized Raman lines have perpendicular polarizations, which indicate that the helical isotactic model is also unacceptable since it requires the corresponding IR bands to have parallel dichroism. The experimental observation of the $[p, \sigma]$ type of mode constitutes further confirmatory evidence in support of the presence of the extended syndiotactic chain in ordinary PVC. Additional evidence for the extended syndiotactic structure is obtained from the observation of depolarized Raman lines (not observed in the IR spectrum) near the frequencies predicted through normal coordinate calculations for the A_2 modes.

7.10 CONFORMATION OF THE POLYMER CHAIN

7.10.1 The Solid State

For polymers the vibrational spectrum reveals skeletal modes. These skeletal modes should have frequencies that are sensitive to conformation, since they are highly coupled modes and any change in the chain conformation varies the coupling. For the planar, 2_1, and 3_1 helices, differences in selection rules for Raman and IR spectra allow a determination of conformation, as was demonstrated in the preceding section.

All monosubstituted vinyl helical molecules belong to the symmetry group $C_{2t/u}$, where t is the number of turns in the helix for u number of units necessary to reestablish the original position along the chain. The vibrational frequencies under this symmetry fall into a symmetrical representation, A, and degenerate asymmetrical representations: $E_1, E_2, \ldots, E_n$. The higher E modes are optically inactive, leaving just A and E_1 and E_2 modes. The optically active modes correspond to helix angle values $\psi = 0(A)$, $\psi = \psi(E_1)$, and $\psi = 2\psi(E_2)$.

The frequencies of both A and E modes depend on the helix pitch angle. The normal modes of a helix can be represented by a dispersion curve of frequency plotted versus a phase angle θ. Each helix exhibits a characteristic dispersion curve with similar motions having similar energies. On the dispersion curve A-mode vibrations occur at $\theta = 0$; E-mode vibrations occur at $\theta = \psi$. Thus for a polymer with the same repeat units, differences in the A modes for different helical conformations depend on energy considerations only, since the phase angle is the same. The E mode shifts from one helical form to another depend on the energy differences and on ψ differences corresponding to the different helix angles. Thus it is expected that the E modes should be slightly more sensitive to the changes in conformation.

Some branches of the dispersion curve are insensitive to environment and are the "characteristic" modes corresponding to uncoupled vibrational modes. These modes have very flat dispersion curves, and the frequency of these modes does not depend on the phase angle. In general these modes do not depend on the tightness of the helix. Other branches of the dispersion curve are sensitive to the phase angle, and the frequency of the mode depends on the phase angle. These same modes are sensitive to the helix tightness and exhibit a shift in frequency with different helical forms. So, in general, the observed spectra have modes that have the same frequency position regardless of helix type and modes that have different frequency positions because of the helix form. These latter modes are useful for characterizing the helical conformation of a polymer in the solid state.

Vibrational spectroscopy will not supplant x-ray diffraction for the determination of the conformation of a polymer in the solid state. However, it can be a very useful adjunct when a fiber pattern cannot be obtained because the form is unstable or cannot be satisfactorily oriented. Polybutene is an example where spectroscopy has been of value for the determination of the conformation of the polymer chain in the solid state.

Structural studies by many workers have shown that polybutene-1 exists in at least three crystalline modifications. Form II has a tetragonal unit cell containing 11_3 helices and is obtained by cooling from the melt. Polybutene-1 transforms slowly and irreversibly from form II to form I at room temperature. Form I has a hexagonal unit cell with six 3_1 helices. Form II is prepared by casting a film from a number of solvents such as benzene, carbon tetrachloride, toluene, p-xylene, and decalin. Form III transforms upon heating to form II, and then spontaneously to form I. A x-ray fiber pattern of form III cannot be obtained due to this transformation. Form III is thought to have an orthorhombic unit cell, and recent Raman results suggest a 10_3 helix. The Raman spectra for forms I, II, and III are substantially different. In general, the IR and Raman frequencies agree (36). The Raman lines at 774, 824, 875, and 982 cm^{-1} shift in frequency as a function of helix angle. All these frequencies have contributions from C—C backbone motion according to the normal coordinate analysis. All four are observed as medium to strong bands in all the spectra, except that the 875 cm^{-1} band is missing from the form III spectra.

To determine the conformation of form III from the spectrum, it is necessary to establish the functional relationship between these vibrational modes and changes in helix angle. The normal coordinate analysis suggested a linear relationship between the calculated frequency and helix angle for the limited region between 98° and 120°. This information suggests that the observed frequencies for form I and form II be plotted, and from the observed frequencies of form III an interpolation can be made to determine the helix angle for form III. The linear plots utilizing the A modes of form III at 764, 822, and 966 cm^{-1} produce helix angles of 101.5°, 106.5°, and 109.7°, respectively. If one rejects the 101.5° value based on the statistical 99% confidence interval for forms I and II, a helix angle of 108.1° is predicted, giving form III a 10_3 helix conformation. Three E modes predict a ψ of 106.8°, which gives a 10_3 helix rather than a 7_3 helix (36).

Using this information, one can roughly index the diffraction pattern of form III. Several reflections of $hk3$ and $hk7$ are noted, which would appear strong for a 10_3 helical molecule. Thus, from the observed spectral behavior and the x-ray diffraction data, the assignment of form III to a 10_3 helix seems likely.

7.10.2   Conformation of Polymers in Liquid and Solution

Vibrational spectroscopy can also be used for the study of polymeric structure in the liquid and solution phase. When a polymer is melted, the molecular chain conformation becomes disordered, and the vibrational spectrum is appreciably different from the spectrum of the crystalline state. Many new frequencies appear in the spectrum arising from the new conformations present in the melt, and some of the crystalline frequencies disappear while other modes decrease in intensity. Band splitting is not evident in the molten state, since the vibrational coupling is disturbed upon melting. The vibrational modes are broad and diffuse due to the variety of structures present in the melt. No specific selection rules are expected, since the structures are disordered. The Raman spectra differ from the IR spectra only in terms of the relative intensities of the various modes. The Raman spectrum can be easily obtained in the low frequency region, where the energy differences of the various conformations should be the greatest.

The Raman spectrum of molten PEO, shown in Figure 7.40, should be compared with the spectrum of the solid state, shown in Figure 7.41 (38). The IR spectrum of molten PEO has been obtained, and the spectrum is considerably different from that of the solid state. The vibrational assignments for the molten state are difficult not only because the bands are

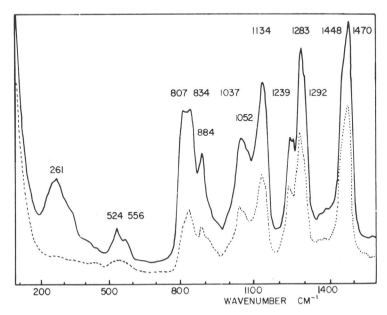

Figure 7.40   Raman spectrum of molten PEG: (———) electric vector of incident beam perpendicular to scattering plane; (− − − −) electric vector of incident beam parallel to scattering plane. (Reprinted by permission of Ref. 38.)

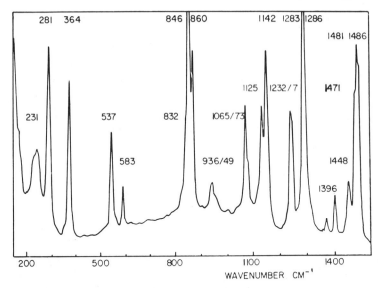

Figure 7.41 Raman spectrum of solid PEG (Carbowax 6000). (Reprinted by permission of Ref. 38.)

weak, badly overlapped, and broad, but also because of the large number of structures with small energy differences which are possible in the melt. In the solid state one structure predominates, and for PEO this conformation is the *TGT*, where the three chain bonds per repeat unit each assume a trans (O—C), a gauche (C—C), and trans (C—O) form. Possible conformations in the molten state are *TGT*, *TGG*, *GGG*, *TTT*, *TTG*, and *GTG*. Statistical weights of the various conformations may be estimated approximately from the abundance ratio $(G + G')/T$ for chain bonds, where G' is the opposite angular notation to G. If the ratios are taken as $\frac{8}{2}$ for the CH_2—CH_2 bonds and $\frac{2}{8}$ for CH_2—O bonds, the statistical weights are calculated as $TGT:1$, $TTT:\frac{1}{4}$, $TGG:\frac{1}{4}$, $TTG:\frac{1}{8}$, $GGG:\frac{1}{64}$, and $GTG:\frac{1}{128}$. The *GTG*, *TGG'*, *GGG'*, and *GG'G* models are discarded, since the probability of finding the *GG'* sequence is negligible because of steric hindrance.

A vibrational analysis of these models has been reported (39). Because none of the conformations is sufficiently prevalent to form helical segments, selection rules do not apply. The calculations were made on helical segments through mathematical necessity, and frequency distributions were calculated in an effort to assign the observed bands. Some of these models have frequency distributions that are sufficiently isolated to be assignable to some of the modes observed in the IR and Raman spectrum. Some of these assignments are listed in Table 7.6 (38). It is observed that some of the modes can be assigned to specific conformations, while other modes apparently have contributions from several conformations. It is apparent that all six of the structures are present in the

Table 7.6 IR and Raman Spectrum of Molten Poly(ethylene glycol) (38)

IR Frequency (cm^{-1})	Raman Frequency (cm^{-1})	Assignment	Form	Models (Tentative)
1485 sh		CH$_2$ scissor	T	TTT, TTG, GTG
1460 m	1470 s	CH$_2$ scissor	G	TGT, TGG, GGG
	1448 sh	CH$_2$ scissor	G	TGG, GGG
1352 m	1352 m	CH$_2$ wag	G	GGG
1326 (m)	1326 w	CH$_2$ wag	T	TTT, TTG, GTG
1296 m	1292 m	CH$_2$ twist	G, T	All
	1283 s	CH$_2$ twist	T	TTT, TTG, GTG
1249 m		CH$_2$ twist	G	TGG, GGG
	1239 m	CH$_2$ twist	T	
1140 sh	1134 s	C—O, C—C	G, T	All
1107 (s)		C—O, C—C, CH$_2$ rock	G, T	All
	1052 m(P)	C—O, C—C	T	TGG
1038 (m)		C—O, C—C, CH$_2$ rock	T	TTT
992 w		C—O, C—C, CH$_2$ rock	T	TTT, TTG
945 (m)		C—O, C—C, CH$_2$ rock	G	TGT
~915	919 sh	CH$_2$ rock, C—O, C—C	G	TGG, GGG
886	884 mw(P)	CH$_2$ rock	G	GGG
855 (m)	—	CH$_2$ rock	G	TGG, GGG
	834 m(P)	CH$_2$ rock	T	TTT, TTG, GTG
~810 (sh)	807 m(P)	CH$_2$ rock	G	TGG
	556 w			
	524 w			
	261 (P)			

melt. Since the vibrational modes have specific extinction coefficients in the IR and specific scattering coefficients in the Raman effect, the relative abundance of the various conformers cannot be calculated. The intensities of the Raman lines at 807 cm^{-1} indicate that the TGG isomer is predominant, with the number of trans ethylene units (TTT, TTG, GTG) also large.

Raman spectroscopic studies of polymer solutions are of interest primarily to relate the conformation of the polymer chain to other solution properties. In many cases the structure of the polymer changes upon dissolution or undergoes transformation with changes in the pH, ionic strength, or salt content of the solution. The ideal solvent for Raman spectroscopy is water, since the scattering of water is very weak except at 1650 cm^{-1} and 3600 cm^{-1}. The previous studies of aqueous solutions have been handicapped, as very few physical techniques can be used. Raman spectroscopy is water, since the scattering of water is very weak except obtained.

The Raman spectra of PEG in aqueous and chloroform solutions are shown in Figure 7.42 (38). Comparison of the Raman spectrum in aqueous

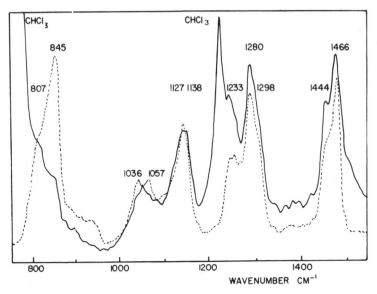

Figure 7.42 Raman spectra of PEG in aqueous solution (– – – –) and in chloroform (———). (Reprinted by permission of Ref. 38.)

solution with that of the melt indicates that the changes on dissolution in water are considerably less dramatic than observed on melting the polymer. On the other hand, the spectrum of PEG in chloroform is very similar to the spectrum of the melt. In aqueous solution the halfwidths of the lines are considerably narrower than those in the molten state, indicating that fewer energy states are available to the molecule in aqueous solution than in the melt. The shifts in frequencies from the solid state are not appreciable. In aqueous solution the Raman lines occur at 884, 846, and 807 cm^{-1} in the methylene rocking region, with the 846 cm^{-1} line predominating. Thus the *TGT* methylene structure is more prevalent than the *TGG*, *TTG*, and *GGG* structures. This indicates that the change in structure of the backbone upon dissolution in water is not as complete as observed in the melt or dissolution in chloroform. This result is in agreement with IR and NMR results (40).

7.11 SUMMARY

The discussion given in this chapter should serve to indicate the popularity and utility of vibrational spectroscopy, both IR and Raman, for the study of the chemical composition and structure of polymer molecules. Over 200 papers are published each year in this field, further illustrating the wide variety of applications. Many more applications will appear as the instrumental techniques improve and our theoretical understanding increases.

REFERENCES

1. W. S. Potts, *Chemical Infrared Spectroscopy*, Vol. I, *Techniques*, John Wiley & Sons, New York, 1964.

2. N. B. Colthup, L. H. Daly, and S. E. Wiberley, *Introduction to Infrared and Raman Spectroscopy*, Academic Press, New York, 1964.

3. J. A. Koningstein, *Introduction to the Theory of the Raman Effect*, D. Reidel Co., Dordredt, Holland, 1972.

4. M. M. Coleman, P. C. Painter and J. L. Koenig, *J. Raman Spectrosc.* 5, 417 (1976).

5. E. B. Wilson, J. C. Decius, and P. C. Cross, *Molecular Vibrations*, McGraw-Hill, New York, 1955.

6. J. R. Ferraro and J. S. Zromek, *Introductory Group Theory and Its Application to Molecular Structure*, Plenum Press, New York, 1975.

7. J. L. Koenig, *Appl. Spectroscopy Rev.*, 4, 233 (1971).

8. *American Laboratory*, 68 (Nov. 1973).

9. C. J. Henniker, *Infrared Spectroscopy of Industrial Polymers*, Academic Press, London, 1967.

10. J. Haslam, H. A. Willis, and M. Squirrell, *Identification and Analysis of Plastics*, 2nd ed., Ileffe, London, 1972.

11. M. Zeller, *Perkin Elmer Infrared Bulletin 35*.

12. J. Higgins and P. Miller in *Laboratory Methods in Infrared Spectroscopy*, 2nd ed., Miller and Stace, Heyden, Philadelphia, 1972.

13. N. J. Harrick, *Internal Reflection Spectroscopy*, Wiley-Interscience, Publishers, New York, 1967.

14. P. R. Griffiths, *Chemical Infrared Fourier Transform Spectroscopy*, John Wiley & Sons, New York, 1975.

15. J. L. Koenig, *Appl. Spectrosc.*, 29, 293 (1975).

16. L. S. Bellamy, *The Infrared Spectra of Complex Molecules*, Chapman & Hall, London, 1975.

17. F. R. Dollish and W. G. Fateley, *Characteristic Raman Frequencies of Organic Compounds*, John Wiley & Sons, New York, 1974.

18. J. G. Grasselli and W. M. Ritchey, *Atlas of Spectral Data and Physical Constants for Organic Compounds*, Vol. I, CRC Press, Cleveland, Ohio, 1975.

19. S. W. Cornell and J. L. Koenig, *Rubber Chem. Technol.* 43, 322 (1970).

20. S. W. Cornell and J. L. Koenig, *Rubber Chem. Technol.* 43, 313 (1970).

21. M. M. Coleman, J. R. Shelton, and J. L. Koenig, *Rubber Chem. Technol.*, 45, 173 (1972).

22. A. C. Angood and J. L. Koenig, *J. Macromol. Sci. (Phys.)*, B3, 321 (1969).

23. D. L. Tabb, J. J. Sevick, and J. L. Koenig, *J. Polym. Sci.*, 13, 815 (1975).

24. D. S. Eiley, *Anal. Chem.*, 40, 894 (1968).

25. R. Zbinden, *Infrared Spectra of High Polymers*, Academic Press, New York, 1964.

26. R. G. Snyder and J. H. Schachtschneider, *Spectrachem. Acta*, 19, 85 (1963).

27. R. G. Snyder, *J. Chem. Phys.*, 27, 1316 (1967).

28. R. G. Snyder, *J. Chem. Phys.*, 27, 969 (1967).

29. G. Bucci and T. Sermonazzi, *J. Polym. Sci.*, 7C, 203 (1964).

30. G. Turrell, *Infrared and Raman Spectra of Crystals*, Academic Press, New York, 1972, Chapter 3, p. 60.

31. M. Tasumi, personal communication.

32. J. L. Koenig, L. D'Esposito, and M. K. Antoon, *J. Appl. Spectrosc.*, 31, 292 (1977).

33. A. C. Angood and J. L. Koenig, *Macromolecules*, **2**, 37 (1969).

34. M. Meeks and J. L. Koenig, *J. Polym. Sci.*, *A-2*, **9**, 719 (1971).

35. D. Druesedow and J. L. Koenig, *J. Polym. Sci.*, **7A**, 1075 (1969).

36. S. W. Cornell and J. L. Koenig, *J. Polym. Sci.*, **7A**, 1965 (1969).

37. J. P. Luongo and R. Salovey, *J. Polym. Sci.*, *A-2*, **4**, 997 (1966).

38. A. C. Angood and J. L. Koenig, *J. Polym. Sci.*, **A28**, 1787 (1970).

39. H. Matsuura and T. Miyazawa, *J. Polym. Sci.*, *A-2*, **7**, 1735 (1969).

40. K. Liu and J. L. Parsons, *Macromolecules*, **2**, 529 (1969).

41. R. G. Snyder, *J. Chem. Phys.*, **47**, 1316 (1967).

8
Nuclear Magnetic Resonance of Polymer Chains

8.1 INTRODUCTION

Nuclear magnetic resonance (NMR) spectroscopy is perhaps the most powerful technique available for studying the environment of individual atoms within a polymer chain. Other spectroscopic techniques must be interpreted empirically and sometimes these qualitative interpretations are equivocal. NMR data yield qualitative as well as quantitative information that is most often unequivocal.

The phenomenon of NMR was first observed in 1946 by Purcell at Harvard and Block at Stanford. A three-line spectrum of the protons of ethyl alcohol in 1951 demonstrated the potential for data on the chemical nature of protons through chemical shifts. High resolution NMR spectra of polymers were first reported by Bovey and Tiers (1). Subsequently, a flood of applications was discovered as the nature and chemical environment of individual protons were examined. The increase in magnetic field generated further applications, as longer sequences could be detected with the increase in resolution and signal. High resolution pulsed ^{13}C NMR of polymers was first reported by Schaefer (2), and a new surge of interest in NMR resulted. For ^{13}C NMR triad and pentad compositional sequences could be observed as well as configurational sequences. Duch and Grant (3) were the first to obtain ^{13}C NMR spectra of bulk polymers. Later Schaefer (4) used magic angle spinning to obtain ^{13}C magnetic resonance on solid polymers of a variety of types. This observation has generated a new wave of interest, and it appears we are on the threshold of exciting new results using NMR.

Bovey (5) has produced a most complete and well written book on the

NMR spectroscopy of polymers. This book is highly recommended for an intensive discussion of the spectra of particular systems and a discussion of applications. A most up-to-date and complete review of the use of ^{13}C NMR for polymers has been written by Schaefer (6). For an in-depth presentation of high resolution NMR in solids, the monograph by Haeberlen (7) certainly details the theory and practice, but is limited in applications—no polymers are mentioned.

8.2 BASIC THEORY OF NMR (8)

8.2.1 Nuclear Spin

With NMR we are using a nucleus to investigate molecules. Atomic nuclei (like ^{1}H, ^{13}C, and ^{19}F) possess magnetic moments and can serve as magnetic probes because the local magnetic field for a particular nucleus depends on its chemical environment, dipolar interactions of the other nuclei, shielding by the electrons, and intermolecular field effects.

When a nucleus with a magnetic moment is placed in a magnetic field H_0, it aligns itself with the field and has an energy of $-\mu H_0$, where μ is the magnetic moment. The nucleus can have $2I + 1$ distinct energy states in the magnetic field, where I is the spin quantum number. These levels are labeled by the value of the magnetic quantum number m, where $m = I\,(I-1)\cdots -I$. This splitting of energy levels is termed *nuclear Zeeman splitting* by analogy with the Zeeman splitting of electronic levels. These Zeeman levels are equally spaced with an energy of $\mu H_0/I$, as shown in Figure 8.1. The basis of the NMR experiment is to supply energy of the appropriate frequency to induce transitions between these levels. The frequency required is

$$h\nu_0 = \frac{\mu H_0}{I} = \gamma \hbar H_0,$$

where γ is the gyromagnetic ratio. For the proton $\mu = 1.42 \times 10^{-23}$ erg/G and at a field of 23.5 kg, the frequency is 100 MHz, whereas for ^{13}C, ν_0 is 25.2 MHz, which can be produced by radiofrequency techniques.

Nuclei with spins of $\frac{1}{2}$, such as protons, actually precess about the applied magnetic field direction with a frequency called the *Larmor frequency* $(\omega_0 = 2\pi\nu_0)$ and, in order to induce them to flip, a second magnetic field H_1 must be applied at right angles to H_0. When this second field rotates at the precession frequency of the nuclei, a transfer of energy occurs from the lower energy to higher energy state and we observe the resonance phenomenon.

The two primary nuclei of interest to polymer chemists are ^{13}C and ^{1}H, which share some common characteristics but differ in some very important ways. First, the gyromagnetic ratio γ of ^{13}C nuclei (^{12}C has no

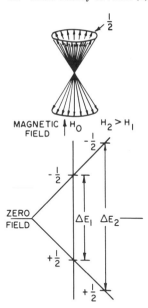

Figure 8.1 Nuclear orientation and energy levels of nuclei in a magnetic field for a nuclear spin number of $\frac{1}{2}$.

nuclear spin) is $\frac{1}{4}$ of the ^{1}H nuclei. Since the sensitivity of the nucleus is proportional to γ^3, ^{13}C gives $\frac{1}{64}$ the signal of the ^{1}H nuclei. Second, the natural abundance of ^{13}C is 1.1%, further lowering the sensitivity to a level 6000 times lower than the proton NMR experiment. Although the ^{13}C NMR experiment is much less sensitive with respect to the resonance signal, there are other factors that make the effort worthwhile.

8.2.2 Spin-Lattice Relaxation

Unlike the other forms of spectroscopy, the differences between the upper and lower states in NMR are very small. The relative numbers are given by

$$\frac{N_{+1/2}}{N_{-1/2}} = 1 + \frac{2\mu H_0}{kT}.$$

For a field of 14,100 G, $2\mu H_0$ is only 0.01 cal or, stated another, more meaningful way, $2\mu H_0/kT$ is less than 10^{-5}. Thus at ordinary temperatures, according to the Boltzmann distribution law, only a very small excess of nuclei are in the lower energy state at thermal equilibrium. At 25°C for every 1 million protons in the upper state, there are 1 million and 6 in the lower. In the presence of a magnetic field H_1 a transfer of energy occurs from lower to upper state. Since the difference is initially so small, saturation can occur, resulting in no resonance signal unless the system can relax sufficiently rapidly from the upper energy state to the lower energy state. Fortunately, such a relaxation process is made available by the molecular motion of the molecules, which are referred to as the *lattice* irrespective of whether the system is liquid or solid. The motion of these neighboring nuclei with respect to the observed nucleus produces fluctuating local magnetic fields that have components in the direction of H_1 and induce emission or relaxation from the higher to lower energy state. The time required for the difference between excess spin population and its equilibrium to be reduced by the factor e is called T_1, or the spin-lattice relaxation time. After 1 T_1, $1/e$ (=36.8%) of the initial magnetization remains, and after $5T_1$ 99% has occurred. It has been found that T_1 varies inversely with η/T, where η is the viscosity, indicating that narrow lines are obtainable from low viscosity solutions. Narrow lines approaching 10^{-4} G are possible if the magnetic field is uniform.

Detailed theoretical and experimental discussions of the spin-lattice relaxation times for ^{1}H (9) and ^{13}C (6) nuclei are available.

8.2.3   Chemical Shift for ^{1}H and ^{13}C Nuclei

In detecting nuclear magnetic resonance, the radiofrequency (rf) radiation is usually held at a fixed frequency while the applied magnetic field is varied linearly with time until the resonance condition is reached. Absorption is recorded as intensity versus magnetic field. The gyromagnetic ratio γ is characteristic of each type of nucleus, thus all protons have the same γ. This means that all protons in a magnetic field H_0 absorb energy at the same frequency. However, it was discovered very early that the three types of protons in ethyl alcohol resonate at slightly different frequencies and that the intensities of the peaks are proportional to the numbers of protons of each type (10). The origin of these different proton resonances is the cloud of electrons about each proton. The electrons have magnetic moments and, as a result of interaction between these electrons and applied magnetic field H_0, an opposing secondary magnetic field directly proportional to H_0 is produced. Thus the actual magnetic field at the sites of the nuclei is no longer equal to H_0 but is

$$H_{\text{loc}} = H_0(1 - \delta),$$

where δ is the shielding constant. The shielding constant δ is a function of the electron distribution around the nucleus and thus is a measure of the chemical environment at the site of the molecule in which the nucleus resides. Protons nearly devoid of electrons, as in carboxyl groups, are subject to little shielding and appear at very low fields. Protons with high shielding appear at high fields. Extremes in chemical environment cause shifts to vary over a range of about 20 parts per million (ppm).

Because nuclear shielding is proportional to the applied magnetic field, the spacing between peaks (or groups of peaks) corresponding to different types of nuclei is proportional to the magnetic field. Thus there is no natural fundamental scale unit; the energies of transitions between quantum levels are proportional to the laboratory field. There is also no natural zero or reference. For practical purposes these difficulties are evaded by the following practices:

1. Using ppm relative change in H_0 as the scale unit.
2. Using an arbitrary reference substance dissolved in the sample and referring all displacements in resonance, called *chemical shifts*, to this "internal" reference.

The use of a dimensionless scale unit has the great advantage that the chemical shift values so expressed are independent of the value of H_0 of any particular spectrometer, and so a statement of chemical shift does not have to be accompanied by a statement of the frequency (or field) employed. The standard internal reference substance for proton spectra is tetramethylsilane (TMS). On this scale tetramethylsilane is assigned a value of zero since its protons are more shielded than those of nearly all other organic compounds. The chemical shift is normalized to the frequency used and is reported as shift in ppm from the reference line. The great majority of proton chemical shifts fall between 0 and 10 on this scale. Figure 8.2 shows the chemical shifts for a number of types of protons in organic structures and functional groups (11).

In ^{13}C NMR the known carbon chemical shifts cover a range of ca. 600 ppm. The carbonyl compounds appear at low field and the methyl carbons at high field. According to convention, in ^{13}C NMR enriched carbon disulfide is the internal standard, but more recently interest has been turning to using TMS as well. A general ^{13}C chemical shift chart (12) is shown in Figure 8.3.

Although theoretical treatments of the chemical shift factors have been attempted, these treatments have not been satisfactory. For the ^{13}C shifts additive empirical equations have been developed for the presence of attached or nearby carbon atoms (13).

For paraffins the chemical shift is calculated assuming that the carbon of interest is an alkyl-substituted methane molecule. First, a constant is used which nearly corresponds to that of methane (-1.87 ppm). Then chemical shift contributions are added for each carbon up to five carbons adjacent to the specified carbon. These contributions are described by constants α for the first bonded carbon, β for the second carbon, which is two bonds away, γ for the third, δ for the fourth, and ϵ for the fifth. If there are two α, β, or other carbons, the constant is multiplied by the appropriate number of adjacent carbons of the type. Values of α through ε were obtained by Grant and Paul from a regression analysis of a series of paraffins. The results are shown in Table 8.1. For branched alkanes upfield shifts are observed for branched carbons and carbons next to branches, so corrective terms are required to account for chemical shifts of carbons associated with branching. Quaternary, tertiary, secondary, and primary carbon atoms are designated by 4°, 3°, 2°, and 1°, respectively. In the corrective terms the

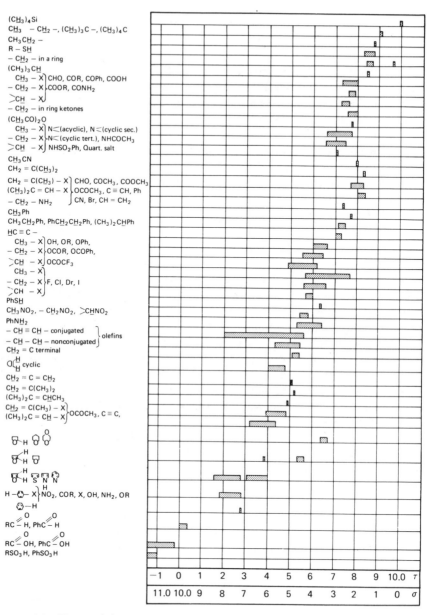

Figure 8.2 Characteristic NMR spectral positions for hydrogen in organic structures. (Reprinted by permission of Ref. 12.)

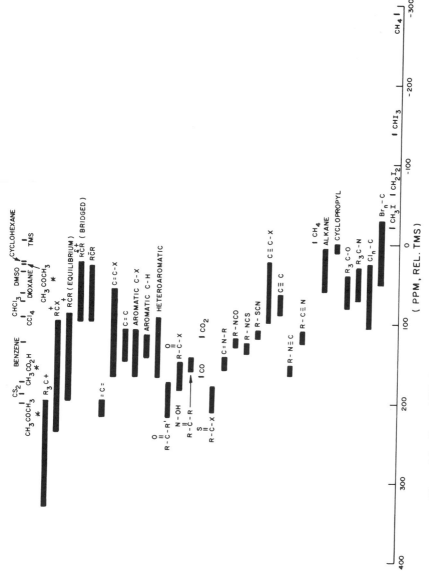

Figure 8.3 General ^{13}C NMR chemical shift chart. (Reprinted by permission of Ref. 12.)

Table 8.1   Grant and Paul Parameter Values of α through ε^a for Paraffins

α	8.61 ± 0.18 ppm	$3°(2°)$	-2.65 ± 0.08 ppm
β	9.78 ± 0.16	$2°(3°)$	-2.45 ± 0.17
γ	-2.88 ± 0.10	$1°(3°)$	-1.40 ± 0.38
δ	0.37 ± 0.14	$\delta(CH_4)_{Rsub.}{}^b$	-1.87 ppm
ε	0.06 ± 0.13	Number of observations	56

^{a} Measured from polymers in 1,2,4-trichlorobenzene solutions at 125 °C.
^{b} With respect to an internal TMS standard.

carbon of interest is tested first and the adjacently bonded carbon is placed in parenthesis. Thus the 3° (2°) term gives the corrective contribution to the methine carbon chemical shift associated with an adjacent methylene group. Table 8.1 shows the chemical shift parameters for a series of ethylene-1-olefin copolymers and hydrogenated polybutadiene (14). This equation can be used to predict chemical shifts with a standard error of 0.8 ppm. Consider calculating the ^{13}C chemical shift of the methyl, methylene, and methine carbons for polypropylene. For the methine the shift contributions are as follows:

The additivity relationship is given by

$$\delta_{CH} = 3\alpha + 2\beta + 4\gamma + 2\delta + 4\varepsilon + 2[3°(2°)] + (-1.87)$$
$$= 3(8.61) + 2(9.78) + 4(-2.88) + 2(0.37) + 4(0.06) + 2(-2.65)$$
$$= 27.68 \text{ ppm (26–29 ppm observed)}.$$

In a similar fashion, for methylene and methyl the relationship is

$$\delta_{CH_2} = 2\alpha + 4\beta + 2\gamma + 4\delta + 2\varepsilon + 2[2°(3°)] + (-1.87)$$
$$= 45.41 \text{ ppm (44–47 ppm observed)};$$

$$\delta_{CH_3} = \alpha + 2\beta + 2\gamma + 4\delta + 2\varepsilon + 1°(3°) + (1.87)$$
$$= 20.74 \text{ ppm (20–22 ppm observed)}.$$

For atactic polypropylene the configurational contribution adds additional splittings of 1–2 ppm in the vicinity of each carbon, but these contributions have not been included.

8.2.4 Dipole-Dipole Interaction

Whereas the dipole-dipole relaxation avoids saturation, it also modifies the magnetic environment of the observed nucleus in the field and must be dealt with in the NMR experiment. This factor induces a local field arising from the magnetic moments of neighboring nuclei in fixed positions. If another nucleus is present at a distance R, the magnitude of the field can range to $\pm 2\mu/R^3$. The $\pm$ sign corresponds to the fact that the local field may add or subtract from H_0. For a proton at $1\,\text{Å}$ this is a range of $57\,\text{G}$. Actually, most rigid lattice solids show a range of $10\,\text{G}$ due to the Gaussian distribution of magnetic interaction. Thus the nuclei do not experience the same uniform magnetic field H_0, but have a broad range that will produce broad resonances of $10\,\text{G}$.

In solids this dipolar effect can be used to obtain molecular geometries about the separation distance R and orientation θ of the dipoles. The separation depends on the orientation of the crystal in the magnetic field.

The equation expressing the energy dependance of this effect is

$$\Delta E = 2\mu[H_0 \pm \tfrac{3}{2}\mu R^{-3}(3\cos^2\theta - 1)],$$

where θ is the angle between the line joining the nuclei and the direction of H_0; under proper conditions both R and θ can be determined. When molecular motion takes place, the variables R and θ are a function of time. For organic systems R is fixed and θ varies with time. For this case the time averaged local field is

$$H_{\text{loc}} = \mu R^{-3} T_2^{-1} \int_0^{T_2} (3\cos^2\theta - 1)dt,$$

where T_2 is the time that the nucleus resides in a given spin state. When θ varies rapidly over all values, this time average can be replaced by a space average:

$$H_{\text{loc}} = \mu R^{-3} \int_0^\pi (3\cos^2\theta - 1)\sin\theta\, d\theta = 0.$$

Thus the net effect of the neighboring magnetic nuclei is lost, and the line width narrows by a factor of 10^4–10^5 compared to its rigid lattice value. So when the motion of the nuclei relating to each other is sufficiently high, there is a spin exchange or flip-flop within the nuclei magnetic energy states, with no overall change of energy of the system but a shortening of the lifetime of each. Hence T_2 is a correlation time and is an inverse measure of the broadening of the spectral line. If the motion is sufficient, the magnetic-dipole interaction of nuclei is neutralized, and the resonance line width ΔH is sharply reduced. From the presence of steps on the $\Delta H(T)$ curve one can judge the onset and character of motion of polymer chains or their parts. The second moment of the line shape is frequently used:

$$\Delta H^2 = \frac{\int_{-\infty}^{+\infty} f(H)(H - H_0)^2 \, dH}{\int_{-\infty}^{+\infty} f(H) \, dH},$$

where $f(H)$ is the absorption line contour, and $(H - H_0)$ is the deviation from the line center (the denominator has been added for normalization). Broad line NMR measurements have proven to be useful in the study of molecular motions of polymers (9).

Techniques (to be discussed later) have been devised for the direct observation of ^{13}C nuclei in solids, and these observations can be used to deduce atomic positions in solids (17, 18). The effects of the nuclear magnetic dipole-dipole interactions are directly specified by interatomic distances and orientations. The chemical shift tensor of a nucleus in a molecule in the solid state can yield information about the local electronic environment of the nucleus and its local site symmetry. The resonance position ν_{res} is given by

$$2\pi\nu_{\mathrm{res}} = \gamma_c H_0 (1 - \lambda_x^2 \sigma_{xx} - \lambda_y^2 \sigma_{yy} - \lambda_z^2 \sigma_{zz}),$$

where σ_{ii} are the principal elements of the chemical shift tensor (CST) and λ_i are the direction cosines between H_0 and the ith principal CST direction. For a methylene group in a long chain hydrocarbon the principal axes must coincide with (1) a vector parallel to the line joining two protons on the same CH_2 unit (σ_{11}), (2) the bisector of the H—C—H bond in the plane of these atoms (σ_{22}), and (3) the chain direction (σ_{33}). Each of these chemical shift tensors have different shielding values, as shown in Table 8.2 (16) for the interior methylene group of a long chain hydrocarbon. The chain axis

Table 8.2   Principal Values of the Chemical Shift Tensors for the Interior Methylene Unit in n-Eicosane and Polyethylene

Tensor Element Direction of Shielding		Magnetitude of Shielding (ppm)	
		—CH_2—[a] (eicosane)	—CH_2—[b] (PE)
Proton-proton vector	σ_{11}	142.6 ± 2.0	143.3
CH_2 angle bisector	σ_{22}	154.6 ± 2.0	157.0
Chain axis	σ_{33}	175.6 ± 2.0	180.0
σ_{ave}		157.6 ± 2.0	160.4
σ (liquid or solution)		163.2	163.5

[a]   D. L. Vander Hart, *J. Chem. Phys.*, **64**, 830 (1976).
[b]   J. Vikina and J. S. Waugh, *Proc. Natl. Acad. Sci.*, **71**, 5062 (1974).

direction corresponds to the direction of the least shielded component. The direction cosines can also be calculated, and the principal axes compared with those determined by crystallography. Observe that the principal values of the CST for the interior methylene group of eicosane differ from those of linear polyethylene, particularly in the chain direction. Observe also that the average shielding (157.6 ppm) for the solid eicosane is different from the liquid (163.2 ppm). This difference is too large to be discounted based on susceptibility arguments, so one must conclude that the chemical shift tensor is dependent to some extent on crystal packing forces such as conformational restraints, slight changes in bond angle caused by packing forces, and anisotropic crystalline force fields. Therefore it is important to observe that the crystal habit influences the chemical shift tensor in the solid, while the isotropic shifts in the liquid state are identical (16).

8.2.5 Electron Coupled Spin-Spin Interaction

In addition to the shielding, electrons produce additional effects due to the magnetic moments of neighboring nuclei. Since this part of the coupling occurs via chemical bonds, it is independent of spatial orientation. This interaction survives the effect of molecular motions. Thus a nucleus with a single neighbor having spin $\frac{1}{2}$ can see its neighbor either in $+\frac{1}{2}$ or $-\frac{1}{2}$ states; consequently its absorption peak splits into two lines.

Assume that we have a molecule such as ethyl chloride. We expect that the methylene group will give rise to a resonance pattern having a different chemical shift than the methyl group. If the methyl group is coupled to the methylene group, what resonance pattern is expected for each type of proton? To answer this question we must view with some detail the effect of the methylene group on the methyl group. One of the protons on the methylene group can be oriented either with or against the field. This is true also for the other proton. Therefore any given methylene group can display three possible total orientations of their spins, as shown in Figure 8.4. In one case they are both oriented with the field, in two cases they are oriented in opposite directions, and in one case they are both against the field. If they were both oriented with the field, they would generate totally a small magnetic field that would add to the applied magnetic field from our sweep. Thus the methyl protons, interacting with the methylene protons both oriented with the field, would come into resonance at a slightly lower field than would all others. If one spin were oriented in each direction, then the small magnetic field produced would be canceled by the other, so that there would be a net contribution of zero to the applied magnetic field. Therefore methyl protons whose adjoining methylene protons had this configuration of spin orientations would give rise to a resonance line slightly upfield from the former line and actually at the true location of the chemical shift for this group. The final orientation,

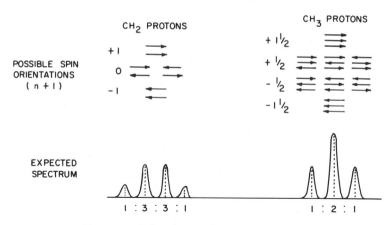

Figure 8.4 Spin-spin coupling for ethyl chloride.

in which both methylene protons are aligned against the magnetic field, would then cause a slightly greater application of the applied magnetic field so that the methyl resonance line having a methylene group whose protons are aligned against the field would give rise to resonance at a slightly higher field location. Since there are a total of three orientations, the methyl group should give rise to a resonance pattern having three lines. The spacing between lines is called the *spin-spin coupling constant J.* Since there are four specific orientations within this methylene group and each one is present statistically in equal amounts, each orientation gives rise to 25% of the resonance of the total methyls. Since two of the orientations are identical in magnetic effect, that species represents a situation where the resonance line arising from this group represents 50% of the total resonance line of the methyls. It follows, therefore, that the three lines have an intensity distribution of 25, 50, and 25%, or $1:2:1$ in an intensity or area ratio. Note that electron coupled spin-spin interactions between members of an equivalent group do not give rise to multiplet splitting.

In a similar manner one can predict the pattern for the methylene group being spin-coupled with the methyl. Here again there are a total of four possible orientations, which give rise to four peaks, and the distribution of the area within the groups is $1:3:3:1$, following the same reasoning as before. Since these magnetic effects are transmitted through the chemical bonds, the spin-spin interaction is the same for methyl and methylene groups. The coupling constant should be the same for each (7 Hz for methyl methylene) and characteristic of the type of interacting groups. It is also obvious that this type of magnetic effect should be entirely independent of the applied magnetic field and, therefore, the coupling constant should be independent of field strength, as is observed.

If a nucleus has n sufficiently close, equivalently coupled neighbors, its resonance splits into $2nI + 1$ peaks, corresponding to the $2nI + 1$

possible spin states of the neighboring group of spins. Intensities of the peaks are given by the number of equivalent arrangements of spins for each spin state, and are therefore proportional to the coefficients of the binomial expansion. Hence there is means of determining the number of equivalent nuclei on neighboring carbon atoms and also a means of determining what group is attached to the carbon group in question. Finally, the intensity of the observed resonance line is always directly proportional to the concentration of nuclei. Therefore areas under resonance multiplets are directly proportional to the concentration of the species.

Let us apply these considerations to steric isomers of vinyl polymers. Consider a molecule of the type

$$\begin{array}{c} M \\ | \\ X-C-Y \\ | \\ M \end{array}$$

If the groups X and Y both have a plane or axis of symmetry (e.g., phenyl, methyl, halogen), the M groups are magnetically equivalent, and one resonance line is observed, as no spin-spin splitting occurs. If one of the groups has no symmetry element, the groups are nonequivalent. An AB quartet of lines results from each proton exhibiting a chemical shift that is split due to interaction with the other proton. When the M groups are equivalent, they are termed *homosteric*; when they are nonequivalent, they are termed *heterosteric* (5). Consider a molecule of the same type as above, where $X = Y = CR_1R_2R_3$, or

$$\begin{array}{c} R_1 \quad M \quad R_1 \\ | \quad | \quad | \\ R_2-C-C-C-R_2 \\ | \quad | \quad | \\ R_3 \quad M \quad R_3 \end{array}$$

It is apparent that we have two similar asymmetric centers.

If X and Y have the same handedness, either a d or l isomer results. These enantiomers are diastereomeric, but the CM_2 group is homosteric. If X and Y are of opposite handedness, it is the meso diastereomer and the CM_2 group is heterosteric. Studies of racemic and meso 2,4-disubstituted pentanes confirm these considerations; for example, with the 2,4-diphenylpentanes, the methylene groups differ by 0.21 ppm in chemical shift in the meso isomer, but are equivalent in the racemic isomer. The extension of these results to polymers is apparent. Note that a methylene group in an isotactic chain can be regarded as the center unit in a meso dyad of monomer units having the same configuration (thus heterosteric) and exhibiting an AB quartet of lines. Similarly, for racemic dyads in syndiotactic chains, the methylene dyads are homosteric, so a single resonance line is observed. These considerations were the basis of the first NMR measurements of the steric configuration or stereoregularity of poly(methyl methacrylate) (1). As the magnetic field becomes higher and the signal is enhanced, it is possible to see effects of longer sequences including triad, tetrad, pentad. (These are discussed in Chapter 10).

Unfortunately, life is not entirely simple. The spin-spin mechanisms just discussed and the simple way of predicting the number of lines to be observed apply only when the spin-spin interactions are first order, which represents about 90% of the observed cases. As the chemical shift between two coupling groups becomes smaller and smaller, there is a mixing of the energy states and, in the extreme, two changes take place. One is that the line intensity relationships no longer hold as it did for first order, and second, the total number of lines observed no longer follows first order predictions. The first is illustrated in Figure 8.5, in which the coupling constant is constant and the chemical shift is changed. This represents the spectrum obtained for two nonequivalent nuclei spin-coupling with each other, and one would expect on the basis of first order predictions that each of these would give rise to a doublet, each line being of equal intensity. The upper trace represents the situation observed when the chemical shift is only 10 times greater than the coupling constant. Here two doublets are seen, except that the two inner lines have slightly greater intensity than the two outer lines. If the chemical shift is decreased so that it is only 5 times the magnitude of the coupling constant, the second trace is observed, and the group of four lines almost represents a quartet. The

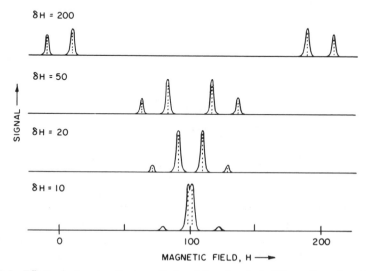

Figure 8.5 Effect of changing the magnitude of the chemical shift on the spin-spin interactions for the resonance lines.

doublet asymmetry is becoming greater. The third trace shows the situation observed when the chemical shift is 2 times the magnitude of the coupling constant; one sees a pattern very similar to a quartet expected from spin interaction with the methyl group. However, the key here is that the intensity distribution is not $1:3:3:1$. Finally, the last trace represents the situation where both the coupling constant and the chemical shift are of the same order of magnitude. Now one sees almost a single line in the center and two very weak satellites on each side. If the chemical shifts were exactly the same, these two species would indeed give rise to a single line and the spin interaction would not be observed. Even in the final trace depicted here, if resolution is not good and the noise is not kept to an absolute minimum, one may only observe a single line here. Thus a general rule of thumb is that first order predictions break down when the ratio J/δ becomes greater than one-tenth. The intensity changes within a resonance multiplet as the chemical shift differences decrease relative to the coupling constant.

Now let us consider a system that has greater than two spins and observe how the number of observable lines may vary beyond first order prediction. In the two spin case we have just observed that the minimum number of lines to be observed is 1, which occurs only when the nuclei are equivalent, and the maximum number that can be observed when they are nonequivalent is 4. The situation now changes somewhat when we go to three spins. If they are all equivalent, only 1 resonance line is observed; but if they are each nonequivalent, then first order predictions would yield 12 lines. However, theoretically, if the coupling constant is somewhat large in comparison to the chemical shift, then there is a possibility of mixing of the spin states and 15 rather than 12 lines is the theoretical maximum. There are 3 combination lines, which brings the total to a maximum of 15. Table 8.3 shows the maximum number of lines that are theoretically observable as a function of the number of nonequivalent nuclei.

In addition, the J couplings may be positive or negative, so the spectra can be quite complex. By way of illustration, if one considers a five

Table 8.3 Spins and Lines

No. of Spins	Maximum No. of Lines
2	4
3	15
4	56
5	210
6	792
7	3,003
8	11,440

spin system such as many normal olefins have, there may be 210 resonance lines. Furthermore, these 210 lines are no longer spaced in the simple manner predicted by first order coupling. Thus to determine the coupling constants and chemical shifts in these situations, one is forced to utilize computer techniques. A number of suitable programs for this purpose have been devised. Values of J and δ can be used to calculate a theoretical spectrum or, alternatively, given the line positions and intensities, the program calculates a solution.

For carbon there can be coupling between two ^{13}C nuclei, coupling between a ^{13}C nucleus and a proton, and coupling between a ^{13}C nucleus and other magnetic nuclei. ^{13}C—^{13}C coupling is not usually observed because of the low probability (1 in 10,000) of having two ^{13}C nuclei in the same molecule. Due to the wide band decoupling of protons in ^{13}C NMR, the ^{13}C—H coupling constants are not obtained. Therefore in most organic systems the ^{13}C NMR resonances are singlets characteristic of the resonating nuclei (15). Molecules containing magnetic nuclei other than ^{1}H and ^{13}C can give rise to spin-spin coupling that is observable in ^{13}C NMR.

8.3 INSTRUMENTATION FOR NMR

8.3.1 Field Sweep NMR (20)

NMR is one type of absorption spectroscopy and requires the same basic components, namely, an energy source that is a monochromatic electromagnetic wave in the radiofrequency region, a sample, and an electronic detector-recorder system. The only unusual requirement is a homogeneous magnetic field—homogeneous to the extent of 1 part in 10^7 with good time stability. There are basically three kinds of magnets: permanent magnets, electromagnets, and superconducting magnets. The permanent magnets are limited to about 7 kG, whereas the electromagnets can reach 13 kG, and the superconducting magnets produce fields of the order of 100 kG. High fields are desirable, since the signal and the chemical shift are a function of field but the separations of the multiplets are independent of the field and, as a result, simplification of the spectra can result. The major problem is the homogeneity of the field, and various techniques are used to improve the homogeneity, including shimming and sample spinning. Sample spinning at 10–20 rps is commonly used for this purpose.

It is nceessary for the rf field to be stable to 1 part in 10^9. This is accomplished by modulation coils that are powered by a field modulator. By using an external lock system with two sample chambers, the field/frequency drift rate can be maintained to within less than 1 part in 10^8 per hr. Even more precise control can be achieved by using an internal lock system with the reference substance contained in the sample tube, and modern instruments are so designed, as illustrated in Figure 8.6. This

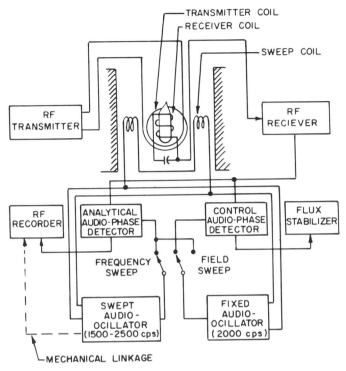

Figure 8.6 Block diagram of a cross-coil NMR spectrometer with internal lock (Varian HA–100).

instrument can generate the spectrum by sweeping either the magnetic field strength or the rf frequency.

Since the line width is 10^{-4} G, or 1 part in 10^4, and saturation must be avoided, one has two alternatives for detecting the signal; rapid sweep rates, which naturally result in low resolution, or low power settings of the rf field, which result in low signal intensities.

Signal enhancement can be accomplished by rapidly sweeping through the spectrum many times in succession, beginning each new sweep at exactly the same point in the spectrum and then summing up the traces. For this purpose one can use a multichannel pulse height analyzer— essentially a small computer that can remember several hundred traces. A "computer of average transients" (CAT) is quite helpful for this purpose.

Mode selection of the signal is accomplished by phase sensitive detection. To do this the amplified signal is compared to the signal from the rf transmitter; by electronic means one may select the component of the probe signal in phase or out of phase with H_1.

The most modern instruments have at least three independent irradiation channels, allowing simultaneous observation, locking, and decoupling of the nuclear species. Two receiving channels are derived from or locked in one master source, preventing frequency drifts. Wide gap (>1 in.) magnet systems utilize universal probes and large diameter tubes. Finally, interfacing is accomplished to a laboratory computer to enhance control and data processing.

8.3.2   Experimental NMR of ^{13}C Nuclei

As previously indicated, the inherent resonance of the ^{13}C nuclei is 5720-fold lower in sensitivity than for protons. However a number of instrumental factors have been utilized to generate a measurable ^{13}C signal (21). First, large sample tubes up to 12 mm in diameter can be used for ^{13}C NMR, and this provides a signal enhancement factor of 11. Second, proton decoupling techniques are used to collapse the multiplets arising from ^{13}C—^{1}H dipolar effects to singlets, giving an increase in sensitivity of 2–15 times.

A by-product of proton irradiation is the nuclear Overhouser enhancement (NOE). Since the ^{13}C nuclei are relaxed from their excited state by the protons, the increase in the Boltzman population of the protons by the irradiation of the protons allows the ^{13}C nuclei to change their own energy level populations. Thus rf energy can be absorbed by the ^{13}C nuclei as a result of the larger population in the lower energy level. The theoretical NOE enhancement is 2.988 times the total signal in the absence of ^{1}H irradiation (22). Finally, time averaging can be used to give a gain of 20–30. But most important, Fourier transform (FT) techniques (to be discussed) give an improvement of 10–100. Hence the instrumental factors allow for from 1000 to 50,000 enhancement of the ^{13}C signal, making the ^{13}C resonance observable even in natural abundance.

8.3.3   FT NMR (23)

The limitation of field sweep or continuous wave (CW) NMR is that only one frequency is being observed at a time. What is desired is a method of observing all of the frequencies all of the time, since for ^{13}C NMR with a chemical range of 5000 Hz, each frequency is observed only 1/5000 of the time. Multiplex methods give substantial improvement. Pulse methods have been developed which excite all of the precession frequencies and the sample responds by absorbing its own frequencies. These pulse methods give an enhancement proportional to $(F/D)^{1/2}$, where F is the chemical shift range and D is the duration of the pulse. These pulse methods require high rf power for a short period. The detector sees a free induction decay (FID) as the pulse decays. With a single frequency of resonance, the FID is an exponentially decaying sine wave, as shown in Figure 8.7 (24). For many resonances the FID is more complicated and the Fourier transformation must be taken to obtain the resonance frequencies.

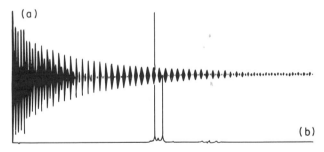

Figure 8.7 Free induction decay (*a*) and frequency domain spectrum (*b*) of $(CH_2=CH)_4Si$ (the small triple is benzene–d_4 solvent). (Reprinted by permission of Ref. 24.)

FT NMR requires a high power amplifier to generate a high flux, rf gates to form the pulse, a data acquisition system capable of accepting data at rates of 10,000 measurements/sec, and processing equipment to carry out the mathematical operations.

8.3.4 High Resolution ^{13}C NMR of Solids

Magic Angle Spinning of Solids in NMR. For liquids and solutions high resolution measurements of isotropic chemical shifts and *J* couplings are possible because rapid molecular motion incoherently but effectively averages out certain line broadening interactions. In the solid state molecules are fixed in specific orientations, so the nucleus experiences not only the applied field but also dipolar interactions. These interactions vary randomly in time and from point to point, causing a broadening of the resonance line shape. It is now understood that this breadth is due primarily to residual chemical shift anisotropy and dipole–dipole interactions. For dilute spins like ^{13}C, the dipolar broadening can be removed by decoupling, using intense rf irradiation at the proton resonance frequency. Nevertheless, a substantial broadening remains because of the anisotropy of the chemical shift in the solid. This line broadening is eliminated by a process called magic angle spinning (MAS) which involves rapid rotation of the sample at an angle of 54.7° relative to the external magnetic field. The criterion for achieving complete narrowing is that the rotation rate, v_R, be greater than the magnitude of the chemical shift anisotropy ($\Delta\sigma$) which means rotational rates of 3–5 kHz. The effect of magic angle spinning is shown for the spectrum of solid poly(methyl methacrylate) in Figure 8.8 (25).

Sample spinning removes any rotationally nonvariant interaction in solids. The line width goes from a broad Gaussian for a rigid solid to a simple exponential decay with a time constant T_Ω, which is expected to increase proportionally to Ω^2, where Ω is the frequency of rotation.

In general the sample must be rotated at a frequency Ω that is large

POLY (METHYL METHACRYLATE)

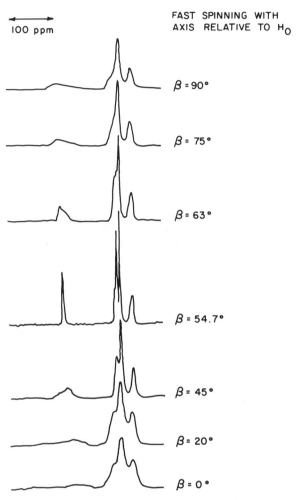

Figure 8.8 Effect of setting angle for fast spinning relative to H_0 on the ^{13}C NMR spectrum of poly(methyl methacrylate). (Reprinted by permission of Ref. 25.)

compared to the line width:

$$\frac{\Omega}{\sigma} \gg 1,$$

where σ is the root mean square second moment of the resonance line shape. This means that for ^{13}C NMR of solids the spinning conditions must be large compared to the 50 kHz ^{13}C linewidth. These speeds of rotation are impossibly high.

However, for ^{13}C spectra, for which the severely dipolar broadened

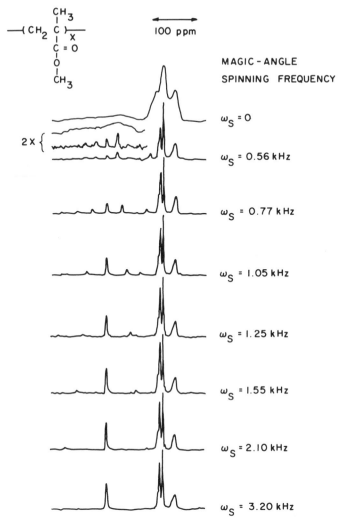

Figure 8.9 Effect of magic angle spinning frequency on the ^{13}C NMR spectrum of poly(methyl methacrylate). (Reprinted by permission of Ref. 25.)

^1H—^1H dipole fluctuations can be removed by decoupling, slow spinning rates can produce narrowing of the ^{13}C lines (25). The effect of a magic angle spinning frequency is shown in Figure 8.9 (29). Side bands do appear in experiments with slow spinning, but these spinning side bands reflect the chemical shift anisotropy observed in a nonspinning experiment and may be of value.

Cross-polarization NMR of Solids (17, 18). Another innovative approach to obtaining ^{13}C NMR spectra of solids has been termed *proton-enhanced* *nuclear double resonance*, or cross-polarization. As previously indicated, the dipolar interaction of the proton spins can be eliminated by strong resonant irradiation of the protons. This is known as *decoupling*. In the cross-polarization experiment a transfer of spin polarization from the protons to the ^{13}C nuclei is accomplished. Imagine applying rf fields at resonance for the ^{1}H or ^{13}C spin systems. Each system in its respective rotating frame sees a static field about which the spin precesses with the Larmor angular frequencies ω_{HH} and ω_{HC}. Under normal conditions the systems have no dynamic effect on each other because of their widely differing resonance frequencies. When ω_{HH} and ω_{HC} are adjusted into resonance, there is a rapid exchange of polarization or energy. After the enhanced ^{13}C magnetization is obtained from the richer ^{1}H, the ^{13}C irradiation is turned off while the ^{1}H radiation is maintained. The ^{1}H irradiation now serves to decouple the ^{1}H and ^{13}C systems while the ^{13}C spin undergoes a high resolution decay for a time T_{2C}^* when the contact is broken. Many of these decays are averaged to increase the signal-to-noise ratio and then Fourier transformed to give a high resolution spectrum. This basic form of the proton-enhanced nuclear double resonance experiment is shown in Figure 8.10. The H spins are interlocked with the ^{13}C spin field. During this process the ^{13}C magnetization builds up during the contact time τ and then undergoes a high resolution decay for a time T_{2C}^* when contact is broken. An enhancement of between 5 and 10 is experimentally observed. This process may be repeated until the proton polarization is depleted. The protons repolarize with sufficient time in the laboratory field and this entire process may be repeated. In such a cross-polarization experiment the proton T_1 rather than the much longer ^{13}C T_1 determines the allowable

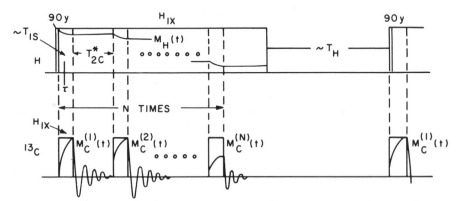

Figure 8.10 Basic form of proton-enhanced nuclear double resonance. After spin-locking the ^{1}H spins, the ^{13}C spin rf field is turned on, satisfying the Hartmann-Hahn condition of resonance. The ^{13}C magnetization builds up during the contact time and then undergoes a high resolution decay for a time when the contact is broken. This process may be repeated until the ^{1}H polarization is depleted. The entire procedure may be repeated after the ^{1}H spins repolarize in the laboratory field.

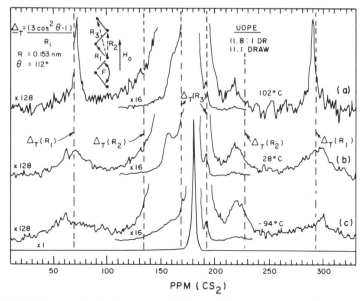

Figure 8.11 ^{13}C cross-polarization spectra of oriented linear polyethylene with an 11.8:1 draw ratio. The draw direction is parallel to the magnetic field. The three pairs of ^{13}C–^{13}C dipolar satellite positions, based on $R_1 = 0.153$ nm and $\theta = 112°$, are given by the dashed lines. (Reprinted by permission of Ref. 26.)

repetition rate. Actually, by repeating the sequence a gain in the power signal-to-noise ratio of the order of 4 is typically achieved per pulse.

Using these ^{13}C—^{1}H cross-polarization techniques for solids, one can use NMR to develop structural information by following the ^{13}C magnetization caused by nearby ^{1}H dipolar fields.

Figure 8.11 shows the ^{13}C spectra of oriented linear polyethylene (26). At low amplifications only one line is observed at 181 ppm relative to CS_2, corresponding to the methylene chemical shift tensor when the chain axis is parallel to H_0. At higher amplification ($\times 16$) significant intensity arises about 40 ppm downfield from the main resonance. This intensity arises from chains that are not fully aligned by the drawing process. At high amplification ($\times 128$) satellites occur due to the ^{13}C—^{13}C dipolar interaction. Calculations show that these lines arise from pairs of carbons on the same chain which are one, two, and three bonds separated. These resonances are quite broad, but sharpen at higher temperatures due to rotational averaging of θ about the chain axis direction. The carbon-carbon distance is calculated from the measurement of these satellite separations (Δ) by the equation

$$\Delta = 1.5\gamma_e^2\hbar^2(1 - 3\cos^2\theta_i)R_i^3,$$

where R is the distance between the ^{13}C nuclei and is exactly the c dimension of the orthorhombic unit cell.

8.4 EXPERIMENTAL TECHNIQUES (5)

8.4.1 Sample Preparation

To obtain narrow lines for polymer chains in solution, the motional narrowing arising from the segmental motion of the polymer chain must be such that

$$\frac{1}{t_c} \geq \delta\omega,$$

where $\delta\omega$ is the linewidth in units of angular frequency, and t_c is the average time required for two protons to rotate around each other by one radian. Thus, for polymers exhibiting a linewidth of 10 G, the t_c for the segmental motion of the polymer chain becomes less than ca. 4×10^{-6} sec.

The rate of spin-lattice relaxation is proportional to the viscosity of the solution and inversely proportional to the temperature. Solvents are selected that yield low viscosity.

Further, the most obvious approach to circumvent this problem is to dilute the solution until a satisfactory T_1 is observed. But, as dilution increases, signal intensity decreases; thus one must achieve a balance in dilution. Another solution to this general problem is to increase the temperature of the sample and thereby decrease the viscosity. Again one is seriously limited by the nature of the solvent, its boiling point, the solubility of the polymer in the solvent as a function of the temperature, and the strict requirement that no thermal gradients can be subjected to the magnetic pole pieces. The latter effect causes inhomogeneities in the field, thus reducing the resolution. Also, as temperature is increased, the density decreases within the sample volume and, therefore, a slight reduction in signal intensity occurs with increase in temperature. Consequently, one must first find the optimum solvent, concentration, and temperature for the study of a particular polymer. Finally, in the preparation of the solutions extreme care must be taken to eliminate any insoluble or gel material from the sample tube, as these particles induce inhomogeneities in the field and line broadening. A millipore filter membrane on the hypodermic needle is recommended.

8.4.2 Spin Decoupling

Spin decoupling is the elimination of spin-spin couplings by irradiation at a second radiofrequency. These experiments are often called *double resonance experiments*. When two nonequivalent nuclei, say A and X, are coupled, the spectrum is split into a characteristic multiplet structure. If (while the spectrum of nucleus A is observed) a second radiofrequency is applied at the resonant frequency of X, it induces transitions among the spin states of X. If these transitions occur rapidly enough, the multiplet structure collapses. This effect is very useful in a complex spectrum to simplify the spectrum. In ^{13}C NMR the "white noise" decoupling of all of the protons gives rise to singlets (as previously discussed). Additionally, the double resonance decoupling proves unequivocally that the two nuclei are coupled, thus allowing a chemical group identification. Selective decoupling is done by successively irradiating with a single proton frequency each of the proton resonances and observing the effects in the ^{13}C spectrum. Irradiation of directly bonded protons leads to collapse of the larger proton-induced splittings. The net result is to give a singlet for only the directly bonded carbons while the resonances of other carbon nuclei are only slightly affected. It is then a simple matter to identify a ^{13}C singlet with specific proton multiplets.

A useful technique for ^{13}C NMR assignments is off-resonance decoupling. In this process the sample is irradiated strongly at a single frequency several hundred hertz from the region of proton resonance frequencies. In the ^{13}C spectrum all the splittings are reduced by different amounts and all long range couplings are eliminated. Those carbon nuclei whose directly bonded protons precess at a rate closest to the decoupling frequency have the smallest splittings while those farthest away have the largest splittings. Thus one can identify several peaks with one experiment. An example of this off-resonance method is shown in Figure 8.12 for

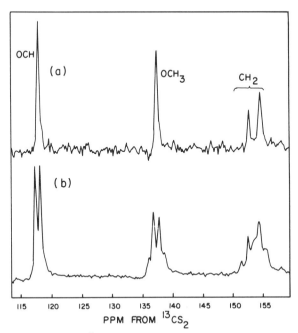

Figure 8.12 Natural abundance ^{13}C NMR spectra of poly(vinyl methyl ether) (25.04 MHz), 50% (w/v) solution in chlorobenzene at 50°C: (a) protons completely decoupled; (b) decoupling "off-resonance." (Reprinted by permission of Ref. 27.)

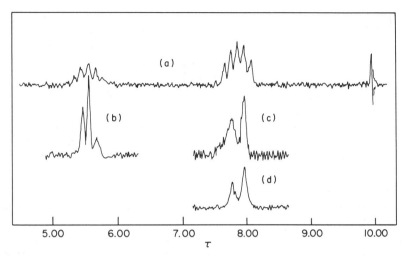

Figure 8.13 Normal and decoupled NMR spectra of poly (vinyl chloride), observed in 15% (w/v) solution in chlorobenzene: (*a*) normal spectrum at 170°C; (*b*) α protons upon radiation of β protons (150°C); (*c*)β protons upon irradiation of α protons (150°C); (*d*) poly-α-d_1-vinyl chloride (150°C). (Reprinted by permission of Ref. 28.)

poly(vinyl methyl ether) (27). As a result of this experiment the chemical shifts of the carbons were assigned as indicated in the figure.

In polymers of monosubstituted monomers, such as vinyl chloride, the vicinal coupling of the α- and β-methylene protons produces a splitting of the observed resonances. On the simplest basis one would expect the α protons, being coupled to four neighboring β protons, to appear as a quintuplet and the β protons, coupled to two α protons, to appear as a triplet. The observed spectrum (Figure 8.13, spectrum *a*) does indeed apparently show a pentuplet centered at δ4.47 ppm (in chlorobenzene solution), but shows a group of five peaks centered at δ ~ 2.1 ppm for the methylene protons (28). The methylene resonance is essentially two overlapping triplets centered at δ2.22 and 2.04 ppm, corresponding to meso and racemic methylene groups, respectively.

This interpretation is confirmed by double resonance. As shown in Figure 8.13 (spectrum *c*), when the β protons are decoupled from the α protons, two peaks are observed, separated by 0.20 ppm. A very similar spectrum is shown by the polymer of α-deuterovinyl chloride (spectrum *d*), ClDC=CH₂ (28).

When the β protons are irradiated, the α-proton resonance shows peaks at 5.48 (δ4.52), 5.59 (δ4.41), and 5.71 (δ4.29 ppm) (Figure 8.13, spectrum *b*). The dependence of the decoupled α-proton spectrum on Δν has been employed to demonstrate that these peaks correspond to syndiotactic, heterotactic, and isotactic triads, respectively, assuming the correctness of the above assignment of the β-proton resonances. The normal α-proton spectrum is actually three overlapping quintuplets.

8.4.3 Isotopic Substitution

For proton NMR the substitution of hydrogen by deuterium represents another means of spectral simplification, for the resonance of deuterium is far removed from that of hydrogen. In addition, the H—D coupling has less than one-sixth the magnitude of the corresponding H—H coupling, and produces no observable multiplicity in polymer spectra. As seen in Figure 8.13, for PVC the deuteropolymer produced the same spectra as double resonance decoupling.

For ^{13}C NMR resonance assignments can be made by selectively enriching a particular position and comparing the spectra with the spectrum obtained from a sample with natural abundance. In Figure 8.14 the ^{13}C NMR spectra are shown for poly(α-methylstyrene) with natural abundance and with specific sites enriched (29).

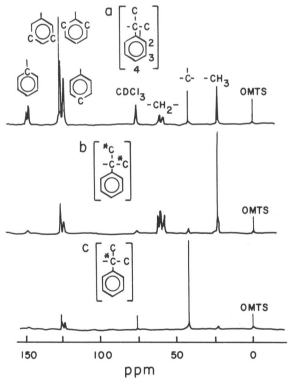

Figure 8.14 ^{13}C NMR spectra of poly(α-methylstyrene): (a) sample with natural abundance; (b, c) ^{13}C enriched samples (10%). The asterisks indicate the positions of enrichment, if not otherwise indicated. The ^{13}C frequency is 67.88. (Reprinted by permission of Ref. 29.)

8.4.4   Use of Model Compounds

To determine the structures of polymers from NMR, model compounds are often used to assist with the band assignments. In most cases the model compounds are low molecular weight compounds that can be prepared and fractionated into known structures. By analogy, when the NMR spectra are similar, the structures are similar. Bovey used model compounds in his initial study of polystyrene (1) to make proper resonance assignments. The spectral peak positions of the polymer and model compounds are compared directly. For the study of the stereochemical configuration of poly-trifluorochloroethylene (30), the model compouunds meso and dl-$CF_2ClCFClCF_2CFClCF_2Cl$ were prepared, and their NMR spectra were compared, as shown in Figure 8.15. The spectrum includes only the resonances of the CF_2 and CFCl groups; the CF_2Cl end groups appear as a triplet at lower field, and are not included because there are no corresponding resonances in the polymer spectrum. The meso form corresponds to the isotactic configuration, whereas the racemic form corresponds to the syndiotactic configuration.

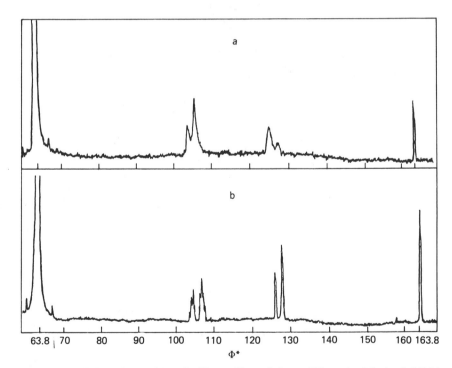

Figure 8.15   NMR spectra: (*a*) polytrifluorochloroethylene, 100 mg in 0.5 ml of 3,3'-bis-trifluoromethylbiphenyl at 150°C; (*b*) meso and racemic $CF_2ClCFClCF_2CFClCF_2Cl$, 10% solution in 3,3'-bistrifluoromethylbiphenyl at 150°C. (Reprinted by permission of Ref. 30.)

8.4.5 Effects of High Magnetic Field

As H_0 increases, there is a directly proportionate increase in the chemical shift differences between different nuclei, but the increase in field does not change the magnitude of the spin-spin J coupling. Thus there is a simplification of the spectra—in many cases going from complex, second order spectra to simpler groupings of lines in well separated multiplets. An excellent example of this effect comes from the ^1H NMR spectra of polypropylene at various fields. The NMR of polypropylene is complex not only because of the large number of protons but also because of the small chemical shift differences between the α, β, and methyl protons. Figure 8.16 shows the spectra of isotactic and syndiotactic propylene (31). Eight broad peaks, not all clearly visible, centered at 8.50τ ($\delta 1.50$ ppm), are given by the α proton, split by three methyl and four methylene protons. To determine the tacticity of polypropylene, the doublets of the methyl group at 9.13τ ($\delta 0.87$ ppm) for the isotactic polymer and 9.18τ ($\delta 0.82$ ppm) for the

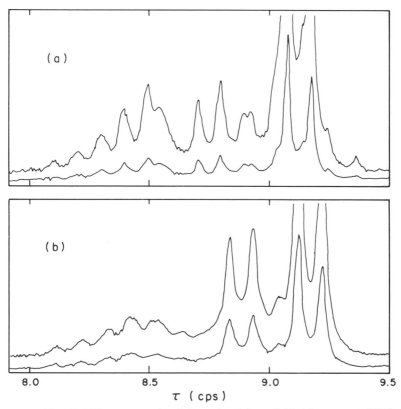

Figure 8.16 The 60 MHz spectra of poly(propylene) in o-dichlorobenzene at 150°C: (a) isotactic; (b) syndiotactic. (Reprinted by permission of Ref. 31.)

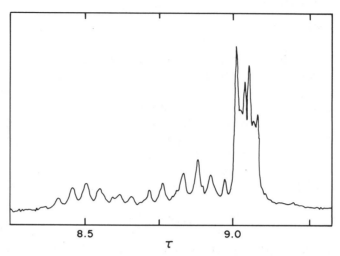

Figure 8.17 The 100 MHz spectrum of stereoblock poly(propylene) in *o*-dichlorobenzene at 150°C. (Reprinted by permission of Ref. 31.)

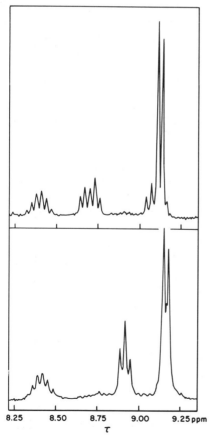

Figure 8.18 The 220 MHz proton spectrum of isotactic (upper) and syndiotactic (lower) poly(propylene). (Reprinted by permission of Ref. 32.)

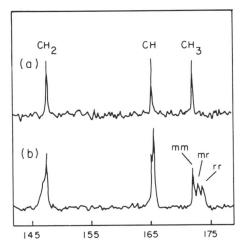

Figure 8.19 ^{13}C NMR spectrum of poly(propylene): (*a*) an isotactic sample, about 5% (w/v) in *o*-dichlorobenzene at 60°C; (*b*) an atactic sample about 30% (w/v) in *o*-dichlorobenzene at 60°C. (Reprinted by permission of Ref. 27.)

syndiotactic spectrum are used. The 100 MHz spectrum of stereoblock polypropylene is shown in Figure 8.17 (31). The better resolution in the ^{1}H 100 MHz spectrum is evident. Observe the heterotactic doublet for the methyl group, which was not observable at 60 MHz. Figure 8.18 shows the 220 MHz proton spectra of isotactic and syndiotactic polypropylene (32). The α, β, and methyl resonances are now clearly separated, the spectrum being nearly first order.

The great simplification produced by ^{13}C NMR resonance is shown in Figure 8.19 (27). For the isotactic polymer there are only three narrow peaks, corresponding to the three types of carbons, methylene, methine, and methyl. The atactic polymer spectrum exhibits triad structures in the methyl resonance, and dyad and tetrad structures in the methylene resonance.

8.4.6 Computer Analysis of NMR Spectra

In many spectra the effects of chemical shift and spin coupling give rise to a complex pattern of lines resulting from the merging of individual multiplets, which then have few features of regularity. One is then faced with the problem of interpreting such spectra, assigning each line to a definite transition, and finally extracting numerical values of the chemical shifts and spin coupling constants. Numerical methods have been developed for calculating the NMR spectra for different models. Comparison of the results of these calculations with the observed spectra allows structural differentiation. This technique was used early in the study of polymers.

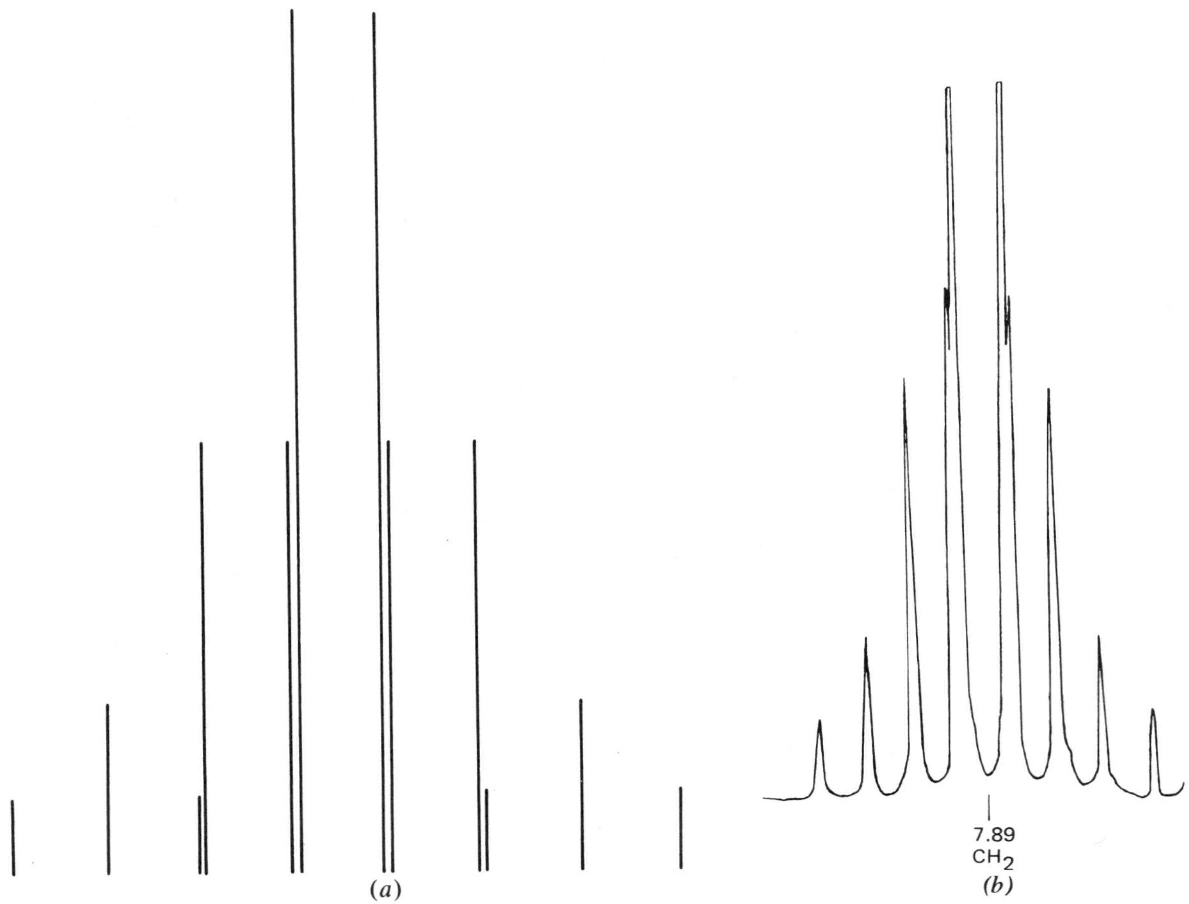

7.89
CH₂
(b)

(a)

Figure 8.20 Decoupled methylene resonance in meso 2,4-dichloropentane: (a) calculated; (b) experimental. (Reprinted by permission of Ref. 33.)

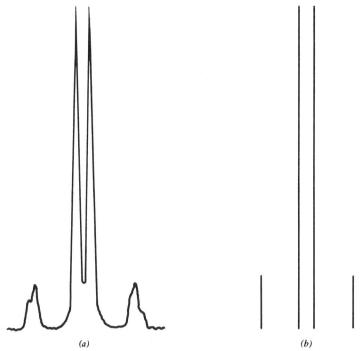

(a)
(b)

Figure 8.21 Decoupled methylene resonance in racemic 2,4-dichloropentane: (*a*) observed; (*b*) calculated. (Reprinted by permission of Ref. 33.)

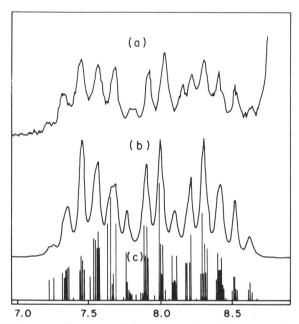

Figure 8.22 NMR spectra of isotactic poly(isopropyl acrylate): (*a*) observed in chlorobenzene at 150°C; (*b*) calculated spectrum with 2.4 Hz linewidth; (*c*) calculated "stick" spectrum. (Reprinted by permission of Ref. 34.)

For example, the methylene resonances of PVC are complex, and to sort the resonances Satoh calculated the methylene resonances for meso and racemic 2,4-dichloropentane (33). For the meso form there are four interacting protons, which he assumed was an ABX_2 spin system:

$$
\begin{array}{ccccccc}
 & H_X & & H_A & & H_X & \\
 & | & & | & & | & \\
CH_3 & -C & -C & -C & -CH_3 \\
 & | & & | & & | & \\
 & Cl & & H_B & & Cl & \\
\end{array}
$$

Here there are six possible spin coupling constants and two chemical shifts. It is assumed (1) that the chemical shift difference of the methylene protons is 0.26τ, (2) equal values for the spin coupling constants between the methylene and methine are equal to 7.1 Hz, and (3) the coupling between the two chemically different methylene protons is 14.5 Hz. The calculated spectrum for the methylene units is compared with the experimental spectrum in Figure 8.20 (33). For the racemic compound a similar calculation yielded the results shown in Figure 8.21 when decoupling of the α protons is accounted for.

When several chemical shifts and many spin-coupling constants exist, the spectra can become very complex, and with the multiple structures present in polymers, these complexities might be expected. For these cases computer analysis is extremely variable. The spectrum of isotactic poly(isopropyl acrylate) has been subjected to such a computer analysis (34). The results are shown in Figure 8.22, with (*a*) the observed spectrum, (*b*) the calculated spectrum with 2.5 Hz linewidth, and (*c*) the calculated "stick" spectrum. The agreement is excellent.

8.5   APPLICATIONS OF NMR TO POLYMERS (5)

8.5.1   Introduction

There are three important parameters in NMR for its application: chemical shifts, spin-spin coupling constants, and areas under resonance peaks. First, all protons and carbons are observable, and the chemical shifts are extremely sensitive to local structural differences in the environments of the ^1H or ^{13}C nucleus. When high magnetic fields are used, differences arising several atoms away from the resonating nucleus can be detected. The ease of obtaining structural correlations with model compounds, additivity relations, or numerical calculations facilitate the use of NMR. Finally, in most cases NMR is a self-calibrating system, since the total resonance is proportional to the number of ^1H or ^{13}C nuclei; the fraction of each chemical entity is the area of its resonance divided by the total area of the spectrum; thus one avoids the troublesome necessity of establishing reliable extinction coefficients.

8.5.2 Determination of Chemical Functionality

The chemical functionality of a heteropolymer can be measured by NMR. This is valuable, particularly since there are usually difficulties in determining the composition of a heteropolymer system with accuracy and simplicity. To determine the composition by NMR, one first needs to determine the number of different components present. Second, one seeks a unique line for all but one of the components present, as well as knowledge of the number of protons displaying resonance in this unique peak. Finally, one integrates the peaks and calculates the fractional area of

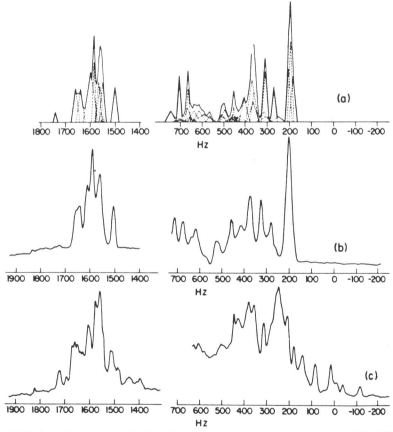

Figure 8.23 Low field and high field portions of the HEW lysozyme spectrum (220 MHz) in D$_2$0: (a) calculated spectrum of denatured protein; (b) observed spectrum at 80°C; (c) observed spectrum at 65°C. The chemical shifts are expressed in hertz, referred to DSS. (Reprinted by permission of Ref. 36.)

the total that is contributed by each unique line. These factors correspond to the composition.

One of the most challenging problems in composition, which will be cited only to indicate the power and sensitivity of NMR, not to suggest the use of NMR for this purpose, is that of proteins. Proteins represent heteropolymers in the ultimate sense—with 22 possible residues, the effects of pH and ionic strength all influence the results. Interaction effects as well as unfolding can be observed. McDonald and Phillips (35) have provided tables of the amino acid chemical shifts in neutral D_2O solution obtained at 220 MHz.

Spectra computed on the basis of known amino acid compositions quite closely fit the experimental results, as shown in Figure 8.23 for lysozyme (36). The agreement is quite reasonable. Comparison of the

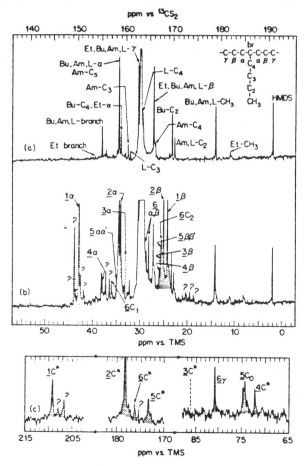

Figure 8.24 ^{13}C NMR spectrum of polyethylene: (*a*) before oxidation, where Bu, Am, and L refer, respectively, to butyl and amyl branches and long chain ends; (*b*, *c*) after being oxidized to 108 ml/g. The oxidized products are identified by numbers as follows:

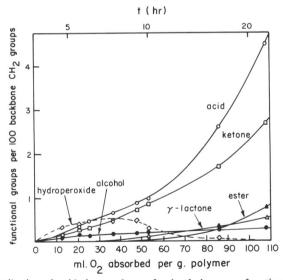

$$\cdots C_{a'}\!-\!C_a\!-\!\overset{\displaystyle O}{\overset{\|}{C}}\!.\!-\!C\!-\!C\cdots \qquad \cdots C_d\!-\!C_a\!-\!\overset{\displaystyle O}{\overset{\|}{C}}\!.\!-\!OH$$

1 = ketones   2 = acids

$$\cdots C_d\!-\!C_a\!-\!\overset{\displaystyle OOH}{\overset{|}{C}}\!.\!-\!C\!-\!C\cdots \qquad \cdots C_d\!-\!C_a\!-\!\overset{\displaystyle OH}{\overset{|}{C}}\!.\!-\!C\!-\!C\cdots$$

3 = secondary hydroperoxides   4 = secondary alcohols

$$\cdots C_d\!-\!C_a\!-\!\overset{\displaystyle O}{\overset{\|}{C}}\!.\!-\!O\!-\!C_0\!\!\left(\!C_{a'}\!-\!C_{\beta'}\cdots\!\right)_{\!2}$$

5 = esters of acid and secondary alcohols

$$\overset{\displaystyle O}{\overset{\|}{C}}\!.\!-\!C_a\!-\!C_{\beta}\!-\!C_{\gamma}\!-\!C_1\!-\!C_2$$

6 = γ-lactones

(Reprinted by permission of Ref. 37.)

native protein, which has never been heated above 65° in D₂O, shows a number of discrete resonances that disappear and new peaks that appear when the protein is heated through the denaturation temperature range to 80°C. Data of this type have been interpreted in terms of a two-state folded-unfolded transition (36).

In the synthetic polymer field ^{13}C NMR has contributed to one of the long-standing problems in the determination of chemical functionality, that is, the oxidation of polyethylene. FT ^{13}C NMR has been used for the

Figure 8.25 Distribution of oxidation products of polyethylene as a function of the extent of oxidation (lower scale, O₂ absorbed; upper scale, time of reaction). The vertical scale refers to the intensity of the backbone CH₂ resonance at 30 ppm; no correction is made for CH₂ groups neighboring the functional groups. (Reprinted by permission of Ref. 37.)

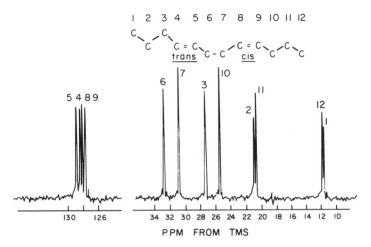

Figure 8.26 ^{13}C FT NMR spectrum of *cis,trans*–4,8–dodecadiene in 50% dioxane. (Reprinted by permission of Ref. 38.)

observation and measurement of the oxidation products of low density polyethylene. In Figure 8.24 the ^{13}C spectra of branched polyethylene are shown before and after thermal oxidation (37). In spectrum (*a*) the principal peak at 30 ppm represents methylene, and the other resonances are identified. In spectra (*b*) and (*c*) a number of groups are observed, including long chain ketones, long chain carboxylic acids, long chain secondary alcohols, long chain secondary hydroperoxides, esters of long chain carboxylic acids with long chain secondary alcohols, and long chain γ-lactones. The quantitative results for the oxidation products as a function of extent of oxidation are shown in Figure 8.25.

Although *cis,trans*-4,8-dodecadiene is not a polymer molecule, its ^{13}C spectrum (Figure 8.26) shows that each carbon atom is clearly distinguished (38). Even the methyl carbons at the ends of the molecule have slightly different electronic environments, as evidenced by the peak separation of 0.14 ppm. This compound is a model compound for the cis-trans linkage in butadiene, and the peaks labeled 6 and 7, representing the methylene carbons for the cis-trans linkage, are separated by 2.4 ppm.

8.5.3 Determination of Steric Configuration in Homopolymers

Of the many significant contributions that NMR has made to our understanding of polymers, perhaps the most important is the measurement of steric configuration or stereoregularity of polymers. The NMR results completely revolutionized the thinking concerning isomerism and led to an understanding of the stereospecific polymerization mechanism. Poly(methyl methacrylate), which was first studied, proved to be a most convenient starting point, since no vicinal coupling by the main chain

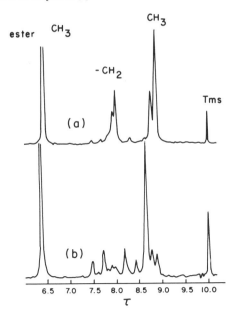

Figure 8.27 The 60 MHz spectra of 15% (w/v) solutions in chlorobenzene of poly(methyl methacrylate) prepared with a free radical initiator (*a*) and an anionic initiator (PhMgBr) (*b*). The ester methyl resonance appears near 6.5τ, the α-methylene protons appear near 8.0τ and the β-methyl protons give three peaks near 8.5 and 9.0τ. (Reprinted by permission of Ref. 40.)

protons complicates the spectrum. Figure 8.27 shows the 60 MHz spectra of (*a*) predominantly syndiotactic and (*b*) predominantly isotactic poly(methyl methacrylate) (40). The ester methyl resonance appears near 6.5τ (δ3.5 ppm), the β-methylene protons appear near 8.0τ (δ2.0 ppm), and the α-methyl protons give three peaks between 8.5 (δ1.5 ppm) and 9.0τ (δ1.0 ppm). As predicted from the preceding discussion in this chapter, the methylene resonance for the syndiotactic polymer should be a singlet, as is observed, although it is somewhat broadened by residual isotactic resonances and complicated by additional fine structure associated with higher sequences such as tetrads. The methylene resonance for the isotactic polymer is predominantly the expected AB quartet ($J = 14.9$ Hz) with some additional syndiotactic resonance. These results are absolute measures of the polymer's predominant configuration—no other supporting evidence is required.

The methyl group (α substituent) shifts have been commonly found to vary appreciably with the relative configurations of the nearest neighboring monomer unit. When three peaks are observed, they are associated with the three possible species of α substitutents in a polymer that is not stereochemically pure: those on the central monomer units of isotactic, syndiotactic, and heterotactic triads of monomer units.

A number of polymer systems have been examined since poly(methyl methacrylate) with the same general technique. When the backbone methine couples, the backbone methylene can often be isolated by deuterium labeling, substitution on the adjacent carbons, or spin decoupling. A single resonance band is expected from syndiotactic (racemic) methylene, a quartet (AB) for the isotactic (meso), and a combination of the two for heterotactic. The syndiotactic line usually appears in the center of the four-line isotactic multiplet. The chemical shift resonance of the methyls from heterotactic triads are flanked by those arising from syndio- and isotactic triads. However, when three separate resonances are observed, the low field methyl resonance may represent either the isotactic or the syndiotactic structures. For example, PMMA, PVAC, and PP all have the resonance for the isotactic triad at low field with the syndiotactic at high field, whereas PA, PVME, PVA, and PVC have the syndiotactic triad at low field and the isotactic at high field. Differences in this line ordering arise from the band anisotropy of the functional groups affecting the protons on neighboring carbons. The ordering isotactic, heterotactic, syndiotactic (with increasing field) is attributed to a functional group that causes a shielding effect, and the reverse order arises if the substituted

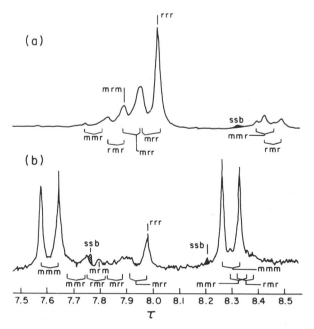

Figure 8.28 (a) The β-methylene proton spectrum of predominantly syndiotactic poly(methyl methacrylate) (polymer I) observed at 220 MHz (about 10% w/v solutions in chlorobenzene at 135°C); (b) the β-methylene proton spectrum of predominantly isotactic poly(methyl methacrylate) (polymer II), same conditions as (a). In these spectra the vertical markers for the doublet components of *mmm*, *mmr*, *rmr*, and *mrr* quartets are joined by slanting lines that intersect at the chemical shift; *ssb* designates spinning side band. (Reprinted by permission of Ref. 5.)

functional group causes a deshielding effect (39). For a new polymer these considerations can be used, as well as other nonspectroscopic considerations.

With increasing strength of the magnetic field, differences in chemical shift may allow the resolution of tetrad sequences of monomer units appearing as fine structure on the meso (m) and racemic (r) resonances. There are six possible tetrads of monomer units. The m resonance should be resolved into three tetrad resonances, all heterosteric, giving 6 different chemical shifts. The r resonance should be split into 2 homosteric resonances and 1 (2 chemical shifts) heterosteric resonance. Thus a total of 10 observable methylene chemical shifts could be observed in an atactic polymer. The methylene spectrum of predominantly syndiotactic PMMA at 220 MHz is shown in Figure 8.28 with the designation of the tetrad

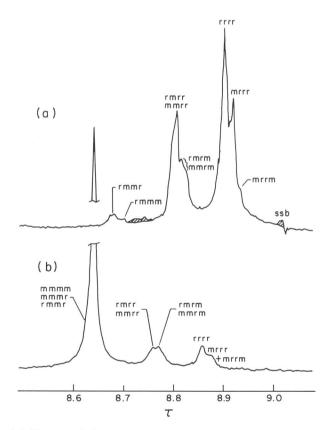

Figure 8.29 (*a*) The α-methyl proton spectrum of predominantly syndiotactic poly(methyl methacrylate) (Polymer I) observed at 220 MHz (about 10% w/v solutions in chlorobenzene at 135°C); (*b*) the α-methyl proton spectrum of predominantly isotatic poly(methyl methacrylate) (polymer II), same conditions as (*a*); *ssb* designates spinning side band. (Reprinted by permission of Ref. 5.)

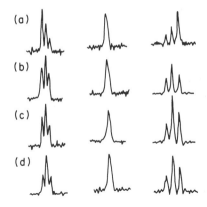

Figure 8.30 ^{13}C NMR spectra of the nitrile carbons (left), the methylene carbons (center), and the methine carbons (right) of dimethylsulfoxide solutions of poly(acrylonitriles) prepared by (*a*) irradiation of a urea-canal complex, (*b*) an organometallic catalyst, and (*c*) a free radical catalyst. (*d*) Spectra of the free radical generated material in dimethylformamide. (Reprinted by permission of Ref. 41.)

assignments for the various resonance lines. These assignments are made on the basis that the proton shieldings are predictable and systematic, and the relative intensities agree with the expected numbers of tetrads. Note that the *mmm* tetrad resonance, which is hardly discernible in the syndiotactic polymer, is the dominant feature of the isotactic polymer.

In a similar fashion the fine structure on the α-methyl spectra of the *mm*, *mr*, and *rr* triad peaks splits into further peaks associated with the pentad sequence at high magnetic field. These spectra and assignments are shown in Figure 8.29 (5).

Extensions of these arguments to longer sequences reveal why the use of ^{1}H resonance even at high fields is limited. The problem arises because the resonances may be broad and difficult to detect because of a wide variety of long-range, configurationally dependent chemical shifts. This long-range dependence comes as a result of strong correlations between conformations of neighboring dyads, which are then propagated from dyad to dyad over sequences of many dyads. The net result is a broad line caused by a dispersion of chemical shifts.

Many of the problems encountered in proton resonance are not a bother for ^{13}C NMR, but new problems arise. ^{13}C spins are isolated from dipolar interactions between themselves and the protons are decoupled.

Let us examine polyacrylonitrile—a polymer that has posed a difficult problem for ^{1}H NMR spectra because of the severely overlapped multiplets. The ^{13}C NMR spectra are shown in Figure 8.30 (41). The nitrile carbons are at low fields, the methylene carbons are in the center, and the methine carbons are on the right. Clearly, the lowest field nitrile-carbon line is isotactic, while the highest field line is assigned to the isotactic methine carbon. General characteristics common to the ^{13}C NMR spectra of vinyl polymers are illustrated in Figure 8.31 (42).

For polystyrene the methylene ^{13}C resonances have been resolved to the extent that hexad sequence assignments have been made (44). The methylene and methine region of a ^{13}C NMR spectrum of an amorphous

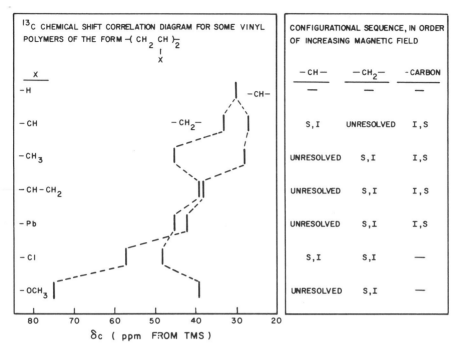

Figure 8.31 A ^{13}C chemical shift correlation diagram for the main chain carbons of some typical vinyl homopolymers. The right-hand side of the diagram shows the effect of steric configuration on relative chemical shifts. "I" and "S" refer to isotactic and syndiotactic dyads for methylene carbons and to isotactic and syndiotactic triads for methine carbons. (Reprinted by permission of Ref. 42.)

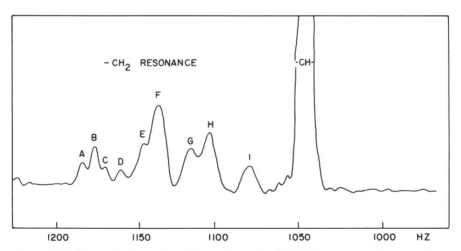

Figure 8.32 The methylene and methine regions of a ^{13}C NMR spectrum of an amorphous polystyrene at 25.2 MHz and 120°C. The hertz values are relative to an internal tetramethylsilane (TMS) standard. (Reprinted by permission of Ref. 44.)

259

Table 8.4 ^{13}C NMR Methylene Assignments for Polystyrene (44)

CH$_2$	(Resonance)	Assignment	Observed	Calc. $P_m = 0.565$
A	46.85	mmrrr^a	0.05 ⎤	0.065 ⎤
B	46.57	mmrrm	0.07 ⎟ 0.21	0.068 ⎟ 0.214
C	46.29	rmrrr	0.04 ⎟	0.040 ⎟
D	45.89	rmrrm^a	0.05 ⎦	0.053 ⎦
E	45.38	rmr	0.10	0.107
F	44.94	mmr	0.28	0.278
G	44.25	mrm	0.13	0.139
H	43.77	mmm	0.19	0.180
I	42.84	rrr	0.09	0.082

^{a} Assignments A and D can be interchanged.

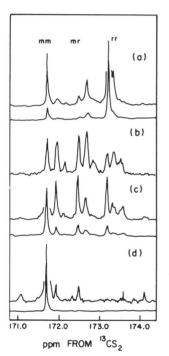

Figure 8.33 ^{13}C NMR spectra of the methyl carbons of solutions of poly(propylene) that are (a) blocklike syndiotactic, (b) atactic, (c) somewhat isotactic, and (d) highly isotactic. Spectra (a), (c), and (d) are shown at two vertical gains. (Reprinted by permission of Ref. 43.)

171.0   172.0   173.0   174.0

ppm FROM ^{13}CS$_2$

polystyrene is shown in Figure 8.32 (44). Here 9 resonances are observed, whereas with just tetrad sequences only 6 resonances would be expected, and a complete hexad sensitivity would produce 20 resonances. In Table 8.4 the assignments are shown. The measurements of relative intensities are compared with the calculated relative intensities for the tetrad and hexad sequences. The agreement is excellent. The assignments are quite different from those expected by analogy with other vinyl polymers. The possibility exists that polystyrene shows anomalous chemical shifts due to mutual shielding effects of an isotropic nature from the aromatic rings.

Thus the possibility exists that the chemical shift of a methylene proton in a syndiotactic dyad in an isotactic polymer depends not only on the stereochemical configuration of the dyad or tetrad in which it is situated, but also on the configurations of many neighboring sequences. Line positions are shown in the left-hand side of Figure 8.31. The right-hand side of the figure shows the effect of steric configuration on relative chemical shifts. The methylene carbon spectra and the spectra of substituted carbons directly bonded to the main chain carbons always appear reversed from one another. Thus the pattern of similarities of the ^{13}C NMR spectra can be used as an aid in their analysis.

Another most difficult polymer for analysis by ^{1}H NMR is polypropylene. Yet by ^{13}C NMR the analysis is quite straightforward (43). Methyl carbon spectra for a series of polypropylenes ranging from blocklike syndiotactic to highly isotactic are shown in Figure 8.33 (43). Spectrum b shows pentad lines resolved nearly to the base line, with indications of heptad structures in higher field lines. In order of increasing field, the pentad lines are tentatively assigned to the sequences *mmmm, mmmr, rmmr, mmrm, mmrr + rmrm* (one line), *rmrr, rrr, mrrr,* and *mrrm*. Observe that each resonance maintains the same relative position regardless of the predominant stereochemistry of the polymer. It appears that configurational analysis by high resolution ^{13}C NMR spectra need not be complicated by considerations of shift effects that vary from polymer to polymer.

The implications of the results of NMR measurements are discussed further in Chapter 10.

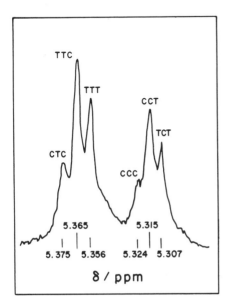

Figure 8.34 Decoupled olefinic methine resonances at 300 MHz of a 1,4-polybutadiene containing cis (C) and trans (T) units. Triad assignments are indicated. (Reprinted by permission of Ref. 45.)

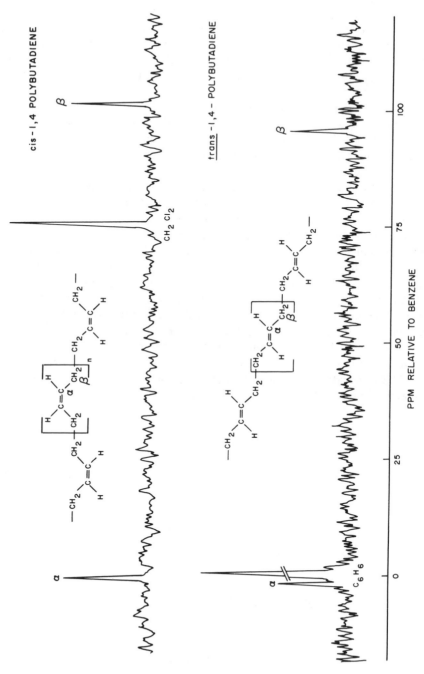

Figure 8.35 ^{13}C NMR spectra of *cis*-1,4-polybutadiene and *trans*-1,4-polybutadiene. (Reprinted by permission of Ref. 46.)

8.5.4 Determination of Structural and Geometric Isomerism in Polymer Chains

When conjugated dienes are polymerized, structures are produced with different chemical types of unsaturation as well as geometrical and configurational isomerism. For example, 1,3-butadiene can give 1,4 or 1,2 enrichment; the 1,4 structure can be cis or trans, and the 1,2 structure can be isotactic or syndiotactic. In addition, the 1,2 units may be linked head-to-head, head-to-tail, or tail-to-tail with one another or with a 1,4 unit. In the proton magnetic resonance spectrum of a polybutadiene, 1,2 units appear as peaks from the olefinic methylene protons at about 4.8τ ($\delta 5.2$ ppm). The trans-1,4 units (1.98τ) ($\delta 8.02$ ppm) are barely distinguishable from the cis-1,4 units (2.03τ) ($\delta 7.97$ ppm). The trans and cis olefinic methine peaks occur at 5.37τ ($\delta 4.63$ ppm) and 5.32τ ($\delta 4.68$ ppm), respectively. When a decoupling experiment is performed at 300 MHz, the olefinic methine resonances of 1,4 units are split into a total of six peaks, as shown in Figure 8.34 (45). In order of increasing field strength, these have been assigned to the olefinic protons of the central butadiene units in cis-trans-cis, trans-trans-cis, trans-trans-trans, cis-cis-cis, cis-cis-trans, and trans-cis-trans triads, respectively (45).

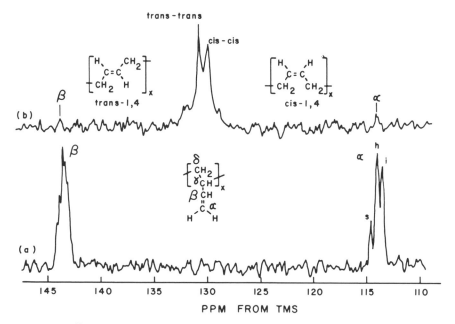

Figure 8.36 ^{13}C NMR spectra of olefinic carbons in polybutadiene (10% in dioxane): (a) 95% 1,2-polybutadiene; (b) 10.5% 1,2 units, 56.3% trans-1,4 units, and 33.2% cis-1,4 units. (Reprinted by permission of Ref. 47.)

The natural abundance ^{13}C NMR spectra of *cis-* and *trans-*1,4-polybutadiene are shown in Figure 8.35 (46). Observe that the β-carbon resonance in the trans unit is 5.4 ppm downfield from that of the cis unit.

In Figure 8.36 (47) the spectra of a butadiene rubber containing 56.3% trans-1,4 units, 33.2% cis-1,4 units, and 10.5% vinyl units is compared with the spectra of a polybutadiene with 95% vinyl groups. The two olefinic carbons are separated by 29.8 ppm, and these are separated by about 15 ppm from any 1,4 olefinic resonance. The α-carbon resonance is clearly separated into iso-, hetero-, and syndiotactic triad resonances. The β carbon is sensitive to pentad resonances.

8.5.5 Determination of Conformation of Polymer Chains

As already indicated, the spin-spin coupling of nuclear moments causes a splitting of the resonance lines. The value of the coupling constant J_{AB} depends on the relative configuration of the interacting protons of group B

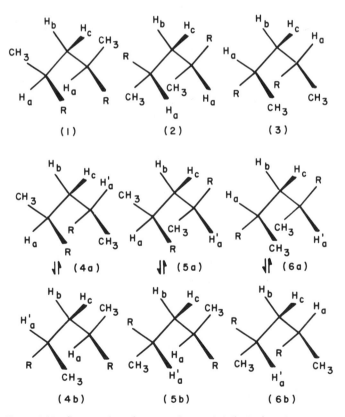

Figure 8.37 Staggered conformers of meso 2,4-disubstituted pentanes.

to the A group. For H_a—C_a—C_b—H_b the coupling constant J_{HH} may be represented in terms of the dihedral angle ϕ (49, 50):

$$J = A + B \cos \phi + C \cos^2 \phi,$$

where A, B, and C are constants having the values of 7, −1, and 5 Hz, respectively (51). Large values of J are predicted for the cis (0°) and trans (180°) configurations; small values result from the gauche (60° and 120°) configurations.

When more than one conformation is available and under conditions of rapid equilibration, the observed vicinal couplings are an average over the conformations present:

$$J_{obs} = \sum_i X_i J_i,$$

where X_i and J_i represent the mole fraction and vicinal coupling of each conformer. For meso 2,4-disubstituted pentanes, there are nine possible conformers (Figure 8.37), including three pairs of equal energy mirror image conformers that are convertible into each other by 120° rotations about the bonds between the α and β carbons. The β-methylene protons H_B and H_C are in different environments in all the conformations. These two protons always have different chemical shifts and have different couplings to the α protons (H_A or H'_A). The following express the couplings assuming equality of all gauche and trans couplings:

$$J_{AB} = X_1 J_g + X_2 J_t + X_3 J_g + X_4 J_g + \frac{X_5(J_g + J_t)}{2} + \frac{X_6(J_g + J_t)}{2},$$

$$J_{AC} = X_1 J_t + X_2 J_g + X_3 J_g + \frac{X_4(J_g + J_t)}{2} + \frac{X_5(J_g + J_t)}{2} + X_6 J_g.$$

It is evident that the J_{AB} and J_{AC} differ substantially if either conformer 1 or 2 is strongly preferred. Similar considerations apply to racemic conformers. If conformer 3 is preferred, the couplings are nearly equal and are of the order of 2–3 Hz. Experimentally, J_{AB} is observed to exceed J_{AC} by as much as 1 Hz for the meso-2,4-disubstituted pentenes except for meso-2-4-diphenyl-pentane, where they are essentially equal. The symmetric conformers cannot be present in significant amounts. Of the asymmetric conformers 4, 5, and 6, apparently conformer 5 is strongly preferred. If only conformer 5 is present, equal vicinal couplings are observed. For polystyrene

$$J_{AB} = J_{AC} = 7.10 \pm 0.10,$$

so it seems reasonable to assume that the TG (conformer 5) is strongly preferred by the polymer (48). This is the segment that gives rise to the 3_1 helical conformation in the crystalline state. The evidence suggests that in solutions of isotactic polystyrene there is a strong predominance of 3_1 helical sequences. However, other conformations must also be present.

It appears that ^{13}C chemical shifts can also be useful in deducing information about the average conformation of polymers. It has been found that the relative ^{13}C chemical shifts of the methylene resonance in meso and racemic 2,4-dichloropentane appear to be dependent on the population of rotational conformations (52). Gauche 1,4 interactions produce a more upfield ^{13}C chemical shift than does a trans-1,4 interaction (53). Thus the relative separation in hertz between the ^{13}C chemical shifts should be a function of the fractional trans or gauche contribution. Hence the relative difference in ^{13}C chemical shifts for the 2,4-dichloropentane is a function of the amount of trans and gauche interactions that a carbon was experiencing in the overall average populations of conformations.

Carman (54) feels that these effects are influencing the ^{13}C NMR spectra of PVC. The methylene carbon in an *rrr* tetrad should experience more trans interaction because the most stable conformation for *rr* is the all-trans one. On the other hand, the *mmm* tetrad should be most shielded because of the gauche interactions arising from the preference for the trans-gauche conformation.

8.5.6 Determination of Copolymer Microstructure

NMR is a particularly useful tool for the study of copolymers, as different chemical shifts can be associated with copolymer sequences. As indicated earlier, copolymer composition measurements are particularly easy and require no external calibration. However, NMR is particularly effective in

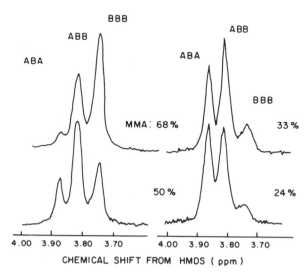

Figure 8.38 Methoxyl resonance in NMR spectra of acrylonitrile-methyl methacrylate copolymer initiated by benzoyl peroxide in tetrahydrofuran at 40°C. (Reprinted by permission of Ref. 55.)

obtaining sequence information. The NMR method is less tedious and probably more accurate than any other analytical method.

Sometimes the spectra are almost elegant in their simplicity. Consider Figure 8.38, which shows the 60 MHz spectra for the methoxy protons of acrylonitrile-methyl methacrylate copolymer of various MMA contents (55). The methoxyl resonance splits into three peaks. The center peak grows with decreases in mole percent of feed monomer of methyl methacrylate (B), while the peak at 3.74 (δ6.26 ppm) decreases. Thus the peaks are assignable to ABA, ABB (and BBA), and BBB triads from the lower to upper fields. The location of the BBB peak corresponds to the methoxyl peak in poly(methyl methacrylate).

When the monomers have no asymmetric centers and produce large deshielding of the protons from neighboring units, the analysis is particularly simple. Copolymers of vinylidene chloride and isobutylene are such examples (56, 57). The spectra of the homopolymers and a copolymer containing 70 mole-% vinylidene chloride are shown in Figure 8.39 (56). The first observation that is apparent is that seven lines are observed, so tetrad sequences are being resolved. The assignments are as follows: (1) AAAA, (2) AAAB, (3) BAAB, (4) AABA, (5) BABA, (6) AABB, (7) AABB. Even at 60 MHz the tetrad peaks are obviously split, so apparently hexad sequences are being observed.

However, if one of the monomers exhibits stereochemical configurational differences, additional lines are predicted because NMR spectroscopy is sensitive to the configurational possibilities of the comonomer sequences. For copolymers of ethylene-vinylchloride these distereomer effects for vinyl chloride have been observed, as shown for the methine resonances in Figure 8.40 (58). Here the six lines are the expected lines for

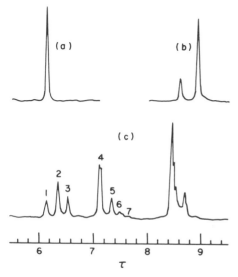

Figure 8.39 The 60 MHz spectra of (a) poly(vinylidene chloride), (b) poly(isobutylene), and (c) vinylidene chloride (A)-isobutylene (B) copolymer (70 mole-% vinylidene chloride). Peaks are identified with tetrad sequences as follows: (1) AAAA, (2) AAAB, (3) BAAB, (4) AABA, (5) BABA, (6) AABB, and (7) BABB. (Reprinted by permission of Ref. 56.)

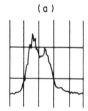

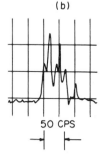

50 CPS

Figure 8.40 The 100 MHz spectra of the α region of an ethylene-vinyl chloride sample: (a) single resonance; (b) double resonance in which the α region is decoupled, revealing α region lines 1–6. (Reprinted by permission of Ref. 58.)

triads where one of the monomers exhibits stereisomeric effects. The assignments for these six lines are shown in Table 8.5.

As might be expected, the use of ¹H NMR for copolymers of polyolefin is limited, as the resonances are severely overlapped, but as the resonances in ¹³C NMR occur over a range of about 40 ppm, problems of this kind should not arise. Figure 8.41 shows the ¹³C NMR spectra of isotactic polybutene and isotactic polypropylene (59). By contrast, Figure 8.42 shows the ¹³C NMR spectra of copolymers of butene-propylene with (a) predominantly BPB, (b) 10% of propylene or blocks, (c) 10% butene as random polymers, (d) 10% butene as blocks (59). The branch methylene group carbon resonances at 27.00, 27.18, and 27.34 ppm are characteristic

Table 8.5 Line Positions and Assignments of Ethylene-Vinyl Chloride in the Methine Region (58)

NMR Line at 100 MHZ	Sequence[a]	Tacticity
4.46	BBB	Syndiotactic
4.35	BBB	Heterotactic
4.23	BBB	Isotactic
4.11	BBA, ABB	Racemic
3.96	BBA, ABB	Meso
3.74	ABA	—

[a] A = ethylene; B = vinyl chloride.

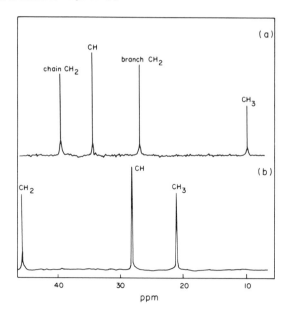

Figure 8.41 ^{13}C NMR spectra: (*a*) poly(butene); (*b*) polypropylene. (Reprinted by permission of Ref. 59.)

of BBB, BBP, and PBP sequences, respectively. Methyl group resonances at 20.96, 20.74, and 20.60 ppm are characteristic of the propylene-centered triad, that is, PPP, PPB, and BPB, respectively. Tetrad sequences for propylene-propylene-centered units BPPB, PPPB, and PPPP are observed from the backbone methylene resonance at 45.86, 45.10, and 46.40 ppm, respectively.

An analysis of the ^{13}C NMR spectrum of copolymer of butadiene-styrene shows the possibility of observing the microstructure of these complicated copolymers (60). The units present are of the type *trans*-1,4-butadiene (*t*), *cis*-1,4-butadiene (*c*), 1,2-butadiene (*ν*), and styrene (*φ*). In addition all units can present themselves by the head or by the tail, that is,

$$\vec{\nu} = —CH—CH_2— \qquad \overleftarrow{\nu} = —CH_2—CH—$$
$$\qquad\quad | \qquad\qquad\qquad\qquad\quad |$$
$$\qquad\quad CH \qquad\qquad\qquad\qquad CH$$
$$\qquad\quad \| \qquad\qquad\qquad\qquad\quad \|$$
$$\qquad\quad CH_2 \qquad\qquad\qquad\qquad CH_2$$

$$\vec{\phi} = —CH—CH_2— \qquad \overleftarrow{\phi} = —CH_2—CH—$$
$$\qquad\quad | \qquad\qquad\qquad\qquad\quad |$$
$$\qquad\quad \phi \qquad\qquad\qquad\qquad\quad \phi$$

The number of possible triads is $6^3 = 216$. However, because of steric hinderance $\vec{\phi}\vec{\nu}$ additions do not seem likely and no $\vec{\nu}\overleftarrow{\nu}$ or $\vec{\phi}\overleftarrow{\phi}$ were present,

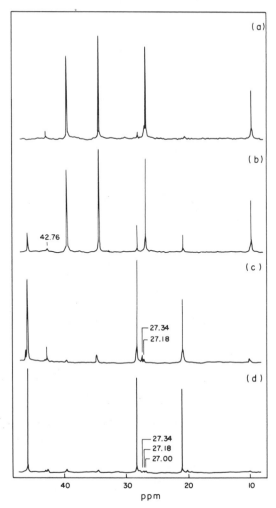

Figure 8.42 ^{13}C NMR spectra of butene propylene copolymers: (*a*) 10 mole-% propylene as isolated units; (*b*) 10 mole-% butene as random copolymer; (*d*) 10 mole-% butene with some butene blocks. (Reprinted by permission of Ref. 59.)

so 152 triads are possible. In all, 30 chemical shift values were observed and their assignments given. Although many of these assignments must be considered tentative, the possibility of obtaining information on the distribution of comonomers in chains of copolymers of the SB type is evident.

REFERENCES

1. F. A. Bovey, G. V. C. Tiers, and G. Filipovich, *J. Polym. Sci.*, **37**, 73 (1959).
2. J. Schaefer, *Macromolecules*, **2**, 210 (1969).
3. M. W. Duch and D. M. Grant, *Macromolecules*, **3**, 165 (1970).

4. J. Schaefer and E. O. Stejskal, *J. Am. Chem. Soc.*, **98:4**, 1031 (1976).

5. F. Bovey, *High Resolution NMR of Macromolecules*, Academic Press, New York, 1972.

6. J. Schaefer, "The Carbon-13 NMR Analysis of Synthetic High Polymers", *Topics in C-13 NMR Spectroscopy*, Vol. 1, George Levy, Ed., 1974.

7. U. Haeberlen, "High Resolution NMR in Solids: Selective Averaging" *Advances in Magnetic Resonance*, J. Waugh, Ed. 1976 Academic Press, Suppl. 1.

8. J. A. Pople, W. G. Schneider, and H. J. Bernstein, *High Resolution Nuclear Magnetic Resonance*, McGraw-Hill, New York, 1959.

9. C. P. Slichter, *Principles of Magnetic Resonance with Examples from Solid State Physics*, Harper & Row, New York, 1967.

10. J. T. Arnold, S. S. Dharmatti, and M. E. Packard, *J. Chem. Phys.*, **19**, 507 (1951).

11. F. Bovey, *High Resolution NMR of Macromolecules*, Academic Press, New York, 1972, Chapter 1.

12. J. Grasselli and W. Ritchey, *Atlas of Spectral Data for Organic Compounds*, Chemical Rubber Co., Cleveland, Ohio, 1975.

13. D. M. Grant and E. G. Paul, *J. Am. Chem. Soc.*, **86**, 2984 (1964).

14. J. C. Randall, *Polymer Sequence Determination—Carbon-13 NMR Method*, Academic Press, New York, 1977.

15. G. C. Levy, *Carbon-13 Nuclear Magnetic Resonance for Organic Chemists*, Wiley-Interscience Publishers, New York, 1972.

16. D. L. VanderHart, *J. Chem. Phys.*, **64**, 830 (1976).

17. A. Pines, M. G. Gibby, and J. S. Waugh, *J. Chem. Phys.*, **56**, 1403 (1971).

18. A. Pines, M. G. Gibby, and J. S. Waugh, *J. Chem. Phys.*, **59**, 569 (1973).

19. J. Vrbina and J. S. Waugh, *Proc. Natl. Acad. Sci.*, **71**, 5062 (1974).

20. F. Bovey, *Nuclear Magnetic Resonance Spectroscopy*, Academic Press, New York, 1969.

21. M. W. Duch and D. M. Grant, *Macromolecules*, **3**, 175 (1970).

22. K. F. Kuhlmann and D. M. Grant, *J. Am. Chem. Soc.*, **90**, 7355 (1968).

23. T. C. Farrar and E. D. Becker, *Pulse and Fourier Transform NMR*, Academic Press, New York, 1971.

24. J. S. Leigh, *Introduction to the Spectroscopy of Biological Polymers*, D. W. Jones, Ed., Academic Press, New York, 1976, Chapter 7.

25. E. O. Stejskal, J. Schaefer, and R. A. McKay, *High-Resolution, Slow-Spinning Magic-Angle Carbon-13 NMR* (in press).

26. D. L. VanderHart, *J. Magn. Resonance*, **24**, 467 (1976).

27. L. F. Johnson, F. A. Heathley, and F. A. Bovey, *Macromolecules*, **3**, 175 (1970).

28. F. A. Bovey, E. W. Anderson, D. C. Douglass, and J. A. Manson, *J. Chem. Phys.*, **39**, 1199 (1963).

29. K. F. Elgert, R. Wicke, B. Stutzel, and W. Ritter, *Polymer*, **16**, 466 (1975).

30. G. V. D. Tiers and F. Bovey, *J. Polym. Sci.*, **1A**, 833 (1963).

31. J. C. Woodbrey, *J. Polym. Sci.*, **B2**, 315 (1964).

32. R. C. Ferguson, *Trans. N.Y. Acad. Sci.*, **29**, 495 (1967).

33. S. Satoh, *J. Polym. Sci.*, **2A**, 5221 (1964).

34. C. Schuerch, W. Fowells, A. Yamada, F. A. Bovey, and F. P. Hood, *J. Am. Chem. Soc.*, **86**, 4882 (1964).

35. C. C. McDonald and W. D. Phillips, *J. Am. Chem. Soc.*, **91**, 1513 (1969).

36. C. C. McDonald, W. D. Phillips, and J. D. Glickson, *J. Am. Chem. Soc.*, **93**, 235 (1971).

37. H. N. Cheng, F. C. Schilling, and F. A. Bovey, *Macromolecules*, **9**, 363 (1976).

38. V. D. Mochel, *J. Polym. Sci.*, **1A**, 1009 (1972).

39. W. Ritchey and F. J. Knoll, *Polym. Lett.*, **4**, 853 (1966).

40. F. A. Bovey and G. V. D. Tiers, *J. Polym. Sci.*, **44**, 173 (1960).

41. Y. Inoue and A. Nishioka, *Polym. J.*, **3**, 149 (1972).

42. J. Schaefer, *Macromolecules*, **5**, 590 (1972).

43. A. Zambelli, D. E. Dormand, A. I. R. Brewster, and F. A. Bovey, *Macromolecules*, **6**, 925 (1973).

44. J. C. Randall, *J. Polym. Sci., Polym. Phys. Ed.*, **13**, 889 (1975).

45. E. R. Santee, V. D. Mochel, and M. Morton, *J. Polym. Sci., Polym. Lett.*, **11**, 453 (1973).

46. M. W. Duch and D. M. Grant, *Macromolecules*, **3**, 165 (1970).

47. V. D. Mochel, *J. Polym. Sci.*, **A1**, 1009 (1972).

48. F. A. Bovey, *Polymer Conformation and Configuration*, Academic Press, New York, 1969.

49. M. Karplus, *J. Chem. Phys.*, **30**, 11 (1950).

50. M. Karplus, *J. Am. Chem. Soc.*, **85**, 2870 (1963).

51. A. Bothner-By, *Adv. Magn. Resonance*, **1**, 195 (1965).

52. C. J. Carman, A. R. Tarpley, Jr., and J. H. Goldstein, *J. Am. Chem. Soc.*, **93**, 2864 (1971).

53. D. E. F. Mooney and P. H. Winson, *Ann. Rev. NMR Spectrosc.*, **2**, 157 (1969).

54. C. J. Carman, *Macromolecules*, **6**, 725 (1973).

55. R. Chujo, H. Ubara, and A. Nishioka, *Polym. J.*, **3**, 670 (1972).

56. J. P. Kinsinger, T. Fischer, and C. W. Wilson, III, *J. Polym. Sci.*, **BA**, 379 (1966).

57. J. P. Kinsinger, T. Fischer, and C. W. Wilson, III, *J. Polym. Sci.*, **B5**, 285 (1967).

58. J. Schaefer, *J. Phys. Chem.*, **70**, 1975 (1966).

59. A. Bunn and M. E. A. Cudby, *Polymer*, **17**, 548 (1976).

60. A. L. Segre, M. Delfini, F. Conti, and A. Boicelli, *Polymer*, **16**, 338 (1975).

9
Chain Isomerism Through Monomer Enchainment

9.1 TERMINOLOGY

In the polymerization of vinyl polymers it is possible for the initiating molecules to attack the monomer, producing two different active species:

$$R^* + CXY{=}CH_2 \nearrow \genfrac{}{}{0pt}{}{R{-}CH_2{-}CXY^*}{R{-}CXY{-}CH_2^*}$$

The head is defined as the substituted (i.e., the α) carbon, and the tail as the unsubstituted carbon or methylene unit. In general, the active center can be a free radical, carboniumion, or carbanion.

With α,α-disubstituted olefins the same rules apply but with α,β and higher multisubstituted olefins, the system must be arbitrarily defined as to its head and tail.

In the propagation step there are four possible types of enchainment, resulting in three different chain structures, as indicated in Figure 9.1 for polystyrene. In the notation of the chemist the 1,3-substituted structure corresponds to head-to-tail addition. The 1,2-substituted structure is head-to-head and the 1,4-substituted structure is tail-to-tail enchainment.

This property has been called *positional isomerism*, and more recently the name *orienticity* has been proposed (1). *Autorienticity* has been suggested as the designation of the head-to-head orienticity. The head-to-tail orienticity is called *heteroorientic*, and the tail-to-tail is termed *postorientic*.

Some monomer units have a sense of direction in the chain, so the terminology of orienticity comes quite naturally. For example, with poly-(propylene oxide) (PO) the monomer can add as

$$\begin{array}{cc} -CH{-}CH_2{-}O{-} & -CH_2{-}CH{-}O{-} \\ \mid & \mid \\ CH_3 & CH_3 \\ A & A^* \end{array}$$

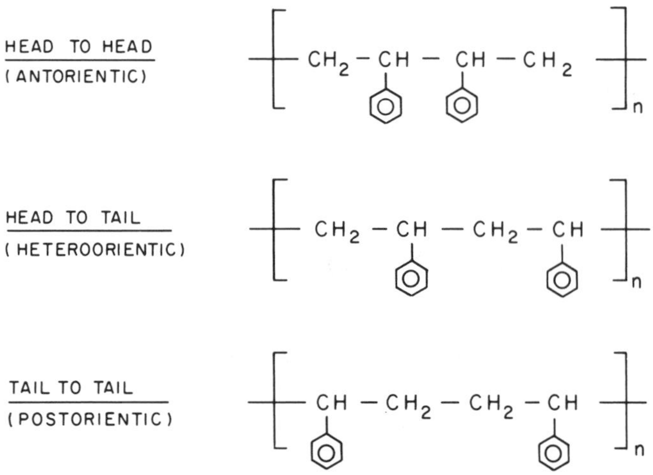

HEAD TO HEAD
(ANTORIENTIC)

$$\left[CH_2-CH-CH-CH_2 \right]_n$$

HEAD TO TAIL
(HETEROORIENTIC)

$$\left[CH_2-CH-CH_2-CH \right]_n$$

TAIL TO TAIL
(POSTORIENTIC)

$$\left[CH-CH_2-CH_2-CH \right]_n$$

Figure 9.1 Possible repeating unit orientations in an α-vinyl polymer chain.

so we have a type of structural isomerism that allows (without consideration of stereochemical differences) four possible ways of combining the monomers. The structural isomer designations of the dimer and trimer sequences of PO are shown in Table 9.1. These sequence variations are further complicated in copolymers of PO.

For copolymers some definitions must be adopted as well. Since the primary copolymers investigated to date have been ethylene-propylene or their fluorinated analogs, we adopt this type of notation. We designate as primary insertion the addition to the CH_2 or methylene group; secondary

Table 9.1 Structural Isomer Designations of the Dimer and Trimer of Polypropylene Glycol[a]

AA*	Tail-tail
AA	Tail-head
A*A*	Head-tail
A*A	Head-head
AAA*	Tail-head, tail-tail
AA*A*	Tail-tail, head-tail
A*AA*	Head-head, tail-tail
AA*A	Tail-tail, head-head
A*AA	Head-head, tail-head
A*A*A	Head-tail, head-head
AAA	Tail-head, tail-head
A*A*A*	Head-tail, head-tail

[a] $A = -CH(CH_3)CH_2O-$;
$A^* = -CH_2CH(CH_3)O-$.

insertion occurs when the addition involves the methine group of propy-lene [—CH(CH$_3$)—]. Ethylene must always add by primary insertion, but propylene can add either way.

Since polymerization of polydienes can occur by several routes, the number of ways of arranging the monomer units is larger than for vinyl monomers. When vinyl polymerization occurs across either double bond (1,2, or 3,4), the structure

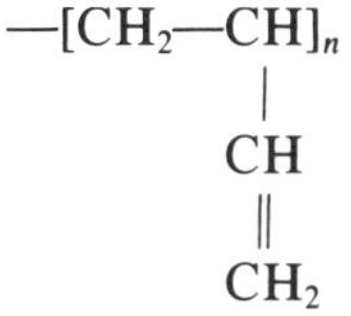

$$—[CH_2—CH]_n$$
$$|$$
$$CH$$
$$\|$$
$$CH_2$$

has the possibilities of either head-to-head or head-to-tail. When addition occurs across atoms 1 and 4 for polybutadiene (1,4), the resulting structure

$$[—CH_2—CH=CH—CH_2—]_n$$

has no difference between head-to-head and head-to-tail arrangements. But with isoprene the methyl group may be on either side of the double bond, hence head-to-head and head-to-tail arrangements are not identical and these linkages do occur in polyisoprene. Additionally, both 1,4 and 3,4 addition occur simultaneously, yielding:

Structure	Linkage
	4,1-1,4 head-to-head
	1,4-4,1 tail-to-tail
	4,1-3,4 head-to-head
	1,4-4,3 head-to-tail

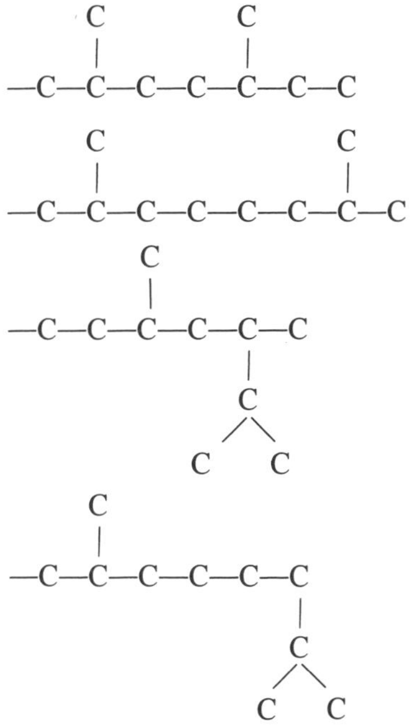

The head-to-tail units of both 1,4 and 3,4 also occur.

9.2 GENERAL CONSIDERATIONS

Polymers prepared by bond-opening polymerizations of α-olefins, vinyl monomers, and acrylic monomers have mainly head-to-tail linkages in the polymer chains. For some of these cases the presence of head-to-head linkages by a radical polymerization has been assumed to occur by a termination reaction caused by radical recombination. Polymers are known where reverse monomer addition is responsible for up to 30% of head-to-head linkages in the polymer chain. Most of the monomers that have detectable amounts of head-to-head addition have a halogen substituent, such as chlorine or fluorine, attached to the double bond. Examples are vinyl chloride, vinyl fluoride, vinylidene fluoride, and chloroprene. The amount of head-to-head structure can be varied in some cases, with higher polymerization temperature usually giving higher reverse monomer addition in the polymer chain.

The orienticity of the monomer in the polymerization is dependent on two processes:

1. The relative energy requirements of the transitions states.
2. The relative stabilities of the products.

The initiation mode generating an active center at the head of the monomer is favored by both processes. That is, the steric requirements in the transition state are lessened, since the substitutent is one carbon removed from the attacking species. Additionally, the head has the substitutent in position to stabilize the product by resonance or induction, while the tail has no effects of this kind. Consequently, the initiation step strongly favors the formation of the active site on the head of the repeating unit, and the propagating step proceeds predominantly by head-to-tail polymerization or heteroorienticity.

Let us compare the free radical polymerization of styrene and vinyl fluoride in this respect. For styrene the benzyl radical is favored by resonance stabilization in the amount of 20 to 25 kcal relative to the tail. With vinyl fluoride very little stabilization is afforded by the fluoride atom. In the transition state the activation energy of the formation of the benzyl radical is lower by ca. 8–10 kcal, since the phenyl rings are far apart. For vinyl fluoride there is some preference for the radical on the head end due to electrostatic repulsion. Consequently, for styrene the free radical polymerization is exclusively head-to-tail, while for vinyl fluoride head-to-tail polymerization is dominant at lower temperatures, but addition is nearly random head-to-head and head-to-tail at higher polymerization temperatures.

Resonance and inductive stabilization effects and steric effects favor heteroorientic addition in nearly all types of polymerizations of monovinyl monomers.

For dienes the 1,2 or 3,4 addition process favors the head-to-tail arrangement because of the stabilizing influence of the vinyl side group of the growing chain. In the 1,4 polymerization of isoprene the preference is for head-to-tail addition in spite of the relatively small stabilizing effect of the methyl on the adjacent carbon. Reversals of monomer units occur with much greater frequency in the dienes than in the simple vinyl systems.

9.3   PREPARATION OF MODEL HEAD-TO-HEAD AND TAIL-TO-TAIL POLYMERS

It is expected that pure head-to-head and tail-to-tail polymers would have different and possibly useful new combinations of properties compared to head-to-tail polymers, so efforts have been made to prepare pure model systems. The methods used fall generally into one of two types: (1) preparation of appropriate one-to-one alternating copolymers, and (2) chemical conversion of one homopolymer into a head-to-head or tail-to-tail homopolymer of another type.

Head-to-head polymers have been prepared of polypropylene (2), poly(vinyl chloride) (3), poly(methyl acrylate) (4), poly(methyl cinnanate) (5), and polystyrene (6). Tail-to-tail polypropylene was obtained by copolymerization of ethylene with an excess of 2-butene. Head-to-head poly(methyl methacrylate) was obtained by the alternating copolymerization of ethylene-maleic anhydride and subsequent esterification. Poly-methyl cinnamate with the head-to-head structure utilized the alternating copolymerization of stilbene and maleic anhydride, followed by esterification. Head-to-head poly(vinyl chloride) was obtained by chlorine addition to polybutadiene. For the preparation of head-to-head polystyrene a styrene dimer anion was prepared and subsequently polymerized.

In general, the head-to-head polymers have higher T_g's than their head-to-tail analogs, as might be expected from the steric interaction of the substitutents adjacent to each other. Additionally, the thermal decomposition behavior of these two types of polymers is considerably different. However, with only the preparation of an isolated number of polymers to compare, insufficient data are available to draw extensive correlations.

9.4   THEORETICAL MICROSTRUCTURE OF POLYMERS WITH POSITIONAL ISOMERISM (7)

9.4.1   Homopolymers

Consider the polymerization of vinyl polymers where the initiation can attack the monomer to produce the two possible different active species:

$$R^* + CXY{=}CH_2 \underset{\searrow\; R{-}CXY{-}CH_2^*}{\overset{\nearrow\; R{-}CH_2{-}CXY^*}{}}$$

Let R—TH* represent the site R—CH_2—CXY* and R—HT represent the R—CXY—CH_2^* site. We let the probability of the formation of a head site be P_1(TH) and that for the tail P_1(HT). We note that

$$P_1(\text{HT}) + P_1(\text{TH}) = 1,$$

since these are the only two possibilities. In the propagation step we have, using this same notation, the following:

Terminal	Addition	Product	Probability	Conditional Probability
R—TH*	k_1(TH)	R—TH—TH*	P_2(TH—TH)	$P(T/H) = k_1/(k_1 + k_2)$ (9.1)
R—TH*	k_2(HT)	R—TH—HT*	P_2(TH—HT)	$P(H/H) = k_2/(k_1 + k_2)$ (9.2)
R—HT*	k_3(HT)	R—HT—HT*	P_2(HT—HT)	$P(H/T) = k_3/(k_3 + k_4)$ (9.3)
R—HT*	k_4(TH)	R—HT—TH*	P_2(HT—TH)	$P(T/T) = k_4/(k_3 + k_4)$ (9.4)

The conditional probabilities are easily determined as follows:

$$P(\text{T/H}) = \frac{k_1[\text{TH}][\text{R—TH*}]}{k_1[\text{TH}][\text{R—TH*}] + k_2[\text{HT}][\text{R—TH*}]} = \frac{k_1}{k_1 + k_2},$$

where (TH) = (HT), since the directional nature of unreacted monomer cannot be determined.

It is apparent that the initiator probabilities and the dyads are related through the equations

$$P_1(\text{TH}) = P_2(\text{TH—TH}) + P_2(\text{TH—HT}) = P_2(\text{HT—TH}) + P_2(\text{TH—TH}),$$

$$P_2(\text{HT}) = P_2(\text{HT—HT}) + P_2(\text{HT—TH}) = P_2(\text{HT—HT}) + P_2(\text{TH—HT}),$$

which requires the reversibility relationship

$$P_2(\text{HT—TH}) = P_2(\text{TH—HT}). \tag{9.5}$$

This result requires that the amounts of head-to-head and tail-to-tail dyads in any polymer are equal. Additional information can be obtained by substituting in eq. 9.5:

$$P_1(\text{TH})P(\text{H/H}) = P_1(\text{HT})P(\text{T/T}),$$

$$P_1(\text{TH})P(\text{H/H}) = [1 - P_1(\text{TH})]P(\text{T/T}),$$

$$P_1(\text{TH})[P(\text{H/H}) + P(\text{T/T})] = P(\text{T/T}),$$

$$P_1(\text{TH}) = \frac{P(\text{T/T})}{P(\text{H/H}) + P(\text{T/T})}, \tag{9.6}$$

which is the statement of the amount of head-to-tail enchainment in terms of the conditional probabilities. Now we can calculate the fraction of head-to-head dyads:

$$P_2(\text{TH—HT}) = P_1(\text{TH})P(\text{H/H})$$

$$= \frac{P(\text{T/T})P(\text{H/H})}{P(\text{H/H}) + P(\text{T/T})} = P_2(\text{HT—TH}). \qquad (9.7)$$

Since the larger portion of addition is head-to-tail, it is convenient to rearrange by substituting $P(\text{H/H}) = 1 - P(\text{T/H})$, so

$$P_2(\text{TH—HT}) = \frac{P(\text{T/T})[1 - P(\text{T/H})]}{1 - P(\text{T/H}) + P(\text{T/T})}. \qquad (9.8)$$

In general, most vinyl polymers examined have a predominance of head-to-tail enchainment, so

$$P(\text{T/H}) \sim 1, \qquad P_1(\text{TH}) \sim 1, \qquad P_2(\text{TH—TH}) \sim 0.99, \qquad P(\text{T/T}) \sim 1,$$

$$P_2(\text{TH—TH}) = P_1(\text{TH})P(\text{T/H}) = \frac{P(\text{T/T})P(\text{T/H})}{1 - P(\text{T/H}) + P(\text{T/T})} \sim 0.99.$$

The conditional probabilities can be calculated by determining the following ratio of dimer concentrations:

$$\frac{P_2(\text{TH—HT})}{P_2(\text{TH—TH})} = \frac{P(\text{H/H})}{P(\text{T/H})} = \frac{1 - P(\text{T/H})}{P(\text{T/H})}. \qquad (9.9)$$

The triad fractions can be obtained in a similar, albeit a longer, process:

Designation

where ^iS_j is the sequence number designation, where i is the number of monomer units, and j is the number in an arbitrary order of sequences.

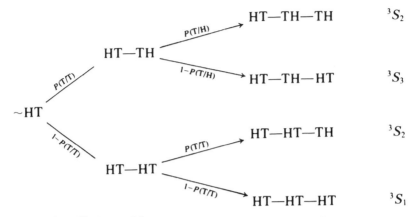

Triad
Designation

Let us simplify by writing

$$p = P(T/H), \qquad q = P(T/T),$$
$$1 - p = P(H/H), \qquad 1 - q = P(H/T).$$

The triad fractions then become

$$P_3(TH—TH—TH) = P(TH)p^2 + P(HT)(1 - q)^2 = P(^3S_1),$$
$$P_3(HT—TH—TH) = P(HT)[q(1 + p - q)] = P(^3S_2),$$
$$P_3(TH—TH—HT) = P(TH)[(1 - p)(1 + p - q)] = P(^3S_4),$$
$$P_3(HT—TH—HT) = P(TH)[(1 - p)q] + P(HT)[q(1 - p)] = P(^3S_3). \qquad (9.10)$$

Since

$$P(TH) = \frac{q}{1 - p + q},$$

$$P(HT) = \frac{1 - p}{1 - p + q},$$

it follows that

$$P_3(^3S_1) = \frac{qp^2 + (1 - q)^2(1 - p)}{1 - p + q},$$

$$P_3(^3S_2) = \frac{(1 - p)q(1 + p - q)}{1 - p + q},$$

$$P_3(^3S_4) = \frac{q(1 - p)(1 + p - q)}{1 - p + q},$$

$$P_3(^3S_3) = \frac{(1 - p)q^2 + q(1 - p)^2}{1 - p + q}. \qquad (9.11)$$

The tetrad probabilities follow in like fashion. Thus one can determine the fundamental parameters from measurements of the dyads, triads, or tetrads.

9.4.2   Theory of Monomer Inversion in Copolymers (8)

A terpolymer model representation is taken for the enchainment orientation involving copolymers. We carry out the analysis for an ethylene-propylene system, but the extension to similar systems is obvious. The ethylene monomer is designated as monomer 1, propylene adding by primary insertion as monomer 2, and by secondary insertion, monomer 3. If penultimate effects are ignored, nine reaction rate constants can be written for addition of three monomers to the three growing ends, and six reactivity ratios can be given. In the monomer mixture the concentrations of the monomer types are

$$(1) = (\text{ethylene}),$$

$$(2) = (3) = (\text{propylene}),$$

since there is no distinction between propylene of type 2 or 3.

There are nine conditional probabilities $P(j/i)$ that the monomer of type j will add to a given chain end of type i:

$$P(j/i) = \frac{k_{ij}[j]}{k_{i1}[e] + k_{i2}[p] + k_{i3}[p]}. \tag{9.12}$$

There are also three necessary relationships between the conditional probability:

$$P(1/i) + P(2/i) + P(3/i) = 1, \quad i = 1, 2, 3.$$

Letting X denote the molar ratio $[e]/[p]$, we have

$$P(2/1) = \frac{1/r_{12}}{X + 1/r_{12} + 1/r_{13}},$$

$$P(3/1) = \frac{1/r_{13}}{X + 1/r_{12} + 1/r_{13}},$$

$$P(3/2) = \frac{1/r_{23}}{X/r_{21} + 1 + 1/r_{23}},$$

$$P(1/2) = \frac{X/r_{21}}{X/r_{21} + 1 + 1/r_{23}},$$

$$P(1/3) = \frac{X/r_{31}}{X/r_{31} + 1/r_{32} + 1},$$

$$P(2/3) = \frac{1/r_{32}}{X/r_{31} + 1/r_{32} + 1}, \tag{9.13}$$

Table 9.2   Numbers of Methylene Sequences of Different Lengths

Length	Occurrence	No. in "Representative Sample"
1	⌊⌊ and ⌋⌋	$s_1 = P(2)P(2/2) + P(3)P(3/3)$
2	⌊⌋ and ⌋_⌊	$s_2 = P(2)P(3/2) + P(3)P(1/3)P(2/1)$
3	⌊_⌊ and ⌋_⌋	$s_3 = P(2)P(1/2)P(1/2) + P(3)P(1/3)P(3/1)$
4	⌊_⌋ and ⌋__⌊	$s_4 = P(2)P(1/2)P(3/1) + P(3)P(1/3)P(1/1)P(2/1)$
5	⌊__⌊, ⌋__⌋	$s_5 = P(1/1)s_3$
6	⌊__⌋, ⌋___⌊	$s_6 = P(1/1)s_4$
$3 + 2n$	⌊⌊, ⌋⌋	$s_{3+2n} = P(1/1)^n s_3$
$4 + 2n$	⌊⌊, ⌋⌋	$s_{4+2n} = P(1/1)^n s_4$
All	⌊⌋, ⌋⌊	$\Sigma s = P(2) + P(3) - P(3)P(2/3)$

where $r_{ij} = k_{ii}/k_{ij}$. When the molar ratio and the six independent probabilities are known, it is possible to solve for the six reactivity ratios.

It is useful to consider this system in terms of the numbers of methylene sequences of different lengths rather than dyads or triads, since IR and NMR spectroscopy are used for the characterization. Table 9.2 shows the various lengths of the methylene sequences for the various types of addition, as well as the number calculated using the conditional probabilities, where $P(1)$, $P(2)$, and $P(3)$ are the numbers of units of the three monomer types in the polymer sample.

The number of methylene units isolated from each other can be found by adding the probabilities of the occurrence of the addition of type ⌊⌊ (33) with ⌋⌋ (22). Thus

$$S_1 = P(2)P(2/2) + P(3)P(3/3). \tag{9.14}$$

For pairs of methylene units to occur, addition of the type ⌊⌋ and ⌋_⌊ must occur and

$$S_2 = P(2)P(3/2) + P(3)P(1/3)P(2/1). \tag{9.15}$$

A sequence of three methylene units can be found in two ways ⌋_⌋ and ⌊_⌊. so

$$S_3 = P(2)P(1/2)P(2/1) + P(3)P(1/3)P(3/1).$$

The total number of methylene sequences is

$$\sum S = P(2) + P(3) - P(3)P(2/3),$$

where the final term subtracts out the head-to-head additions. Thus it is possible to establish relationships between the methylene sequences and the conditional probabilities for those copolymers exhibiting positional isomerism.

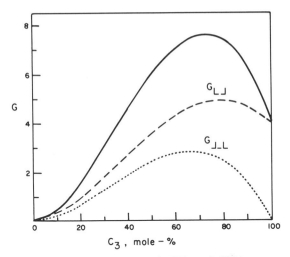

Figure 9.2 Contribution to the total amount of $(CH_2)_2$ of $(CH_2)_2$ groups formed by two tail-to-tail P units $(G_{\llcorner\lrcorner})$ and of $(CH_2)_2$ groups formed by one E unit lying between two head-to-head P units $(G_{\lrcorner\llcorner})$. The three curves correspond to $r_1 r_2 = 1$, $P(3/2) = 0.06$, and $P(1/3) = 0.30$. (Reprinted by permission of Ref. 9.)

A simpler model results when some assumptions are made concerning the conditional probabilities (9). Since head-to-head addition does not occur for the homopolymer of propylene, one can write the conditional probabilities for the inverted unit:

$$P(2/2) = 0, \qquad P(2/3) = 0.$$

Likewise, one can assume that the inversion of the propyl group does not occur when adding to an ethylene unit, so $p(3/1) = 0$. Three parameters remain, namely, $P(3/2)$, which is the probability of entrance of a normal P unit after an inverted P unit, $P(1/3)$, the probability of an E unit after an inverted P unit, and $P(1/2)$, the probability of unit adding to a P unit.

The contribution to the total amount of pairs of methylenes $(CH_2)_2$ that are associated with inverse monomer addition in copolymers of the type just discussed has been calculated (9). Some $(CH_2)_2$ groups are formed by two tail-to-tail propylene units, and some $(CH_2)_2$ groups are formed by one ethylene unit lying between two head-to-head propylene units. The results are shown in Figure 9.2 for a random polymerization process. The probability of head-to-head addition is set at 0.06, and the probability of an ethylene unit adding is 0.30. At high propylene content of the copolymers, the amount of $(CH_2)_2$ can reach almost 8%, while in the homopolymer it is very small.

9.5 METHODS OF DETERMINING THE NATURE OF CHEMICAL LINKAGE BETWEEN MONOMERS

9.5.1 Chemical Methods

The chemical methods for determining the relative amounts of positional isomers depend on differences in chemical reactivity between the three structures:

| 1,3 structure | 1,2 structure | 1,4 structure |

or on the difference in the structure of the products obtained by chain scission.

Different types of reactions have been used which show differences in ease of cyclization of the 1,3 structure relative to the 1,2 structure. For example, with PVC a zinc-catalyzed cyclization occurs for the head-to-tail structure:

head-to-tail

but no similar cyclization occurs for either of the other two structures:

Similarly, condensation reactions of aldehydes can be used such as for poly(vinyl alcohol):

Although these reactions are distinctive in their response, they are not as useful as might be imagined for polymers, since these chemical reactions are insensitive to small amounts of the head-to-head structure because of the "widow" effect. The random reaction of groups along the chain isolates some groups that would ordinarily be available for reaction. This effect is illustrated as follows:

Pure 1,3 structure

Statistical calculations indicate that, if reaction occurs at random for the pure head-to-tail structure, the number fraction of "widowed" or isolated units is $1/e^2 \sim 0.135$, while if the structure has random head-to-head and head-to-tail structure, the limit is $1/e \sim 0.184$. In the usual cases the yields of the reactions are between 84 and 86%, so unless the amount of head-to-head structure is very high, the cyclization process is not very conclusive.

A more sensitive method for chemical analysis of monomer enchainment is identification of the products obtained from chain fragmentation. For example, with styrene destructive distillation can yield

$$\text{1,3 structure} \rightarrow \underset{\displaystyle \phi}{CH_2}-\underset{\displaystyle \phi}{CH_2}-\underset{\underset{\displaystyle \phi}{|}}{\overset{\overset{\displaystyle H}{|}}{C}}-\underset{\displaystyle \phi}{CH_2}-CH_2$$

$$\text{1,2 structure} \rightarrow \underset{\displaystyle \phi}{CH_2}-\underset{\displaystyle \phi}{CH_2}$$

For polystyrene no product has been obtained with the phenyl ring on adjacent carbons, so no 1,2 structure is present in polystyrene.

Another chemical scission method with good sensitivity is based on the chemical reaction of the 1,2 structure. For PVA the following reactions occur with periodic acid:

$$-CH_2-\underset{\underset{\displaystyle OH}{|}}{CH}-\underset{\underset{\displaystyle OH}{|}}{CH}-CH_2- \xrightarrow{\ HIO_4\ } -CH_2CHO + \overset{\overset{\displaystyle O}{\|}}{CH}-CH_2-\underset{\underset{\displaystyle OH}{|}}{CH}-$$

1,2 structure

$$-CH_2-\underset{\underset{\displaystyle OH}{|}}{CH}-CH_2-\underset{\underset{\displaystyle OH}{|}}{CH}- \xrightarrow{\ HIO_4\ } \text{no reaction}$$

Thus the increase in the number of molecules is directly proportional to the number of 1,2 structures in the polymer chain. A measurement of the change in molecular weight allows calculation of the number of 1,2 structures. For poly(vinyl acetate) (precursor to PVA) at 25°, the results are that $p \sim q = 0.99$, and at 100°C, $p \sim 0.98$ (10).

Another example of chain scission leading to results indicative of chain isomerization is the ozone degradation of poly(propylene oxide) (11). Consider the chain

$$\overset{a}{-O}-CH_2\underset{\underset{\displaystyle CH_3}{|}}{CH}-\overset{b}{O}-CH_2-\underset{\underset{\displaystyle CH_3}{|}}{CH}-\overset{c\ \ \ a}{O}-\underset{\underset{\displaystyle CH_3}{|}}{CH}-CH_2-\overset{b}{O}-CH_2-\underset{\underset{\displaystyle CH_3}{|}}{CH}-\overset{c}{O}-$$

If chain scission occurs at the points *aa*, one obtains the product of head-to-tail addition:

$$HOCH_2CH—O—CH_2—CH—OH$$
$$\qquad | \qquad\qquad\qquad |$$
$$\qquad CH_3 \qquad\qquad CH_3$$
$$I$$

whereas if scission occurs at the points *bb*, one obtains

$$HO—CH_2—CH—O—CH—CH_2OH$$
$$\qquad\qquad | \qquad\quad |$$
$$\qquad\qquad CH_3 \quad CH_3$$
$$II$$

which is the head-to-head product. Finally, scission at the points *cc* gives

$$HOCH—CH_2—O—CH_2—CH—OH$$
$$\quad | \qquad\qquad\qquad\qquad |$$
$$\quad CH_3 \qquad\qquad\qquad CH_3$$
$$III$$

from the tail-to-tail structure. These three products can be identified and separated on a vapor phase chromatographic column. The results show that compounds II and III are formed to the extent of 10–30%, indicating the presence of substantial monomer inversion in poly(propylene oxide) (11).

It has also been found possible to obtain information about the directional isomeric distribution in poly(propylene oxide) (PO) by hydrolysis of PO with subsequent GLPC column separation of the glycol ethers (12). The dimers and trimers represent runs of two and three PO units in the chain. The chromatograms of commercial poly(propylene glycol)s are shown in Figure 9.3. The mass spectra of the four dimer fractions show relative intensities of the peak at m/e 103 (parent ion minus —CH_2OH) of 0.6, 14.2, 15.6, and 16.2, respectively. Since the dimer AA* has no terminal —CH_2OH group, it is assigned to the first fraction (in order of elution).

NMR spectral analysis of the second fraction established that it is the sum of the AA and A*A* dimers. The dimer fractions 3 and 4 are the stereo pairs (*dd, ll, dl, ld*) of A*A, respectively. Mass spectroscopy and NMR spectroscopy revealed that trimer fractions 1 and 5 arise from AAA*, AA*A* and A*AA, A*A*A, respectively. Trimer fraction 2 is assigned to one stereo pair of A*AA*, AA*A; fraction 3 to AAA, A*A*A*; and fraction 4 to the sum of one stereo pair each of A*AA*, AA*A, and A*AA, A*A*A. In commercial acid-catalyzed poly(propylene glycol) the amounts of both the dimer and trimer are nonrandom (12). For the dimers the amount of dimer AA* is expected to be 25%, but 36% is found.

Similar analysis has been performed on propylene oxide-sulfur dioxide copolymers (13). The primary structure of these chains consists of sulfite linkages connecting the monomer units. The head-to-head, head-to-tail isomers in closed runs of PO of length two and three were determined

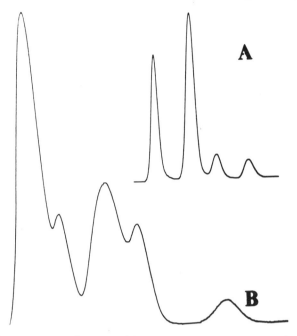

Figure 9.3 Chromatograms of commercial polypropylene glycols: (*A*) dimer; (*B*) trimer. (Reprinted by permission of Ref. 12.)

by chromatographic analysis of the corresponding di- and tripropylene glycol ethers. The relative concentration of the dimer isomers AA*, AA(A*A*), and A*A were 0.24, 0.56, and 0.20, respectively, indicating completely random distributions in the closed runs of two PO units. Similar results were observed for the triad distributions (13).

9.5.2 Physical Methods

NMR Spectroscopy. Consider the simplest example, that of a disubstituted vinyl monomer. The methylene protons in the H—T structure have slightly different chemical shifts from the T—T structure. In both structures the protons are equivalent, so no spin-spin splitting is observable. In the ^{1}H NMR spectrum of poly(vinylidene chloride) only a single proton resonance at $\tau = 6.14$ ppm is observed (14). Consequently, no monomer inversion occurs in the homopolymerization of vinylidene chloride. The ^{1}H NMR spectrum of poly(vinylidene fluoride) (PVF$_2$) gives quite a different picture, which is unfortunately complicated by ^{19}F coupling to ^{1}H. After ^{19}F spin decoupling, the T—T protons occur at $\tau = 7.76$ ppm, while the H—T protons occur at $\tau = 7.08$ ppm (14).

The structure of PVF$_2$ becomes much easier to understand using the ^{19}F NMR techniques, since the ^{19}F chemical shifts are one or more orders

Table 9.3 Monomer Sequences and Carbon Designations

Sequence Notation	Monomer Sequence	Designation	Carbon Sequence H-Centered	Carbon Sequence T-Centered	Designation
Two-monomer			*Three-carbon*		
1 1	HT—HT	$^2S_1$	TH*T	HT*H	$^2S_1 + {}^2S'_1$
1 −1	HT—HT	$^2S_2$		HT*H	$2\,^2S_2$
−1 1	TH—HT	$^2S_3$	T H*T		$2\,^2S_3$
−1 −1	TH—TH	$^2S'_1({}^2S_4)$	T H*T	HT*H	$^2S_1 + {}^2S'_1$
Three-monomer			*Five-carbon*		
1 1 1	HT—HT—HT	$^3S_1$	HT H*TH	THT T*HT	$^3S_1 + {}^3S'_1$
1 1 −1	HT—HT—TH	$^3S_2$	HT H*TT	THT T*TH	$^3S_2 + {}^3S'_2$
1 −1 1	HT—TH—HT	$^3S_3$	TT H*HT	HT T*HH	$^3S_3 + {}^3S'_3$
−1 1 1	TH—HT—HT	$^3S_4$	TH H*TH	HHT T*HT	$^3S_4 + {}^3S'_4$
Four-monomer			*Seven-carbon*		
1 1 1 1	HT—HT—HT—HT	$^4S_1$	THT H*THT	HTH T*HTH	$^4S_1 + {}^4S'_1$
1 1 1 −1	HT—HT—HT—TH	$^4S_2$	THT H*TTH	HTH T*HTT	$^4S_2 + {}^4S'_2$
1 1 −1 1	HT—HT—TH—HT	$^4S_3$		HTH T*THT	$^4S_3 + {}^4S'_3$
1 −1 1 1	HT—TH—HT—HT	$^4S_4$	HTT H*HTH	THH T*THT	$^4S_4 + {}^4S'_4$
1 1 1 1	TH—HT—HT—HT	$^4S_5$	HHT H*HTH	THH T*THT	$^4S_5 + {}^4S'_5$
−1 1 1 1	HT—TH—HT—TH	$^4S_6$			$2\,^4S_6$
1 1 −1 1	TH—HT—HT—TH	$^4S_7$			$2\,^4S_7$
−1 1 1 1	HT—TH—HT—HT	$^4S_8$	HTT H*HTT	THH T*HTT	$^4S_8 + {}^4S'_8$
1 1 −1 1	TH—HT—TH—HT	$^4S_9$	HHT H*TTH	THH T*HTT	$2\,^4S_9$
−1 1 1 1	TH—TH—HT—HT	$^4S_{10}$	THT H*HTH	THH T*THH	$2\,^4S_{10}$

of magnitude greater than the chemical shifts of ^1H spectra. For an unsymmetrical monomer that has a distinguishable head and tail, the two-monomer sequences can be represented as shown in Table 9.3, where ^iS_j is the sequence number designation, i being the number of monomer units and j the number of the sequence in an arbitrary order of sequences (25). Since we cannot specify the chain direction $^2S_1 = ^2S_4$, only three two-monomer sequences are distinguishable. Sequence $^2S_4$ can be designated $^2S_1'$, indicating the transpose (reverse) of $^2S_1$. These sequences produce three different carbon sequences based on the $^2S_1$, $^2S_2$, and $^2S_3$ sequences, since the $^2S_4 = ^2S_1'$.

The sequence representation is visualized more easily by a notation where we let 1 represent HT and -1 represent TH. Using this notation, we can generate $^iS_j'$ by multiplying the numerical sequence by -1 and reversing the order. In NMR the triad sequences are resolved when the chemical shift of a carbon is affected measurably by the substitutents on the carbons α and β on it. Thus the five-carbon sequences are unique and determine the resonances. Table 9.3 shows the five-carbon sequences generated by the four three-monomer sequences. Tetrad effects are observed if the γ as well as the α and β substitutents affect the chemical shift. The 10 four-monomer sequences generate 8 H-centered and 8 T-centered seven-carbon sequences, as shown in Table 9.3. The ability to observe these different sequences depends on the strength of the magnetic field, and work has been reported for ^{19}F NMR at 40, 56.4, 94.1, and 188 MHz for PVF_2 (25). The spectra of the three lower fields show only four resolved lines that have been assigned based on the three-monomer approximation, while the 188 MHz spectrum has eight resonances that have been assigned based on the four-monomer sequences. The 40 MHz ^{19}F NMR spectrum of PVF_2, shown in Figure 9.4, consists of four observable resonances, with three of the bands being small and of nearly equal intensity, and one being quite strong. These four peaks are from chemical shift differences arising not only from nearest neighbors but next nearest neighbors surrounding a particular $-CF_2-$ group; they are assigned in Figure 9.4. When chemical shifts based on compounds of low molecular weight are used and additivity is assumed, these sequence assignments are confirmed by calculation of the expected shifts. When a $-CH_2-$ group is substituted for a $-CF_2-$ group in either the first or second neighbor position, the ^{19}F shift of the central $-CF_2-$ group is affected in a regular fashion. The second neighbor groups affect the chemical shifts about one-fourth as much as the first neighbors (2.5 ppm versus 9.55 ppm). Further evidence for these resonance assignments arise from the ^1H spin decoupling of the ^{19}F NMR spectrum. Peak A is a singlet upon ^1H decoupling, indicating that the peak corresponds to a $-CF_2-$ group flanked on both sides by $-CH_2-$ groups. Similar arguments are made for the fine structure of the remaining peaks.

An interesting observation has been made concerning the fact that peak D has an intensity nearly equal to peaks B or C. Peak D only exists

Table 9.4 Assignments of 188 MHz ^{19}F NMR of Poly(vinylidene fluoride) (25)

Sequence Designation	Seven-Carbon Sequence	δ (ppm)[a] Observed	δ (ppm)[a] Calculated	Relative Intensity	Sequence Probability $P(\text{T/H}) = 0.948$
$^4S_1$	HTH T* HTH	−91.6	−91.6	0.804	0.805
$^4S_2$	HTH T HTT	−92.2	−92.2	0.035	0.045
$^4S_5$	THH T HTH	−94.8	−95.0	0.049	0.045
$^4S_8$	THH T HTT	−96.0	−97.0	0.001	0.005
$^4S_3$	HTH T THH	−113.6	−113.6	0.045	0.045
$^4S_6$	HTH T THT	−114.1	−114.2	0.005	0.005
$^4S_9$	THH T THH	−115.6	−114.3	0.009	0.005
$^4S_3$	THH T THT	−116.0	−116.2	0.045	0.045

[a] δ (CFCl$_3$, internal) = 0; the minus sign indicates lower frequency (higher magnetic field). T = CF$_2$; H = CH$_2$.

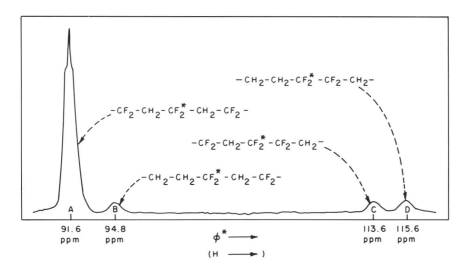

Figure 9.4 ^{19}F NMR spectrum of polyvinylidene fluoride. (Reprinted by permission of Ref. 15.)

when H—H addition is immediately followed by T—T. Since the intensities are nearly the same, only one monomer is inverted; the subsequent unit adds to restore the chain to the preferred configuration. The ^{19}F NMR results indicate that ca. 5–6% of the monomeric additions are inverted (15).

The 188 MHz ^{19}F spectrum of PVF$_2$ has eight resonances whose chemical shifts, assignments, relative intensities, and four-monomer sequence probabilities are shown in Table 9.4. The assignments are calculated using an additivity relationship for the seven-carbon sequences. The observed intensities are compared with calculated intensities corresponding to a head-to-tail addition probability of $P(\text{T/H}) = 0.948$. The

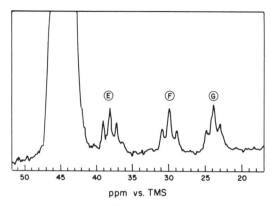

Figure 9.5 The 25 MHz spectrum of the CH$_2$ carbons of poly(vinylidene fluoride) observed at 90°C using a 30% (w/v) solution in ethylene carbonate at 90°C. The very large resonance centered at ca. 45 ppm is that of the normal head-to-tail sequences. The "defect" resonances E, F, and G are associated with the "reversed" monomer sequences. (Reprinted by permisssion of Ref. 16.)

calculated intensities are in agreement with the observed intensities, indicating Bernoullian polymerization statistics. Ziegler-Natta catalyst–initiated PVF$_2$ has a non-Bernoullian distribution and is therefore formed by a non free radical mechanism.

The ^{13}C NMR spectra of PVF$_2$ yield results similar to the ^{19}F NMR spectra (16), and in fact the spectra of the methylene carbons are remarkably like the ^{19}F spectra. The 25 MHz spectrum of the methylene carbons of PVF$_2$ is shown in Figure 9.5. The resonance at 45 ppm is the normal head-to-tail sequences. The inverted carbons show well resolved multiplets and are all more shielded than the principal resonance. In terms of the chain structure the resonances are assigned as follows:

$$
\begin{array}{ccc}
\text{E} & \text{G} & \text{F} \\
\downarrow & \downarrow & \downarrow
\end{array}
$$

—CH$_2$CF$_2$CH$_2$—CF$_2$CH$_2$CF$_2$CF$_2$CH$_2$CH$_2$CF$_2$CH$_2$CF$_2$—
H—H   T—T

which is in order of increasing shielding. The assignments involve the use of a deshielding inductive effect by fluorines on neighboring carbons (12 ppm) and an opposing shielding effect for fluorines on next nearest carbons (6.5 ppm).

When a monosubstituted vinyl monomer (CH$_2$=CHX) is polymerized, the NMR spectra are complicated by stereoisomeric effects and problems are associated with the determination by NMR of the amount of inverted monomer units. The ^{1}H NMR spectra are quite complicated, even with decoupling experiments, and have yielded little information about the

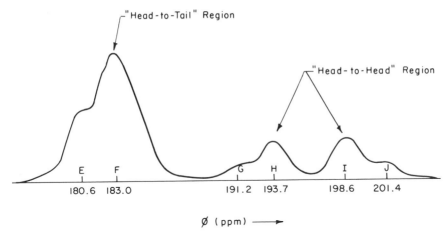

Figure 9.6 ^{19}F NMR spectrum of poly(vinyl fluoride). Lines E and F are associated with the normal head-to-tail sequences, and lines G, H, I, and J with the head-to-head region. (Reprinted by permission of Ref. 15.)

positional isomerism in the homopolymer except to indicate that if monomer inversion exists it is very small. In the case of poly(vinyl fluoride) (PVF), it is not clear from the ^{1}H NMR spectra whether the H—H structure is present. Spectral irregularities, possibly associated with inversion, are observed. On the other hand, ^{19}F NMR spectroscopy is quite useful, as shown in Figure 9.6 (15), where six chemically shifted peaks are observed over a range of 20.8 ppm. The relative intensities suggest that the peaks designated as E and F are the H—T components. The relative intensities of the upfield peaks (G, H, I, J) arise from the H—H and T—T structures. Differences in stereoconfigurations of the PVF account for neither the number of peaks (only three are expected, isotactic, heterotactic, and syndiotactic triads) nor the magnitude of the chemical shifts (a total spread of 5 ppm is expected for stereoconfigurational triads). Combining the stereochemical variations with the monomer inversion process predicts 11 resonances. The H—T should have three lines, T—T should have two lines, the H—H has four possible triad configurations, and two resonances occur for the structure with an H—H followed immediately by a T—T. The expected spacings between these lines are only 1.0–2.5 ppm, so they are badly overlapped, hence the observation of only six resonances. For PVF the monomer inversion occurs once for every six monomers added. The fraction of H—H polymerization in PVF varies linearly with polymerization temperature in the range of 25 to 180°C, increasing from 26 to 32% in the interval (15).

The large chemical shift range of ^{13}C NMR and lack of spin-spin coupling make the detection of positional isomers particularly easy. For vinyl monomers $CH_2{=}CR_1R_2$, such as α-methylstyrene, dimer dianions can be prepared with the tail-to-tail structure:

$$
\begin{array}{ccc}
R_1 & & R_1 \\
| & & | \\
-C-CH_2-CH_2-C- \\
| & & | \\
R_2 & & R_2
\end{array}
$$

Using normal dibromoalkanes $Br(CH_2)_nBr$, $n = 3–10$, as linking agents and α-methylstyrene as the monomer, copolymers can be prepared with the structure

$$
\left[
\begin{array}{ccc}
Me & & Me \\
| & & | \\
-C-CH_2-CH_2-C-(CH_2)_n- \\
| & & | \\
\phi & & \phi
\end{array}
\right]_m
$$

but other structures including H—T can occur (17). For these systems the ^{13}C NMR spectra of the α-methylstyrene-alkane dimers can be obtained to aid in the assignment of the resonances in the copolymers (17). The ^{13}C NMR spectrum of the copolymers shows that all the major peaks in the spectra can be predicted from the dimer chemical shifts having the T—T structure, suggesting that this T—T structure is the predominant structure in the copolymers. However, the regularity of the structure decreases as the length of the alkane chain increases, the long dibromoalkanes giving a considerable number of H—T linkages. The ^{13}C NMR resonances of the quaternary carbons in the T—T structure occur at 40.6 ppm, and for the H—T structure they occur at 41.5 ppm.

In spite of the limited amount of monomer inversion in homopolymers (due primarily to steric effects), copolymers exhibit considerably different results.

In an ^{1}H NMR study of vinyl chloride (VC)-vinylidene chloride (VDC) copolymers, it was found that the chain structure contained head-to-head or tail-to-tail VDC sequences. In the copolymers a singlet at 6.27 ppm appeared but was not observed in either homopolymer. This resonance is sharp, so there are no proton-proton splittings: it is assigned to the methylene protons from VDC that are polymerized tail-to-tail. The ratio of this resonance at 6.27 ppm and the resonance at 6.14 ppm found in the VDC homopolymer increases as the VC concentration increases, which indicates that VC is in some way involved (16). The following propagation steps for the copolymerization were suggested:

$$R-CCl_2\cdot + CH_2{=}CHCl \rightarrow R-CCl_2-CH_2-CHCl\cdot$$

$$R-CCl_2-CH_2-CHCl\cdot + CCl_2{=}CH \rightarrow R-CCl_2-CH_2-CHCl-CCl_2-CH_2\cdot$$

$$R—CCl_2—CH_2—CHCl—CCl_2—CH_2\cdot + CH_2{=}CCl_2 \rightarrow$$

$$R—CCl_2—CH_2—CHCl—CCl_2—CH_2—CH_2—CCl_2$$

$$H—T \qquad H—H \qquad T—T$$

In the second equation steric effects are not as pronounced with a VC unit, and head-to-head addition may occur. Now tail-to-tail addition of a VCD unit results in the new resonance at 6.27 ppm.

^{13}C NMR spectroscopy has also contributed to our understanding of the monomer sequence distribution in ethylene-propylene elastomers which exhibit monomer inversion. Figure 9.7 shows the spectrum of a 45% propylene PPDM rubber. Table 9.5 gives the assignments (8). The nomenclature is such that each secondary carbon has two Greek subscripts indicating its position relative to the nearest tertiary carbons in both directions along the polymer chain. The locational code is shown in Figure 9.8 for the methylenes in a head-to-tail chain, and Figure 9.9 shows the methylene notation for sequences resulting from an inversion of the propylene during polymerization (18). Some of the subscripts are followed by a plus sign, for example, $S_{\gamma\delta}+$, indicating a combined designation for $S_{\gamma\delta}$, $S_{\delta\epsilon}$, $S_{\delta\zeta}$, and so on, which are not resolved spectroscopically. Similarly, each tertiary carbon is denoted by a T with two Greek subscripts showing the positions of the nearest tertiary neighbors. Each primary (methyl) carbon is given the letter P with Greek subscripts that are the same as those for the attached tertiary carbon. The relative areas of the methyl, methylene, and

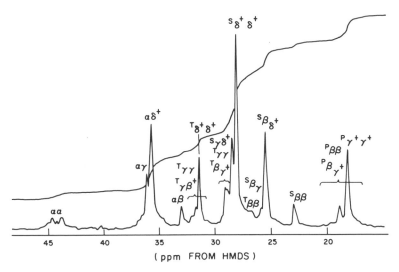

Figure 9.7   Pulsed FT ^{13}C NMR spectrum of a 45 wt-% propylene EPDM rubber. The lettering, defined in Figures 9.8 and 9.9, denotes the relative position of each carbon to its tertiary carbon neighbors. The spectrum was obtained at 120°C from a trichlorobenzene solution. (Reprinted by permission of Ref. 8.)

Table 9.5 ^{13}C Chemical Shifts for Ethylene-Propylene Rubbers

Species	^{13}C NMR Shift[a]	Occurrence
$S_{\alpha\alpha}$	44.6–43.7	Methylene sequence length 1
$S_{\alpha\beta}$	32.9	Two in each sequence length 2
$S_{\alpha\gamma}$	[b]35.9, 36.4	Two in each sequence length 3
$S_{\alpha\delta^+}$	35.5	Two in each sequence length $M > 3$
$S_{\beta\beta}$	22.7	Methylene sequence length 3
$S_{\beta\gamma}$	25.8	Two in each sequence length 4
$S_{\beta\delta^+}$	25.4	Two in each sequence length $M > 4$
$S_{\gamma\gamma}$	28.8	Methylene sequence length 5
$S_{\gamma\delta^+}$	28.4	Two in each sequence length $M > 5$
$S_{\delta^+\delta^+}$	28.0	$M - 6$ in each sequence length $M > 6$
$T_{\alpha\beta^+}$	[c](38.9)	Two for each ⅃⅃
$T_{\beta\beta}$	27.0	⌊⌊⌊ and ⅃⅃⅃
$T_{\beta\gamma^+}$	[b]29.0, 28.8	_⌊⌊, _⅃⅃, ⌊⌊_, ⌊⌊⌊, ⌊⅃⅃, ⅃⅃_
$T_{\gamma\gamma}$	31.6	⌊⅃_⌊ and ⅃_⌊⅃
$T_{\gamma\delta^+}$	31.6	_ _⌊⅃, _⅃_⌊, ⌊_⌊⅃, ⌊⅃_ _, ⌊⅃_⅃, ⅃_⌊_
$T_{\delta^+\delta^+}$	31.3	_ _⌊_, _⅃_ _, _⅃_⅃, ⌊_⌊_
$P_{\alpha\beta^+}$	[c](22.9)	Attached to $T_{\alpha\beta^+}$(⅃⅃)
$P_{\beta\beta}$	[b]19.8, 19.6, 18.9, 18.7	Attached to $T_{\beta\beta}$ (⌊⌊⌊ and ⅃⅃⅃)
$P_{\beta\gamma^+}$	18.7	Attached to $T_{\beta\gamma^+}$ (_⅃⅃, ⌊⌊_, etc.)
$P_{\gamma^+\gamma^+}$	18.2	Attached to $T_{\gamma\gamma}$, $T_{\delta\delta^+}$, and $T_{\delta^+\delta^+}$

[a] Downfield from internal hexamethyldisiloxane (HMDS) at 120°C in OCDB or TCB.
[b] Stereostructure produces nonequivalent chemical shifts.
[c] Not detected; predicted chemical shift is shown in parentheses.

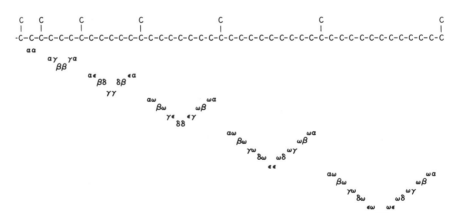

Figure 9.8 Description of methylene carbons in normal ethylene-propylene sequences (i.e., head-to-tail propylene addition). The Greek letters under each carbon stand for the number of bonds separating a methylene carbon from its nearest neighbor tertiary carbons. (Reprinted by permission of Ref.18.)

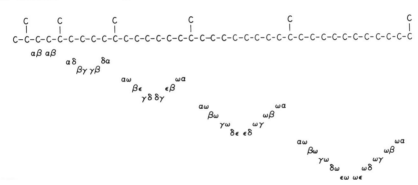

Figure 9.9 Description of methylene carbons in ethylene propylene sequences containing an inverted propylene monomer. The Greek letters are used as in Figure 9.8. (Reprinted by permission of Ref. 18.)

methine carbon resonances can be used to calculate ethylene-propylene composition. The areas assigned to the methine carbon resonances provide enough data to measure some monomer sequences up to pentads. Inversion occurs because resonances arising from even-numbered methylene sequences, lengths 2 and 4, are observed. The probabilities that give the best fit of the ^{13}C spectra of ethylene-propylene rubber are shown in Table 9.6. Although the composition derived from the NMR spectra agree with

Table 9.6 **Probabilities from Five-Parameter Model that Best Fit the ^{13}C NMR Spectra of Ethylene-Propylene Rubber**

	EP1	EP2	EP3
$p(1/1)$	0.708 ± 0.018	0.604 ± 0.111	0.465 ± 0.240
$p(2/1)$	0.290 ± 0.017	0.361 ± 0.169	0.530 ± 0.092
$p(3/1)$	0.002 ± 0.028	0.035 ± 0.235	0.006 ± 0.294
$p(1/2)$	0.857 ± 1.580	0.696 ± 0.668	0.608 ± 0.263
$p(2/2)$	0.095 ± 1.655	0.099 ± 0.789	0.258 ± 0.353
$p(3/2)$	0.048 ± 0.077	0.205 ± 0.291	0.134 ± 0.133
$p(1/3)$	0.931 ± 29.376	0.776 ± 1.991	0.310 ± 0.723
$p(2/3)^b$	0.0	0.0	0.0
$p(3/3)$	0.069 ± 29.376	0.224 ± 1.991	0.690 ± 0.723
Mole-% C_3	25.3 ± 0.8	35.5 ± 5.4	51.0 ± 6.9
$r_1 r_2$	0.39 ± 0.04	0.60 ± 0.44	0.82 ± 0.37
$r_1 r_2(\chi)^a$	0.40	0.64	0.57

^{a} $r_1 r_2(\chi)$ as calculated from the tertiary carbon areas after the conditional probability model has accounted for peak areas and determined the contiguous contribution to $T_{\gamma\delta^+}$.

^{b} $p(2/3)$ is set to zero because of the absence of $T_{\alpha\beta^+}$.

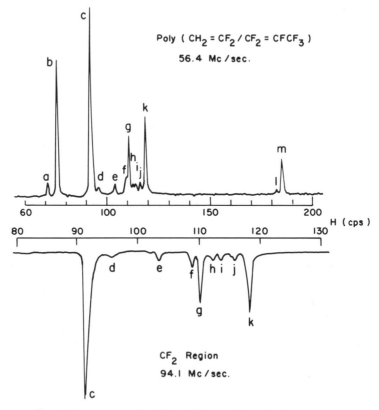

Figure 9.10 ^{19}F NMR spectrum of a $CH_2 = CF_2$ and $CF_2 = CFCF_2$ copolymer 80 mole-% vinylidene fluoride. Upper spectrum at 56.4 MHz. Lower spectrum: CF_2 at 94.1 MHz. The internal reference was CCl_2F. (Reprinted by permission of Ref. 19.)

other analytical methods, the limitation of the method is that one cannot accurately determine the percent of propylene inversion. One can only conclude that in the copolymers ranging from 20 to 60 wt-% propylene, between 10 and 40% of the total propylene present is inverted (8).

Another example of the determination of monomer inversion in copolymers is found for vinylidene fluoride-hexafluoropropylene (19). Hexafluoropropylene does not homopolymerize under the conditions employed in synthesizing the copolymers. The ^{19}F NMR spectra of a copolymer with 80 mole-% vinylidene fluoride at 56.4 MHz and at 94.1 MHz are shown in Figure 9.10, with the assignments given in Table 9.7, where $H = —CH_2—$, $T = —CF_2—$, and $X = —CFCF_3$. Since the hexafluoropropylene does not homopolymerize, no dyads containing two hexafluoropropylene units occur.

Polyisoprenes can be polymerized into four types of isomeric structures that can be arranged by head and tail linkages. The ^{13}C NMR

Table 9.7 ^{19}F Resonances of Poly (CH$_2$=CF$_2$-CF$_2$=CFCF$_3$)

Resonance	φ (ppm)[a]	Group Assignment[b]
a	71.4	H T X T H (CF$_3$)
b	75.9	T T X H T (CF$_3$)
c	91.9	T H T H T
d	95.7	H H T H T
c	103.7	T H T X T
f	109.4	T H T T X
g	110.7	
h	113.0	X H T T X
i	114.0	T H T T H
j	116.3	H T T H H
k	118.9	H T T X H
l	182.3	H T X T H (CF)
m	184.9	T T X H T (CF)

[a] Upfield from CCl$_3$F internal; acetone solution.
[b] H = —CH$_2$—, T = —CF$_2$—, X = —CF(CF$_3$)—.

resonances are complicated, but after hydrogenation cis-1,4 and trans-1,4 become similar:

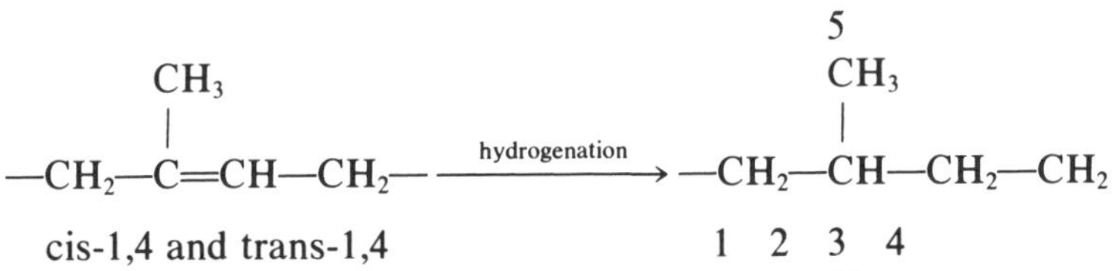

Signals at 34.41 and 27.42 ppm are assigned to the C$_1$ and C$_4$ atoms of structure (I) in head-to-head (4,1-1,4) and tail-to-tail (1,4-4,1) linkages. Polymers having 4,1-3,4 (head-to-head) and 1,4-4,3 (tail-to-tail) linkages have characteristic signals around 38.1 and 25.00 ppm, respectively. The amount of each of these dyads varies with the nature of the polymerization process (20).

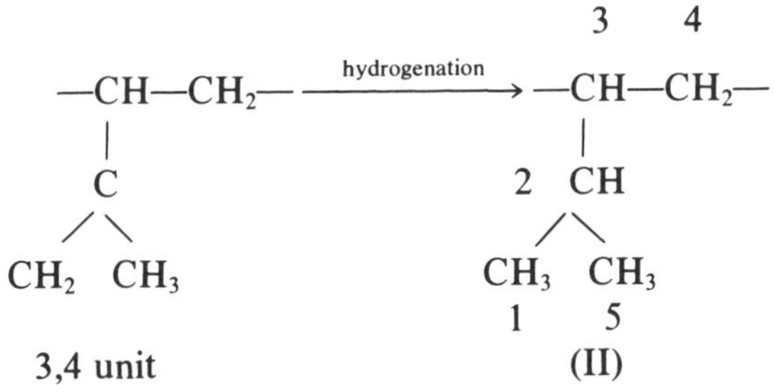

IR Spectroscopy of Monomer Enchainment Orientation. IR spectroscopy can be useful in detecting the presence of monomer reversals in the polymer chain when the head-to-head, tail-to-tail structures give rise to frequencies that can be associated with these structures. When a vinyl monomer is used, it is usually possible to distinguish between the isolated methylene of the head-to-tail structure and the pair of methylenes of the tail-to-tail structure. With polypropylene the isolated methylene absorbs at $812 \, cm^{-1}$ and the pair at $752 \, cm^{-1}$. The head-to-head (adjacent methyl groups) structure of polypropylene gives rise to a band at $1120 \, cm^{-1}$, which can also be observed.

The IR method has been used for fluorocarbon polymers. For copolymers of 1,1-difluoroethylene and tetrafluoroethylene, the spectra of the 1,3 and 1,4 structures are sufficiently different to allow detection. The 1,4 absorbs at $2870 \, cm^{-1}$, whereas the 1,3 and 1,2 absorb at $2777 \, cm^{-1}$. A model compound, obtained by copolymerizing ethylene with the monomer, was used to verify the band assignments. However, no head-to-head polymerization was found for these systems (21).

The most detailed study has been made on ethylene-propylene copolymers in conjunction with an IR study of the sequence distribution in this system. Origin of the $(CH_2)_n$ bands can be visualized from the following hypothetical polymer structure:

tail-to-head PP

An IR spectrum of the methylene rocking region is shown in Figure 9.11 (22). A method has been devised for evaluating the distribution of methylene sequences and in turn the distribution of methylene sequences and further in turn the distribution of monomeric units by means of a three-point measurement and a set of three simultaneous equations (22). The absorptivities were obtained from model systems: n-nonadecane (five or more contiguous methylenes), hydrogenated natural rubber ($n = 3$), and atactic polypropylene containing some head-to-head units ($n = 2$). The following set of simultaneous equations was used:

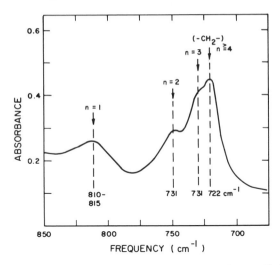

Figure 9.11   IR spectrum of an ethylene-propylene copolymer (54.3 wt-% C_2) in the methylene rocking region of the spectrum. (Reprinted by permission of Ref. 23.)

$$\frac{A_{752}}{t} = A_2 + 0.20A_3 + 0.12A_5,$$

$$\frac{A_{733}}{t} = 0.1A_2 + A_3 + 0.35A_5,$$

$$\frac{A_{722}}{t} = 0.50A_3 + A_5.$$

The thickness t was determined from the intensity of the $4310\,cm^{-1}$ combination band. The three unknown absorbances per unit thickness, A_2, A_3, and A_5, were converted to the concentration of methylene groups by applying —CH_2— absorptivities determined from model compounds. Absorptivities of $0.10, 0.15, 0.20$, and 0.25×10^4 ml/mole · cm were obtained for CH_2 sequences of 1, 2, 3, and 5 or more from the following model compounds, respectively: atactic polypropylene with head-to-tail linking, atactic polypropylene with a definite amount of head-to-head linking, hydrogenated natural rubber or squalane (2, 6, 10, 14-tetramethylpentadecane), and linear C_{10}–C_{19} hydrocarbons (22).

Drushel and co-workers (23) attempted to resolve the seriously overlapped bands for the methylene sequences in the rocking region by digitizing the IR spectra and applying curve-fitting techniques on the computer. Spectra were digitized by means of a general purpose data-acquisition system constructed from standard components. The digitized, smoothed, and interpolated spectra were mathematically resolved on the computer using an iterative least squares method. Methylene group sequence distributions (assuming that the absorptivity of each band is the same)

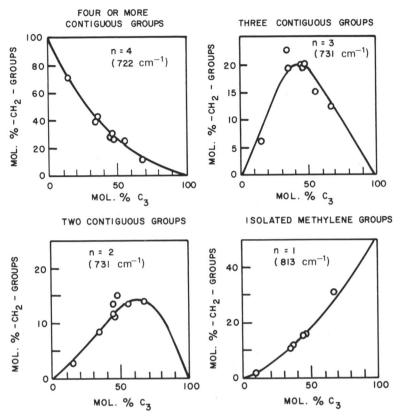

Figure 9.12 Methylene group sequence distribution as calculated from computer-resolved band parameters. (Reprinted by permission of Ref. 23.)

calculated from the computer-resolved band parameters are shown in Figure 9.12. These results agree with the previous results (23).

Optical Rotary Dispersion. As in low molecular weight compounds, the optical rotation in polymers is connected with optically active electronic transitions in definite chromophoric systems. The optical rotation can be

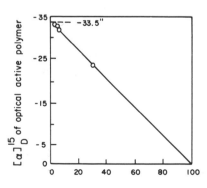

Figure 9.13 Relation between the specific rotatory power of optically active poly(propylene oxide) in benzene and tail-to-tail content (%) of poly(propylene-α-d oxide) by NMR spectra. (Reprinted by permission of Ref. 24.)

modified by mutual interactions among the chromophoric systems in the different monomer units. The decrease in specific rotation of poly(propylene oxide) samples prepared from optically active monomer is caused by the presence of tail-to-tail and head-to-head linkages in the polymers. When D, L-propylene-α-d oxide is polymerized, a linear relationship is observed between the specific rotation of the optically active monomer and the content of tail-to-tail linkage in the poly(propylene-α-d oxide) (Figure 9.13) (24). From the linear relationship the specific rotation of the polymer containing no tail-to-tail linkages is expected to be ca. $-33.5°$.

REFERENCES

1. F. M. Mirabella, Jr., and J. F. Johnson, *J. Macromol. Sci.*, **C12**, 81 (1975).

2. G. Natta, G. Dall'Asta, G. Massanti, I. Pasquon, A. Valvassori, and A. Zambelli, *J. Am. Chem. Soc.*, **83**, 3343 (1961).

3. F. E. Bailey, Jr., J. P. Henry, R. D. Lundberg, and J. M. Whelan, *J. Polym. Sci.*, **2B**, 447 (1964).

4. T. Otsu, S. Aoki, and R. Nakatani, *Makromol. Chem.*, **134**, 331 (1970).

5. T. Tanaka and O. Vogel, *Polymer*, **6**, 522 (1974).

6. D. H. Richards, N. F. Scilly, and F. Williams, *Polymer*, **10**, 603 (1969).

7. P. Meares, *Polymers: Structure and Bulk Properties*, Van Nostrand, London, 1965, Chapter 2.

8. C. J. Carman, R. A. Harrington, and C. E. Wilkes, *Macromolecules*, **10**, 536 (1977).

9. C. Tosi, A. Valvassori, and F. Campelli, *Eur. Polym. J.*, **5**, 575 (1969).

10. H. E. Harris and J. G. Pritchard, *J. Polym. Sci.*, **2**, 3673 (1964).

11. C. C. Price, R. Spector, and A. L. Tumolo, *J. Polym. Sci.*, **A5**, 407 (1967).

12. J. Schaefer, R. J. Katnik, and R. J. Kern, *Macromolecules*, **1**, 101 (1968).

13. J. Schaefer, R. J. Kern, and R. J. Katnik, *Macromolecules*, **1**, 107 (1968).

14. J. L. McClanahan and S. A. Previtera, *J. Polym. Sci.*, **3A**, 3919 (1965).

15. C. W. Wilson, III, and E. R. Santee, Jr., *J. Polym. Sci.*, **8C**, 97 (1965).

16. F. A. Bovey, F. C. Schilling, T. K. Kwei, and H. L. Frisch, *Macromolecules*, **10**, 559 (1977).

17. A. V. Cunliffe, P. E. Fuller, and R. A. Petrick in *Structural Studies of Macromolecules by Spectroscopic Methods*, K. J. Ivin, Ed., John Wiley & Sons, New York, 1976, p. 227.

18. C. E. Wilkes, C. J. Carman, and R. A. Harrington, *J. Polym. Sci.*, 237 (1973).

19. R. C. Ferguson, *Sonderdr. Jahrg.*, **11**, 723 (1965).

20. Y. Tanaka and H. Sato, *Polymer*, **17**, 413 (1976).

21. H. F. White, *J. Polym. Sci.*, **3A**, 309 (1965).

22. G. Bucci and T. Simonazzi, *J. Polym. Sci.*, **7C**, 203 (1964).

23. H. V. Drushel, J. J. Ellerbe, R. C. Cox, and L. H. Love, *Anal. Chem.*, **40**, 370 (1968).

24. N. Oguni, S. Watanabe, M. Maki, and H. Tani, *Macromolecules*, **6**, 195 (1973).

25. R. C. Ferguson and E. G. Brame, Jr., *J. Phys. Chem.*, **83**, 1397 (1979).

10

Chain Isomerism Involving Stereoconfiguration

10.1 INTRODUCTION

Consider the polymerization of an unsymmetrical monomer ($CH_2{=}CHR$) like propylene ($R{=}CH_3$). The monomer unit has two opposite steric configurations, one being the specular image of the other, but they are not superimposable on each other (Figure 10.1). In the polymerization process, both of these configurations can polymerize. In the absence of controlling catalyst systems, the difference in the two configurations is zero and they are equally likely. The polymerization is then determined by the active site of the polymerizing chain. In this case a nearly random placement of the two stereoconfigurations occurs. Stereospecific catalysts have been discovered that allow the polymerization to be controlled such that the relative placement of these two configurations can be steered in a regular manner. Polymers prepared in this manner are usually crystalline.

The position of the R group is determined by the stereochemical configuration of the asymmetric or chiral carbon (CHR) to which it is attached. The designations of the asymmetric centers are arbitrary, that is, either *d* or *l*. However, the relative stereochemical configuration of a given pair of asymmetric centers is important and is a property that is not influenced by conformational changes. The smallest configurational unit of the polymer chain is the stereoconfigurational dyad (Figure 10.2). When the stereoconfigurations of two neighboring asymmetric carbon atoms are the same (*dd* or *ll*), the placement is termed *isotactic*. If two successive asymmetric atoms have opposite stereoconfigurations (*dl*, *ld*), the placement is termed *syndiotactic*. Pairs of like designations *dd* or *ll* are interconvertible by an axis of rotation and are arbitrarily defined in space;

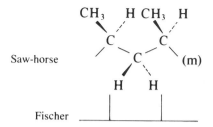

Figure 10.1 Monomer units of propylene having opposite configurations (saw-horse drawing).

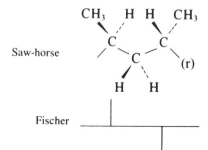

Figure 10.2 Isotactic (m) and syndiotactic (r) dyads in polypropylene [m (meso) and r (racemic)] for the two diastereomeric dyads.

they cannot be changed to unlike pairs (*dl* or *ld*) by any symmetry operation or geometric isomerism. Interconversion between isotactic and syndiotactic dyads in a polymer chain can only occur with breakage of chemical bonds between the pairs.

In isotactically regular polymers the substitutent groups on the asymmetric carbons are oriented similarly either above or below the plane of the extended zig-zag chain of the polymer (Figure 10.3). In syndiotactically regular polymers the substitutent groups on adjacent repeat units are alternately above and below the plane of the extended zig-zag chain of the polymer (Figure 10.4). Depending on the dimensions of the R group, the isotactic pair may be distorted by steric effects, resulting in a helix, but the

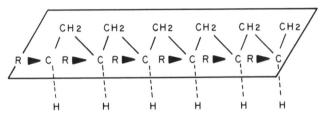

Figure 10.3   Isotactic polymer chain. All chiral carbons have the same chirality.

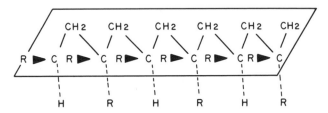

Figure 10.4   Syndiotactic polymer chain. Chiral carbons alternate in chirality.

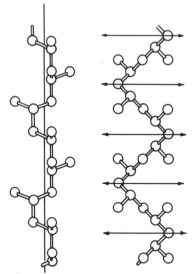

ISOTACTIC   SYNDIOTACTIC   **Figure 10.5**   Helical conformation of isotactic
POLYPROPYLENE   POLYPROPYLENE   polypropylene and of syndiotactic polypropylene.

stereoconfiguration is preserved, as shown in Figure 10.5, for helices of
isotactic and syndiotactic polypropylene. When the isotactic and syndiotactic placements are not regular, the polymer is termed *atactic*. This
property of being isotactic, syndiotactic, or atactic is called *tacticity* or
stereoregularity.

Vinyl monomers that are α,β-disubstituted, such as

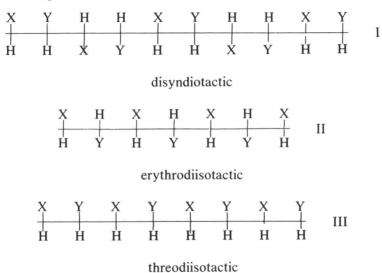

have two different asymmetric carbon atoms and their polymers are termed ditactic polymers. There are four different configurations, since two possibilities exist for each of the carbons. However, the number is restricted by the necessity of retaining the configuration of the monomeric bond between the two carbons. Thus the units are either diisotactic or disyndiotactic. The polymers with the same sequence are termed *erythropolymers* and those with alternating sequences are called *threopolymers*. The three regular configurations that result are:

I

disyndiotactic

II

erythrodiisotactic

III

threodiisotactic

10.2 EFFECT OF STEREOREGULATION

If a polymer is crystallizable, its monomeric units must have configurations that enable them to occupy equivalent positions along the chain axis. Thus the importance of stereoisomerism has long been recognized as a determining factor in the ability of polymers to crystallize. The discovery of stereospecific catalysts allowed the steering and control of the stereoconfiguration and has lead to crystalline polymers. Before the discovery of stereospecific polymerization processes, only a relatively small number of synthetic crystalline polymers were known. As the definition of crystal implies a three dimensional order, a linear polymer that is crystallizable must show regularity in the chain of monomeric units. Only isotactic and syndiotactic structures satisfy this requirement. All monomeric units in a crystal occupy geometrically equivalent positions with respect to the chain axis. Thus a regular succession of units must necessarily follow an n/p-fold helix, where n is the number of repeat units and p is the number of pitches contained within the identity period. Isotactic polymers are the cis stereoisomers of vinyl head-to-tail polymers. The bulky dimensions of the substitutent groups do not permit these polymers to assume a planar conformation. For polypropylene a planar structure would result in a distance of only 2.5 Å between nuclei of the two carbon atoms of successive methyl groups, which is unlikely since the hydrogen atoms of the methyl group would be nearer to each other than allowed by the minimum energy postulate. A suitable accommodation of the methyl groups may be achieved in a helical structure when the successive monomeric units are arranged on a threefold helix (Figure 10.5).

Syndiotactic polymers may be defined as the trans stereoisomers of vinyl head-to-tail polymers. Repetition can be achieved along the chain axis if every two monomeric units are considered ordered or as a helix type associated with twofold axes (Figure 10.5). The bulkiness of the side groups does not hinder a planar or nearly planar conformation of the chain.

Since the primary effect of stereoregularity is to permit preparation of crystallizable polymer systems, the role of stereoregularity in physical, mechanical, and thermal properties is quite profound. In general, the effects are such that correlations of the stereoregular polymers with their irregular analogs is quite impossible, particularly when both the pure syndiotactic and isotactic polymers have been prepared from the same monomer. Due to the increased packing density for the syndiotactic form compared with the isotactic form, general trends can be expected. That is, the melting points and T_g's of the polymers decrease in the order syndiotactic, isotactic, and heterotactic. If one considers the stereoregular

Table 10.1 Fractionation of Raw Polymers Obtained with Different Catalysts Prepared from Titanium Chlorides and Triethyl Aluminum at Temperatures from 35 to 90°C (22)

Fraction	Soluble in	Insoluble in	Crystal-linity (%)	Melting temperature (°C)	Irregularity (%)[a]
I	—	Trichloroethylene	75–85	176	0
II	—	n-Octane	64–68	174–175	0.4–0.8
III	n-Octane	2-Ethyl hexane	60–66	174–175	0.4–0.8
IV	2-Ethyl hexane	n-Heptane	52–64	168–170	2.5–3.4
V	n-Heptane	n-Hexane	41–54	147–159	7.2–12.2
VI	n-Hexane	n-Pentane	25–37	110–135	17.3–27.8
VII	n-Pentane	Ethyl ether	15–27	106–114	26.1–29.5

[a] It is assumed that ideal sterically pure crystalline polymer melts at 176°C.

polymer systems as copolymers of isotactic and syndiotactic placements, the melting points can be used to determine the relative stereoregularity of polymers prepared under slightly different polymerization conditions or fractions extracted from the whole polymer. Usually, the stereoregular forms are less soluble than the irregular forms. In fact, fractional solubility experiments involving different solvents, mixed solvents, and the same solvent at different temperatures have been used to determine tacticity semiquantitatively. Table 10.1 is a summary of the effects for fractionation of a whole polymer prepared with an isospecific catalyst (22).

Stereoregular effects are not limited to α-olefins, but are also important in diolefins as well as polyaldehydes, where it is possible to link the (—O—CHR) monomeric units in a sterically regular way.

10.3 DESIGNATION OF
SEQUENCE TYPES (1)

Direct experimental measurements of configurational sequences in polymers have been made. High resolution NMR spectroscopy has been the principal contributor to our understanding, so the notation used here is based on the NMR perspective.

Molecules having CM_2 groups, where M is an observable nucleus or group, are considered to be *geminally heterosteric* when the M groups can be differentiated through differences in average environment, and are termed *geminally homosteric* when they are equivalent to each other. For a syndiotactic placement of a vinyl monomer, the protons in the methylene group are racemic and have the same average environment, so they are not distinguishable. For a syndiotactic placement or a racemic dyad (*r*), a single line appears in the NMR. For an isotactic placement, or meso dyad (*m*), the protons of each methylene group do not experience the same average environment and have different chemical shifts. The methylene protons are heterosteric and the AB quartet of lines is observed. NMR measurements can distinguish on an absolute basis the isotactic and syndiotactic dyads in some vinyl polymers.

The original NMR observations were made on polymers of methyl methacrylate. Three species of α-methyl groups were detected. These three species were considered to be triads. Thus those monomer units flanked on both sides by units of the same configuration are *isotactic triads* (*i*); those monomer units having units of opposite configuration on both sides are *syndiotactic triads* (*s*); and those monomer units having a unit of the same configuration on one side of opposite configuration as the other side are heterotactic triads (*h*).

As experimental ^{1}H NMR technique has improved, finer line structures have become observable; thus tetrad resonances appear as fine structures on the dyad resonances, and pentad resonances appear as fine

Table 10.2 Stereochemical Sequence Designation and Bernoullian Probabilities

α–Substituent

	Designation	Projection	Bernoullian probability
Triad	Isotactic, mm (i)		P_m^2
	Heterotactic, mr (h)		$2P_m(1-P_m)$
	Syndiotactic, rr (s)		$(1-P_m)^2$
Pentad	$mmmm$ (isotactic)		P_m^4
	$mmmr$		$2P_m^3(1-P_m)$
	$rmmr$		$P_m^2(1-P_m)^2$
	$mmrm$		$2P_m^3(1-P_m)$
	$mmrr$		$2P_m^2(1-P_m)^2$
	$rmrm$ (heterotactic)		$2P_m^2(1-P_m)^2$
	$rmrr$		$2P_m(1-P_m)^3$
	$mrrm$		$P_m^2(1-P_m)^2$
	$rrrm$		$2P_m(1-P_m)^3$
	$rrrr$ (syndiotactic)		$(1-P_m)^4$

β–CM₂

	Designation	Projection	Bernoullian probability
Dyad	meso, m		P_m
	racemic, r		$(1-P_m)$
Tetrad	mmm		P_m^3
	mmr		$2P_m^2(1-P_m)$
	rmr		$P_m(1-P_m)^2$
	mrm		$P_m^2(1-P_m)$
	rrm		$2P_m(1-P_m)^2$
	rrr		$(1-P_m)^3$

structure on the triad resonances. The observation of longer sequences has given increasingly more detailed information about stereoregular polymer chains and their mode of polymerization.

Nomenclature has been suggested for the tetrads by analogy with the aldahexoses, but their use is not common (2). The nomenclature is as follows:

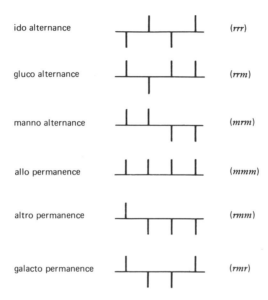

ido alternance		(rrr)
gluco alternance		(rrm)
manno alternance		(mrm)
allo permanence		(mmm)
altro permanence		(rmm)
galacto permanence		(rmr)

It is refreshing that no attempt has been made to name the pentad sequences, which are designated by the combination of dyads making up the pentad. Thus *mmmm* is the all-isotactic-pentad sequence. This designation system has considerable merit in simplicity, and is shown in Table 10.2 for the stereoregular sequences up to the pentads. Extension to longer sequences is obvious.

10.4 GENERAL RELATIONS FOR STEREOREGULAR SEQUENCES (3)

10.4.1 Homopolymers

We have seen that the number of distinguishable sequences increases with length. One can show that the number of distinguishable sequences (N_n) increases as

$$N(n) = 2^{n-2} + 2^{m-1},$$

where $m = n/2$ if n is even and $m = (n-1)/2$ if n is odd. The number of possible types of sequences increases rapidly with length asymptotically as 2^{n-2} as $N \to \infty$ (3).

Therefore for sequence length

$$n = 2 \quad 3 \quad 4 \quad 5 \quad 6 \quad 7 \quad 8,$$

$$N_n = 2 \quad 3 \quad 6 \quad 10 \quad 20 \quad 36 \quad 72.$$

The relative proportions of each sequence length are subject to certain necessary relations. For example,

$$(m) + (r) = 1,$$

$$(mm) + (mr) + (rr) = 1,$$

$$\text{Sum (tetrads)} = 1,$$

$$\text{Sum (all pentads)} = 1,$$

where we have used only the distinguishable n(ads), thus

$$(mr) = P[mr] + P[rm],$$

where the brackets indicate the normalized frequency or analytic fraction. For the triads there is a reversibility relationship:

$$P_3[mr] = P_3[rm].$$

For the tetrads the following reversibility relationships are appropriate:

$$P_4[mmr] = P_4[rmm],$$

$$P_4[mrr] = P_4[rrm],$$

$$P_4[mrm] + P_4[mrr] = P_4[mmr] + P_4[rmr],$$

which generates another necessary relationship between the observables for the tetrad sequences:

$$(mmr) + 2(rmr) = 2(mrm) + (mrr).$$

For the pentads the reversibility relationships are

$$P_5[mmmr] = P_5[rmmm], \qquad P_5[mrrr] = P_5[rrrm],$$

$$P_5[mmrm] = P_5[mrmm], \qquad P_5[rrmm] = P_5[mmrr],$$

$$P_5[mrmr] = P_5[rmrm], \qquad P_5[rmrr] = P_5[rrmr],$$

$$P_5[mmrm] + P_5[mmrr] = P_5[mmmr] + P_5[rmmr],$$

$$P_5[rmrr] + P_5[mmrr] = P_5[mrrr] + P_5[mrrm],$$

from which two additional relationships between the observables can be given:

$$(mmmr) + 2(rmmr) = (mmrm) + (mmrr),$$

$$(mrrr) + 2(mrrm) = (rmrr) + (mmrr).$$

In addition to the relations exhibited between observable n(ads) of the same order, there are relations between observables of different orders. These relations are easily derived by the process of adding to each unit. Thus

$$P_2[m] = P_3[mm] + P_3[mr] = P_3[mm] + P_3[rm],$$

so

$$(m) = (mm) + \tfrac{1}{2}(mr)$$

and

$$(r) = (rr) + \tfrac{1}{2}(mr).$$

Table 10.3 gives all the relationships,

The number average lengths of isotactic and syndiotactic sequences can also be derived in terms of dyads and triads. Let n_m be the isotactic sequence of n units. This isotactic sequence contains $n_m - 1$ isotactic (mm) triads and n_m meso (m) dyads. For the n_r syndiotactic sequence it has $n_r - 1$ syndiotactic (rr) triads and n_r racemic (r) dyads. If $N(n_m)$ is the number of isotactic sequences of length n_m, then there are $\sum_{n_m=1}^{\infty} N(n_m) \times$

Table 10.3   Relationships Between Stereoregular Sequences of Different Orders

Dyad-triad
$$(m) = (mm) + \tfrac{1}{2}(mr)$$
$$(r) = (rr) + \tfrac{1}{2}(mr)$$

Triad-tetrad
$$(mm) = (mmm) + \tfrac{1}{2}(mmr)$$
$$(mr) = (mmr) + 2(rmr) = (mrr) + 2(mrm)$$
$$(rr) = (rrr) + \tfrac{1}{2}(mrr)$$

Tetrad-pentad
$$(mmm) = (mmmm) + \tfrac{1}{2}(mmmr)$$
$$(mmr) = (mmmr) + 2(rmmr) = (mmrm) + (mmrr)$$
$$(mrm) = \tfrac{1}{2}(mmrm) + \tfrac{1}{2}(mrmr)$$
$$(mrr) = 2(mrrm) + (mrrr) = (mmrr) + (rmrr)$$
$$(rmr) = \tfrac{1}{2}(mrmr) + \tfrac{1}{2}(rmrr)$$
$$(rrr) = (rrr) + \tfrac{1}{2}(mrrr)$$

Dyad-tetrad
$$(m) = (mmm) + (mrmr) + \tfrac{1}{2}(mmr) + \tfrac{1}{2}(mrr)$$
$$(r) = (rrr) + (rmr) + \tfrac{1}{2}(mmr) + (mrr)$$

Triad-pentad
$$(mm) = (mmmm) + (mmmr) +)rmmr)$$
$$(mr) = (mmrm) + (mmrr) + (rmmr) + (rmrr)$$
$$(rr) = (rrrr) + (mrrr) + (mrrm)$$

$(n_m - 1)$ isotactic (mm) triads. Similarly, there are $\sum_{n_r=1}^{\infty} N(n_r)(n_r - 1)$ syndiotactic sequences. Finally, the number of isotactic and syndiotactic sequences must be equal:

$$\sum N(n_m) = \sum N(n_r).$$

Therefore

$$\frac{(mm)}{(rr)} = \frac{\sum N(n_m)(n_m - 1)}{\sum N(n_r)(n_r - 1)} = \frac{\sum N(n_m)n_m - \sum N(n_m)}{\sum N(n_r)n_r - \sum N(n_r)} = \frac{\bar{n}_m - 1}{\bar{n}_r - 1},$$

where $\bar{n}_m$ and $\bar{n}_r$ are the number average sequence lengths of the isotactic and syndiotactic sequences, respectively, given by

$$\bar{n}_m = \frac{\sum n_m N(n_m)}{\sum N(n_m)},$$

$$\bar{n}_r = \frac{\sum n_r N(n_r)}{\sum N(n_r)}.$$

It also follows that

$$\frac{(m)}{(r)} = \frac{(\bar{n}_m)}{(\bar{n}_r)}$$

and

$$(m) = \frac{\bar{n}_m}{\bar{n}_m + \bar{n}_r},$$

$$(r) = \frac{\bar{n}_r}{\bar{n}_m + \bar{n}_r}.$$

Therefore

$$\bar{n}_m = \frac{[1 - (mm)/(rr)]}{[1 - (mm)(r)/(rr)(m)]},$$

$$\bar{n}_r = \frac{[1 - (mm)/(rr)]}{[(m)/(r) - (mm)/(rr)]}.$$

The intensity of the heterotactic triad resonance is measure of the block lengths, so

$$(mr)^{-1} = \tfrac{1}{2}(\bar{n}_m + \bar{n}_r) = \bar{n},$$

where $\bar{n}$ is the number average length of all blocks. Combining of the above equations yields

$$\bar{n}_m = \frac{2(m)}{(mr)} = 1 + \frac{2(mm)}{(mr)},$$

$$\bar{n}_r = \frac{2(r)}{(mr)} = 1 + \frac{2(rr)}{(mr)}.$$

The persistence length ρ is given by

$$\rho = \frac{2[m][r]}{[mr]},$$

or

$$\rho = \bar{n}_m(r) = \bar{n}_r(m).$$

The above relations are general for all systems and are particularly useful for assignment purposes and checks of internal consistency.

10.4.2   Copolymer Microstructure with Stereoregular Polymerization (1)

In the previous treatment of copolymerization, (Chapter 3) stereochemistry was ignored. If the stereoconfiguration is considered, the number of structures to be considered increases substantially. The structures for dyads and triads are shown in Table 10.4. There are 6 possible dyad

Table 10.4   Configurational Sequences in Copolymers

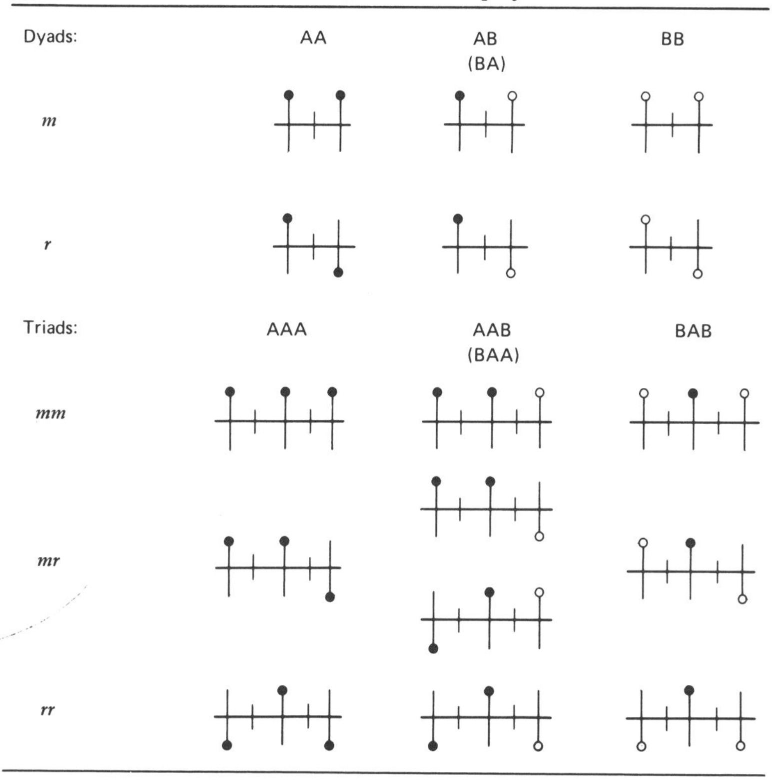

Dyads: AA AB (BA) BB

Triads: AAA AAB (BAA) BAB

+10 other with ● and ○ reversed.

combinations and 20 triads. The number of sequences $N'(n)$ as a function of n is given by (1)

$$N'(n) = s^{2(n-1)} + 2^{n-1};$$

thus

n	2	3	4	5	6
$N'(n)$	6	20	72	272	1056˙

Thus there are many complications.

10.5 MICROSTRUCTURE OF POLYMERS EXHIBITING STEREOREGULARITY (1)

The most convenient statistics for the description of the structure of the stereoregular polymer chains is based on the same principles as the statistics of binary polymerization. Thus the polymerization can be considered simply as a copolymerization of isotactic and syndiotactic additions.

10.5.1 Terminal or Bernoulli Model

For the terminal model of stereopolymerization of a single monomer, four propagation reactions are possible.

End	Add		Final	Conditional Probability
$\sim d^*$	M	k_{dd}	$\sim dd^*$	$P(d/d) = \dfrac{r_i}{r_i + 1}$
$\sim d^*$	M	k_{dl}	$\sim dl^*$	$P(l/d) = \dfrac{1}{r_i + 1}$
$\sim l^*$	M	k_{ll}	$\sim ll^*$	$P(l/l) = \dfrac{r_i}{r_i + 1}$
$\sim l^*$	M	k_{ld}	$\sim ld^*$	$P(d/l) = \dfrac{1}{r_i + 1}$

where

$$r_i = \frac{k_{dd}}{k_{dl}} = \frac{k_{ll}}{k_{ld}}.$$

The fraction of each of these sequences can be calculated in a manner similar to the method used for copolymerization. For the stereoregular dyads that correspond to the composition of copolymers, we have two addition processes yielding the indistinguishable pairs

$$P_2(m) = P_l P(l/l) + P_d P(d/d),$$

$$P_2(r) = P_l P(d/l) + P_d P(l/d),$$

where P_l and P_d are the probabilities of finding an l and d configuration, respectively, on the active site and $P(l/d)$ is the conditional probability of adding an l given a d configuration.

Note that, since only two configurations exist for the active site

$$P_l + P_d = 1,$$

and since the up or down or d or l possibilities are equal,

$$P_l = P_d = 0.5.$$

Thus

$$P_2(m) = 0.5 P(l/l) + 0.5 P(d/d),$$

$$P(l/l) = P(d/d) = P_m,$$

$$P_2(m) = P_m = (m),$$

where (m) is the experimental fraction of isotactic dyads.

The probability of generating a meso or isotactic pair when a new

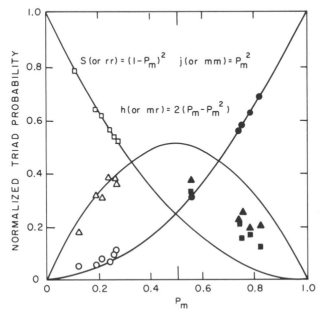

Figure 10.6 Triad fractions for poly(methyl methacrylate). Unfilled symbols represent free radically polymerized PMMA, and shaded symbols, anionically polymerized PMMA.

monomer unit is added to the end of a growing chain is designed as P_m. This assumes that the probability of forming an m or r sequence is independent of the stereoregularity of the terminal pair on the chain. Since a single parameter is involved, the terminal model follows Bernoullian statistics. Therefore the dyad fraction is given by

$$P_2(m) = (m) = P_m,$$
$$P_2(r) = (r) = 1 - P_m.$$

A triad sequence involves two monomer additions and, if the addition does not depend on the stereochemical configuration of the final unit, the two additions are independent and

$$P_3(mm) = (mm) = P_m^2,$$
$$2P_3[mr] = (mr) = 2P_m(1 - P_m),$$
$$P_3(rr) = (rr) = (1 - P_m)^2.$$

A plot of these relations is shown in Figure 10.6, with the solid lines representing the theoretical curve. The data points are discussed later.

Similarly, the tetrad sequences can be calculated, and a plot of these relations is shown in Figure 10.7. Likewise, the pentad sequences are calculated and plotted versus P_m in Figure 10.8.

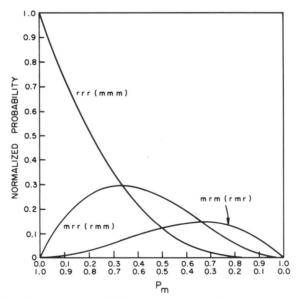

Figure 10.7 Tetrad sequence probabilities as a function of P_m. For *rrr*, *mrr*, and *mrm* the upper P_m scale is used; for *mmm*, *rmm*, and *rmr* the lower P_m scale is used. (Reprinted by permission of Ref. 5.)

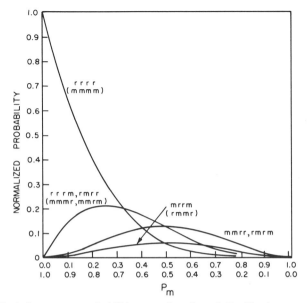

Figure 10.8 Pentad sequence probabilities as a function of P_m. The lower scale should be used for sequences in parentheses. (Reprinted by permission of Ref. 1.)

10.5.2 Penultimate Model

The penultimate model assumes that the terminal pair influences the stereospecific homopolymerization and involves only two independent parameters, following simple Markov statistics. They are

$$P(r/m) = 1 - P(m/m) = u,$$
$$P(m/r) = 1 - P(r/r) = w.$$

These two parameters can be related to the measured sequences by the following type of derivation. Starting with the reversibility relationship,

$$P_3[mr] = P_3[rm],$$

and substituting, we obtain

$$P_2(m)P(r/m) = P_2(r)P(m/r).$$

Since

$$P_2(r) = 1 - P_2(m) = 1 - (m),$$

it follows that

$$(m)P(r/m) = P(m/r) - (m)P(m/r),$$

so for the dyads

$$(m) = \frac{P(m/r)}{P(r/m) + P(m/r)} = \frac{w}{u + w},$$

$$(r) = \frac{u}{u + w}.$$

In a similar manner for the triads,

$$(mm) = \frac{(1 - u)w}{u + w},$$

$$(mr) = \frac{2uw}{u + w},$$

$$(rr) = \frac{u(1 - w)}{u + w}.$$

For the tetrads

$$(mmm) = \frac{(1 - u)^2 w}{u + w},$$

$$(mmr) = \frac{2u(1 - u)w}{u + w},$$

$$(mrm) = \frac{uw^2}{u + w},$$

$$(mrr) = \frac{2uw(1 - w)}{u + w},$$

$$(rmr) = \frac{u^2 w}{u + w},$$

$$(rrr) = \frac{u(1 - w)^2}{u + w}.$$

For pentad sequences

$$(mmmm) = \frac{w(1 - u)^3}{z},$$

$$(mmmr) = \frac{2uw(1 - u)^2}{z},$$

$$(mmrm) = \frac{2uw^2(1 - u)}{z},$$

$$(mmrr) = \frac{2uw(1 - u)(1 - w)}{z},$$

$$(mrmr) = \frac{2u^2 w^2}{z},$$

$$(mrrm) = \frac{uw^2(1-w)}{z},$$

$$(mrrr) = \frac{2uw(1-w)^2}{z},$$

$$(rrrr) = \frac{u(1-w)^3}{z},$$

where $z = u(1-w) + 2uw + w(1-u)$.

10.5.3 Penpenultimate Model

The penpenultimate model has four independent parameters but eight conditional probabilities, and follows second order Markov statistics. For convenience, we adopt the following shorthand notation:

$$P(m/mm) = \alpha, \qquad P(r/mm) = \bar{\alpha},$$
$$P(m/mr) = \beta, \qquad P(r/mr) = \bar{\beta},$$
$$P(m/rm) = \gamma, \qquad P(r/rm) = \bar{\gamma},$$
$$P(m/rr) = \delta, \qquad P(r/rr) = \bar{\delta},$$

where $P(m/mm)$ is the probability of a monomer adding in m fashion to a chain ending in mm, and so on. There are four independent parameters, since

$$\alpha + \bar{\alpha} = 1,$$
$$\beta + \bar{\beta} = 1,$$
$$\gamma + \bar{\gamma} = 1,$$
$$\delta + \bar{\delta} = 1.$$

This model reduces to penultimate when

$$\alpha = \gamma,$$
$$\beta = \delta.$$

The observable dyads through pentad frequencies are

$$(m) = \frac{(\bar{\alpha} + \gamma)\delta}{s},$$

$$(r) = \frac{(\bar{\beta} + \delta)\alpha}{s},$$

$$(mm) = \frac{\gamma\delta}{s},$$

$$(mr) = \frac{2\bar{\alpha}\delta}{s},$$

$$(rr) = \frac{\bar{\alpha}\bar{\beta}}{s},$$

$$(mmm) = \frac{\alpha\gamma\delta}{s},$$

$$(mmr) = \frac{2\bar{\alpha}\gamma\delta}{s},$$

$$(mrm) = \frac{\bar{\alpha}\beta}{s},$$

$$(mrr) = \frac{2\bar{\alpha}\bar{\beta}\delta}{s},$$

$$(rmr) = \frac{\bar{\alpha}\bar{\gamma}\delta}{s},$$

$$(rrr) = \frac{\bar{\alpha}\bar{\beta}\bar{\delta}}{s},$$

$$(mmmm) = \frac{\alpha^2\gamma\delta}{s} = \frac{(mmm)^2}{(mm)},$$

$$(mmmr) = \frac{2\alpha\bar{\alpha}\gamma\delta}{s} = \frac{(mmm)(mmr)}{(mm)},$$

$$(mmrm) = \frac{2\bar{\alpha}\beta\gamma\delta}{s} = \frac{2(mmr)(mrm)}{(mr)},$$

$$(mmrr) = \frac{2\bar{\alpha}\bar{\beta}\gamma\delta}{s} = \frac{(mmr)(mrr)}{(mr)},$$

$$(mrmr) = \frac{2\bar{\alpha}\beta\gamma\bar{\delta}}{s} = \frac{4(mrm)(rmr)}{(mr)},$$

$$(mrrm) = \frac{\bar{\alpha}\beta\delta^2}{s} = \frac{(mrr)^2}{4(rr)},$$

$$(mrrr) = \frac{2\bar{\alpha}\beta\delta\bar{\delta}}{s} = \frac{(mrr)(rrr)}{(rr)},$$

$$(rmmr) = \frac{\bar{\alpha}^2\gamma\delta}{s} = \frac{(mmr)^2}{4(mm)},$$

$$(rmrr) = \frac{2\bar{\alpha}\beta\bar{\gamma}\delta}{s} = \frac{2(mrr)(rmr)}{(mr)},$$

$$(rrrr) = \frac{\overline{\alpha\beta}\delta^2}{s} = \frac{(rrr)^2}{(rr)},$$

where

$$s = \bar{\alpha}\bar{\beta} + 2\bar{\alpha}\delta + \gamma\delta.$$

The conditional probabilities can be calculated from the experimental measurements as follows:

$$\alpha = \frac{(mmm)}{(mm)}, \qquad \bar{\alpha} = \frac{(mmr)}{2(mm)},$$

$$\beta = \frac{2(mrm)}{(mr)}, \qquad \bar{\beta} = \frac{(mrr)}{(mr)},$$

$$\gamma = \frac{(mmr)}{(mr)}, \qquad \bar{\gamma} = \frac{2(rmr)}{(mr)},$$

$$\delta = \frac{(mrr)}{2(rr)}, \qquad \bar{\delta} = \frac{(rrr)}{(rr)}.$$

10.5.4 Two State Propagation or Coleman–Fox Model (4, 5)

In this model two different propagating species are assumed, and each state proceeds by a terminal or Bernoulli-trial process as follows:

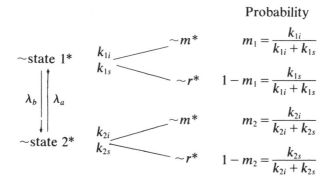

Probability

$$m_1 = \frac{k_{1i}}{k_{1i} + k_{1s}}$$

$$1 - m_1 = \frac{k_{1s}}{k_{1i} + k_{1s}}$$

$$m_2 = \frac{k_{2i}}{k_{2i} + k_{2s}}$$

$$1 - m_2 = \frac{k_{2s}}{k_{2i} + k_{2s}}$$

Let $m_1 = (k_{1i}/k_1)$ be the probability of isotactic placement in state 1 and $m_2 = (k_{2i}/k_2)$ the probability of isotactic placement in state 2, where k_{1i} and k_{2i} are the rate constants for isotactic placement in states 1 and 2, respectively, and $k_1 = k_{1i} + k_{1s}$ and $k_2 = k_{2i} + k_{2s}$ are the total propagation rates in each state. The fraction of the total polymer produced in state 1 is

$$w_1 = \frac{\lambda_a k_1}{\lambda_a k_1 + \lambda_b k_2},$$

where λ_a and λ_b are the rate constants for the transitions state $2 \to 1$ and its reverse, respectively.

From these definitions it can be established that the dyads are

$$(m) = \frac{\lambda_a k_{1i} + \lambda_b k_{2i}}{\lambda_a k_1 + \lambda_b k_2}, \qquad (r) = \frac{\lambda_a k_{1s} + \lambda_b k_{2s}}{\lambda_a k_1 + \lambda_b k_2}.$$

For simplicity let

$$a = \frac{\lambda_a \lambda_b k_1 k_2}{(\lambda_a k_1 + \lambda_b k_2)^2}\left(\frac{k_{1i}}{k_1} - \frac{k_{2i}}{k_2}\right)^2,$$

$$b = \frac{k_{1i} k_{2i}}{k_1 k_2},$$

$$c = \frac{k_{1s} k_{2s}}{k_1 k_2},$$

$$d = \frac{\lambda_a k_1 + \lambda_b k_2}{k_1 k_2},$$

$$x = \frac{[M]}{[M] + d},$$

where [M] is the monomer concentration.

The triad relationships can be shown to be

$$(mm) = (m)^2 + ax,$$

$$(mr) = 2(m)(r) - 2ax,$$

$$(rr) = (r)^2 + ax,$$

Likewise with the tetrad,

$$(mmm) = \frac{(mm)^2 + abx^2}{(m)},$$

$$(mmr) = \frac{(mm)(mr) - 2abx^2}{(m)},$$

$$(mrm) = \frac{(mr)^2 + 4acx^2}{4(r)},$$

$$(mrr) = \frac{(mr)(rr) - 2acx^2}{(r)},$$

$$(rmr) = \frac{(mr)^2 + 4abx^2}{4(m)},$$

$$(rrr) = \frac{(rr)^2 + acx^2}{(r)}.$$

It should be noted that the two state model reduces to first order Markov statistics when only *m* placements are possible in one state and only *r* placements are possible in the other. However, the two state model cannot reduce identically to second order Markov statistics. It may, however, approach the second order Markov statistics very closely.

10.5.5   Stereoconfiguration for Copolymers

It is possible, formally, to describe both the sequences and configurations of the monomer units in copolymers. The term *coisotactic* or *cosyndiotactic* has been used to describe the stereoconfiguration of copolymers (7). Four placement probabilities can be described, σ_{AA}, σ_{AB}, σ_{BA}, and σ_{BB}, where σ_{AB} is the probability that monomer B will add in a coisotactic manner to a growing chain ending in an A unit. The terms σ_{AA} and σ_{BB} can be obtained from the homopolymers polymerized under identical conditions. However, for the configurational parameters, cross propagation, σ_{AB} and σ_{BA}, cannot be independently determined, as they enter as

$$\Sigma_s = \sigma_{AB} + \sigma_{BA},$$
$$\Sigma_P = \sigma_{AB} \cdot \sigma_{BA}.$$

Let us consider the dyad possibilities for the terminal model for coisotactic copolymerization. The number of measured dyads (A_iB) of isotactic configuration is the sum of the two probabilities of the forward and reverse pairs:

$$(A_iB) = P(A_iB) + P(BA_i),$$

where the A_i notation implies isotactic configuration with the neighboring unit. The conditional probabilities are

$$(A_iB) = P(A)P(B/A)\sigma_{AB} + P(B)P(A/B)\sigma_{BA}.$$

From the reversibility relationships we have

$$P(AB) = P(BA),$$
$$P(A)P(B/A) = P(B)P(A/B),$$

so substitution yields

$$(A_iB) = P(A)P(B/A)(\sigma_{AB} + \sigma_{BA})$$
$$= P(A)P(B/A)\Sigma_s.$$

Likewise, for the syndiotactic AB pairs

$$(A_sB) = P(A)P(B/A)(2 - \Sigma_s)$$

Consequently, from the dyads only Σ_s can be obtained.

Table 10.5 **Probabilities for the 20 Magnetically Nonequivalent Triads in a Binary Vinyl Copolymer with Asymmetric α Carbon Atoms in Both Monomer Units, According to Terminal Model Copolymerization Statistics**

	Triad Number	Probability[a]
Isotactic	1	$P(A_iA_iA) = P(A)P(A/A)^2\sigma_{AA}^2$
Triads	2	$P(B_iB_iB) = P(B)P(B/B)^2\sigma_{BB}^2$
	3	$P(A_iA_iB^+) = P(A)P(A/A)P(B/A)\sigma_{AA}\Sigma_s$
	4	$P(A_iB_iB^+) = P(A)P(B/B)P(B/A)\sigma_{BB}\Sigma_s$
	5	$P(A_iB_iA) = P(A)P(A/B)P(B/A)\Sigma_p$
	6	$P(B_iA_iB) = P(A)P(B/A)^2\Sigma_p$
Syndiotactic	7	$P(A_sA_sA) = P(A)P(A/A)^2(1-\sigma_{AA})^2$
Triads	8	$P(B_sB_sB) = P(B)P(B/B)^2(1-\sigma_{BB})^2$
	9	$P(A_sA_sB^+) = P(A)P(A/A)P(B/A)(2-\Sigma_s)(1-\sigma_{AA})$
	10	$P(A_sB_sB^+) = P(A)P(B/B)P(B/A)(2-\Sigma_s)(1-\sigma_{BB})$
	11	$P(A_sB_sA) = P(A)P(A/B)P(B/A)(1-\Sigma_s+\Sigma_p)$
	12	$P(B_sA_sB) = P(A)P(B/A)^2(1-\Sigma_s+\Sigma_p)$
Heterotactic	13	$P(A_iA_sA^+) = 2P(A)P(A/A)^2(1-\sigma_{AA})\sigma_{AA}$
Triads	14	$P(B_iB_sB^+) = 2P(B)P(B/B)^2(1-\sigma_{BB})\sigma_{BB}$
	15	$P(A_iA_sB^+) = P(A)P(A/A)P(B/A)\sigma_{AA}(2-\Sigma_s)$
	16	$P(A_sA_iB^+) = P(A)P(A/A)P(B/A)(1-\sigma_{AA})\Sigma_s$
	17	$P(A_iB_sB^+) = P(A)P(B/B)P(B/A)(1-\sigma_{BB})\Sigma_s$
	18	$P(A_sB_iB^+) = P(A)P(B/B)P(B/A)\sigma_{BB}(2-\Sigma_s)$
	19	$P(A_iB_sA^+) = P(A)P(A/B)P(B/A)(\Sigma_s-2\Sigma_p)$
	20	$P(B_iA_sB^+) = P(A)P(B/A)^2(\Sigma_s-2\Sigma_p)$

[a] $P(B)P(A/B)$ substituted by $P(A)P(B/A)$ in all expressions.

The probabilities for the 20 magnetically nonequivalent triads are shown in Table 10.5 (8) for the terminal model. From this table it can be seen that Σ_P and Σ_s, but not σ_{AB} and σ_{BA}, can be determined from the triad sequences. The basis of the inability to determine σ_{BA} and σ_{AB} is that equal amounts of the two cross-propagation steps are required in a given NMR observable triad.

For the terminal model the triads for the cross-propagation steps can be reduced to two, and only two, independent equations leading to determination of Σ_s and Σ_p:

$$\Sigma_s = \sigma_{AB} + \sigma_{BA} = \frac{(A_iA_iB)}{P(A)P(A/A)P(B/A)\sigma_{AA}}$$

$$= \frac{(A_iB_iB)}{P(A)P(B/B)P(B/A)\sigma_{BB}},$$

$$\Sigma_p = \sigma_{AB} \cdot \sigma_{BA} = \frac{(A_iB_iA)}{P(A)P(A/B)P(B/A)}$$

$$= \frac{(B_iA_iB)}{P(A)P(B/A)^2}.$$

The solution yields

$$\sigma_{AB} \text{ or } \sigma_{BA} = \frac{\Sigma_s \pm \sqrt{(\Sigma_s)^2 - 4\Sigma_p}}{2}$$

From a pair of Σ_s and Σ_p two numbers can be calculated, but they cannot be assigned to one of the two σ_{BA} or σ_{AB}. If the discriminant is zero, $\sigma_{AB} = \sigma_{BA}$.

A determination of Σ_s and Σ_p can be made by using a curve-fitting procedure (8). A straight-line intersection method has also been proposed, but it is limited to NMR triad peak intensities of a homopolymer derived from the copolymer by a chemical reaction proceeding with retention of configuration (9). These conditions are highly restrictive but have been met by methyl methacrylate-metacrylic acid copolymers (9).

10.6 DETERMINATION OF STEREOPOLYMERIZATION MECHANISMS

With the experimental data in hand, one naturally turns to procedures for fitting the data to postulated mechanisms. The problem is simpler than for copolymerization, as the number of parameters is smaller and, in general, the quality of the experimental data (principally from NMR) is better because the monomer ratio does not enter as a variable. But a careful understanding of the limitations of the data arising from resonance overlap, and so forth, and the applicability of the models is required to get meaningful correlations.

Similarly to the statements for fitting of copolymerization models, several observations can be made for fitting stereopolymerization models (3).

From dyad information alone any mechanism can be fitted, but none can be tested. From triad and dyad information a Bernoulli or terminal model can be tested, and Markov models of any order can be fitted. For example, it is entirely possible for the second order Markov (penpenultimate) to produce the same triad proportions as a simple Bernoullian scheme. From tetrad information, a penultimate or first order Markov model can be tested and higher orders fitted. From pentad information, a penpenultimate or second order Markov model can be tested and higher orders fitted. The extension of these statements to higher sequences is obvious.

10.6.1   Terminal or Bernoullian Stereopolymerization Model

The first test of the terminal or Bernoullian model is the fit of the triad data to the parabolic curves of Figure 10.6. The triad results should fall on a single vertical line giving P_m. The triad data for free radical polymerized polymethyl methacrylate on the left side fits the model with $P_m = 0.24$. The dyad and triad data can be used to confirm the Bernoullian statistics, as the following ratios of areas must be satisfied:

$$(mm) = (m)^2,$$

$$(mr) = 2(m)(r),$$

$$(rr) = (r)^2.$$

Another test using the triad results is based on the following calculation of first order Markov conditional probabilities:

$$P(r/m) = \frac{(mr)}{2(mm) + (mr)},$$

$$P(m/r) = \frac{(mr)}{2(rr) + (mr)}.$$

If Bernoullian statistics apply,

$$P(m/r) + P(r/m) = 1.$$

Another criterion for Bernoullian statistics is the persistence ratio

$$\rho = \frac{2(m)(r)}{(mr)} = 1.$$

If $\rho \neq 1$, statistics are not terminal and are nonrandom.

   The final test is to calculate the probability of addition (P_m) and to use this parameter to calculate the expected values of the triad, tetrad, and pentad fractions, testing the internal consistency and the agreement with the experiment. The agreement should be within the experimental error.

10.6.2   Penultimate Stereopolymerization Model

Once triad sequences have been determined by experiment, it becomes possible to decide whether the mechanism governing the propagation is terminal, penultimate, or higher. Again, inspection of the data on anionically polymerized poly(methyl methacrylate) on the right-hand side of Figure 10.6 shows disagreement. No examination of the fractions of low orders such as dyads and triads can rule out the possibility of a higher order mechanism. If only the triad and dyad proportions are observable, the penultimate (first order Markov) model can be tested through the relationships

$$P(r/m) = u = \frac{(rm)}{2(m)} = \frac{(mr)}{2(mm)+(mr)},$$

$$P(m/r) = w = \frac{(mr)}{2(r)} = \frac{(mr)}{2(rr)+(mr)},$$

$$P(m/m) = 1 - u = \frac{(mm)}{(m)} = \frac{2(mm)}{2(mm)+(mr)},$$

$$P(r/r) = 1 - w = \frac{(rr)}{(r)} = \frac{2(rr)}{2(rr)+(mr)},$$

where it is required that $P(m/r) + P(r/r) = 1$, and $P(m/m) + P(r/m) = 1$.
    If

$$P(r/m) = P(r/r) = 1 - P_m$$

and

$$P(m/r) = P(m/m) = P_m,$$

the propagation follows Bernoulli statistics.
    If

$$P(r/m) + P(m/r) \neq 1,$$

first order Markov or higher is suggested.
    With tetrad sequences the terminal and penultimate models require

$$(mmm) = \frac{(mm)^2}{(m)},$$

$$(mmr) = \frac{(mm)(mr)}{(m)},$$

$$(mrm) = \frac{(mr)^2}{4(r)},$$

$$(mrr) = \frac{(mr)(rr)}{(r)},$$

$$(rmr) = \frac{(mr)^2}{4(m)},$$

$$(rrr) = \frac{(rr)^2}{(r)}.$$

If four of these relationships hold, this is sufficient to test the penultimate model. The relationships hold for both Bernoullian and first order Markov statistics and can be used to test deviations from first order statistics.

With pentad statistics the following relationships are consistent with terminal and penultimate models:

$$(mmmm) = \frac{(mmm)^2}{(mm)},$$

$$(mmmr) = \frac{(mmm)(mmr)}{(mm)},$$

$$(mmrm) = \frac{2(mmr)(mrm)}{(mr)},$$

$$(mmrr) = \frac{(mmr)(mrr)}{(mr)},$$

$$(mrmr) = \frac{4(mrm)(rmr)}{(mr)},$$

$$(mrrm) = \frac{(mrr)^2}{4(rr)},$$

$$(mrrr) = \frac{(mrr)(rrr)}{(rr)},$$

$$(rrrr) = \frac{(rrr)^2}{(rr)}.$$

It is necessary to show that six of the relationships hold to establish first order Markov statistics.

10.6.3 Penpenultimate Stereopolymerization Model

When tetrad proportions are observable, the second order Markov model (penpenultimate) can be tested by

$$P(m/mm) = \frac{(mmm)}{(mm)}, \qquad P(m/rm) = \frac{(mmr)}{(mr)},$$

$$P(m/mr) = \frac{2(mrm)}{(mr)}, \qquad P(m/rr) = \frac{(mrr)}{2(rr)}.$$

If the above measurements reveal that

$$P(m/mm) = P(m/rm) = P(m/m) = 1 - P(r/m),$$

$$P(m/mr) = P(m/rr) = P(m/r),$$

$$P(r/rr) = P(r/mr) = P(r/r) = 1 - P(m/r),$$

$$P(r/rm) = P(r/mm) = P(r/m),$$

the tetrad proportions are consistent with first order Markov (or penultimate) statistics.

We have used the data reported by Bovey (1) for poly(methyl methacrylate) polymers, which he has labeled polymers I and II, to illustrate the procedures for testing the data. The results are shown in Table 10.6 for the system that obeys the terminal model or Bernoullian statistics. Table 10.7 gives results for a polymer system that obeys penpenultimate model or second order statistics. Part A of the table shows the dyad and triad results and the appropriate calculations to test the terminal model, which is found to be lacking. The penultimate model can be fit to the dyad and triad data as shown in Part B of the table, while part C uses the tetrad data to test the penultimate model. It is found that the expected agreement between the tetrad data and the dyad and triad results are insufficient to satisfy the penultimate model. In part D of Table 10.7 the conditional probabilities for the penpenultimate model are calculated using the tetrad sequences. The tetrad results are consistent with the pen-

Table 10.6 Model Testing of NMR Data on Stereoregularity

Polymer I	Observed $P_m = 0.24$	Calculated $P_m = 0.24$	Testing (terminal model)
Dyad			
$(m) = 0.24$			
$(r) = 0.76$			
Triad			
$(mm) = 0.04$	0.06	0.20	$(m)^2 = 0.05$
$(mr) = 0.36$	0.36	0.23	$2(m)(r) = 0.36$
$(rr) = 0.60$	0.60	0.23	$(r)^2 = 0.60$
Tetrad			
$(mmm) = 0.00$	0.01		$(mm)/(m) = 0.00$
$(mmr) = 0.07$	0.09		$(mm)(mr)/(m) = 0.06$
$(rmr) = 0.19$	0.20		$(mr)^2/4(m) = 0.14$
$(mrm) = 0.04$	0.04		$(mr)^2/4(r) = 0.04$
$(mrr) = 0.23$	0.23		$(mr)(rr)/(r) = 0.28$
$(rrr) = 0.43$	0.44		$(rr)^2/(r) = 0.47$
Pentad			
$(rmmr) = 0.05$	0.03		$(rmm)^2/4(mm) = 0.03$
$(mmmr) = 0.02$	0.02		—
$\dfrac{(rmrr)}{(mmrr)} = 0.25$	0.27		—
$(rmrm) = 0.06$	0.07		$4(mrm)(rmr)/(mr) = 0.08$
$(mmrm) = 0.02$	0.02		$2(mmr)(mrm)/(mr) = 0.02$
$(rrrr) = 0.40$	0.39		$(rrr)^2/(rr) = 0.31$
$(mrrr) = 0.14$	0.16		$(mrr)(rrr)/(rr) = 0.16$
$(mrrm) = 0.04$	0.03		$(mrr)^2/4(rr) = 0.02$

Table 10.7 Model Testing of NMR Data Stereoregularity for Polymer II

A. Tests of Terminal Model

Dyad	Observed	P_m	
(m)	0.82	0.82	
(r)	0.18	0.18	
Triad			Test
(mm)	0.75	0.86	$(m)^2 = 0.67$
(mr)	0.14	0.822	$(m)(r) = 0.29$
(rr)	0.11	0.67	$(r)^2 = 0.30$

Triad and dyad data not terminal

$P(r/m) = 0.08$ $\qquad\qquad$ $P(m/r) = 0.39$

$\qquad$ $P(r/m) + P(m/r) = 0.45$

$\rho = 0.82$ $\qquad\qquad\qquad$ $\therefore$ Not Bernoullian

B. Fit of Penultimate Model

Triads/Dyad	Triads
$P(r/m) = 0.085$	0.085
$P(m/r) = 0.39$	0.28
$P(m/m) = 0.91$	1.0
$P(r/r) = 0.61$	0.44
$P(m/r) + P(r/r) = 0.72$	$\therefore$ Not penultimate

C. Test of Penultimate Model

Observed	Penultimate Test	Penpenultimate (calculated)
$(mmm) = 0.70$	$(mm)^2/(m) = 0.60$	0.70
$(mmr) = 0.09$	$(mm)(mr)^2/(m) = 0.13$	0.09
$(rmr) = 0.03$	$(mr)^2/4(m) = 0.01$	0.03
$(mrm) = 0.04$	$(mr)^2/4(r) = 0.03$	0.04
$(mrr) = 0.06$	$(mr)(rr)/(m) = 0.02$	0.06
$(rrr) = 0.07$	$(rr)^2/(r) = 0.07$	0.07

Agreement not within experimental error
$\therefore$ Not penultimate

D. Fit of Penpenultimate Model

$$\alpha = \frac{(mmm)}{(mm)} = 0.94 \qquad\qquad \bar{\alpha} = \frac{(mmr)}{2(mm)} = 0.06$$

$$\beta = \frac{2(mrm)}{(mr)} = 0.56 \qquad\qquad \bar{\beta} = \frac{(mrr)}{(mr)} = 0.44$$

$$\gamma = \frac{(mmr)}{(mr)} = 0.63 \qquad\qquad \bar{\gamma} = \frac{2(rmr)}{(mr)} = 0.37$$

$$\delta = \frac{(mrr)}{2(rr)} = 0.28 \qquad\qquad \bar{\delta} = \frac{(rrr)}{(rr)} = 0.64$$

Table 10.7 (*Contd.*)

$\alpha \neq \gamma, \quad \beta \neq \delta$	$\therefore$ Not penultimate
$\beta + \bar{\beta} = 1, \gamma + \bar{\gamma} = 1$ $\alpha + \bar{\alpha} = 1, \delta + \bar{\delta} = 1$	$\therefore$ Penpenultimate

E. Test of Penpenultimate Model and Two Stage Model

	Pentads	Observed	Penpenultimate	Two stage
Peak 1	(*mmmm*) (*mmmr*) (*rmmr*)	0.75	0.77	0.74
Peak 2	(*rmrr*) (*mmrr*)	0.07	0.06	0.07
Peak 3	(*rmrm*) (*mmrm*)	0.07	0.08	0.08
Peak 4	(*rrrr*)	0.07	0.05	0.07
Peak 5	(*mrrr*) (*mrrm*)	0.04	0.05	0.05

penultimate model but again eliminate the penultimate, since the conditional probabilities, which are equal for the penultimate model, are found to be experimentally different. Finally, in part E of Table 10.7 the results for the pentad sequences are used ot test the penpenultimate model and the agreement with experiment is acceptable. In similar fashion, the pentad results are given in part E for the two stage model. The agreement with experiment appears to be at least as good as for the penpenultimate model.

10.6.4 Two Stage Model

To evaluate the applicability of the two stage or Coleman–Fox model, it is necessary to solve a quadratic equation in (m) for the roots m_1 and m_2:

$$m^2\left[\frac{a}{c} + p + (1-p)\frac{b}{c}\right] - m\left[\frac{a}{c} + \frac{b}{c} + p^2\left(1 - \frac{b}{c}\right)\right] + p(1-p)\frac{b}{c} = 0,$$

where a/c, b/c are constants and $p = (m)$. From a determination of m_1 and m_2, w_1 can be calculated:

$$p = w_1 m_1 - (1 - w_1)m_2 = (m).$$

For polymer II

$$\frac{a}{c} = 2.9,$$

$$\frac{b}{c} = 6.7,$$

$$w_1 = 0.81,$$

$$m_1 = 0.96,$$

$$m_2 = 0.21.$$

These parameters indicate that with the two stage model 80% of the polymer is produced under conditions such that isotactic placement is nearly exclusively preferred; the remainder has a probability of isotactic addition characteristic of a free radical or uncompleted anionic chain end.

10.7 METHODS OF DETERMINING STEREOREGULARITY

10.7.1 NMR Measurement of Stereoregularity (1)

A number of polymer systems have been examined by NMR for the determination of the stereoregularity since the elucidation of the micro-structure of poly(methyl methacrylate). As discussed in Chapter 7, if the backbone methylene group can be effectively isolated by spin decoupling or substitution on adjacent carbons, the relative number of tactic placements can be determined. A single resonance band is expected for syndiotactic methylene, a quartet (AB system) for isotactic, and a combination of the two for heterotactic.

Polymers having backbone methyl groups provide another approach to tacticity measurements if spin decoupling can be eliminated. The methyl resonances from heterotactic triads are flanked by those arising from syndiotactic and isotactic triads.

Table 10.8 Line Ordering

Polymer	Structure	CH₃	αCH	βCH₂
	i—h—s ($H_9 \rightarrow$)			
PMMA	$(-C-\underset{\underset{CH_3}{|}}{\overset{\overset{O \,\shortmid\, O^R}{\diagdown\!\diagup}}{\overset{|}{C}}}-)_n$	✓		AB

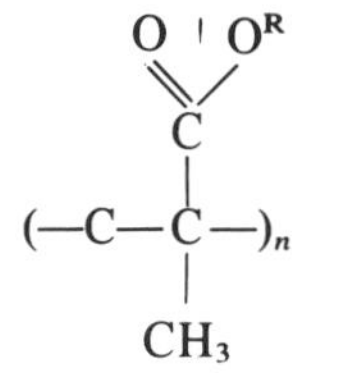

Table 10.8 (*Contd.*)

Polymer	Structure	CH$_3$	αCH	βCH$_2$
PVAc	$(-C-C-)_n$ with O–C(=O)–CH$_3$ side group		✓	
PP	$(-C-C-)_n$ with CH$_3$	✓		AB
	s—h—i ($H_0 \rightarrow$)			
PA	$(-\overset{CH_3}{\underset{}{C}}-O-)_n$	✓		
PVME	$(-C-C-)_n$ with –O–CH$_3$ (O=) side		✓	AB
PVA	$(-C-C-)_n$ with O–H side group		✓	S^b
PVC	$(-\overset{Cl}{\underset{}{C}}-C-)_n$		3^a	2^a
	None			
PMAN	$(-C-\overset{C\equiv N}{\underset{CH_3}{C}}-)_n$	✓		S^b
PAN	$(-C-\overset{C\equiv N}{\underset{}{C}}-)_n$		✓	S^b
PVF	$(-\overset{F}{\underset{}{C}}-C-)_n$		✓	S^b

^{a} Denotes three different chemical shifts for *s*, *h*, and *i* α-CH protons and two bands observed for β-CH$_2$'s when decoupled.

^{b} Denotes equivalence of β protons.

The basic problem is that it has not always been possible to predict when a polymer is susceptible to tacticity studies by NMR. First, the multiplets expected for the isolated methylenes do not always appear. Second, the resonance lines arising from substitution do not always exhibit different chemical shifts. However, it is generally possible to predict the type of resonance pattern to be observed for any given polymer. In general, the observation of tactic polymer systems falls into three groups as shown in Table 10.8 (10). The isotactic bands can appear at low or high field with respect to the syndiotactic bands. Finally, a group exists where the resonances cannot be distinguished. In general, the ordering isotactic, heterotactic, syndiotactic (with increasing field) is attributed to a functional group that causes a shielding effect on adjacent groups, whereas the reverse ordering is characterized by functional groups that cause a deshielding effect. For those polymers in which there is no observed ordering or separation of lines, the functional groups exhibit anisotropic effects that are orthogonal to the backbone of the chain.

For ^{13}C NMR the spectra of solutions of all vinyl polymers are inherently similar (11). If the methylene carbon line of an isotactic dyad is at low field relative to that of a syndiotactic dyad, the same relative positions hold for dyads in all other polymers of similar type. This similarity is related to the absence of a strong dependence of the NMR spectra of carbons in or near the main chain of a polymer on through-space magnetic interactions. But the spectra do depend on the conformation of the carbon backbone. It has been observed that the chemical shifts are determined by the steric configurations of nearest and next nearest neighbors, with more distant neighbors playing a much less important role. Thus configurational analysis of high resolution ^{13}C NMR spectra of stereoregular polymers need not be complicated by considerations of shift effects that vary from polymer to polymer. However, in polymers where structural isomerism can occur, such as poly(propylene oxide), this effect is more important than steric isomerism in determining the relative chemical shifts (11).

Thus, if the methylene carbon line of an isotactic dyad is at low field relative to a syndiotactic dyad for vinyl polymers, the same relative positions hold for dyads in all other polymers of this type. The recognition of this rule allowed assignments of the lowest field peak of nitrile carbon to the isotactic triad for polyacrylonitrile by comparison with previous assignments for polypropylene and polystyrene (12).

The NMR method for determination of tacticity has been extensively reviewed by Bovey (1), and no attempt is made here to reproduce his efforts. However, the results of a recent study on the configuration of MMA/MAA copolymers illustrates the power of the combined techniques of ^1H and ^{13}C NMR (13). Figure 10.9 shows the ^1H NMR spectra of the α-CH$_3$ and β-CH$_2$ resonance regions of an atactic copolymer of MMA and MMA-d_5 in pyridine, for different mole fractions of MMA: trace I, $P(A) =$

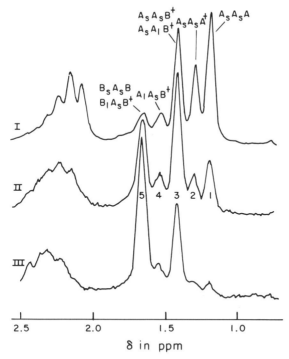

Figure 10.9 ^{1}H NMR spectra of the α-CH$_3$ and β-CH$_2$ resonance regions of atactic MMA MAA-d_5 copolymers in pyridine solutions. Mole fractions of MMA: $P(A)$ decreases from top to bottom; $P(A) = 0.74$ (trace I); $P(A) = 0.53$ (trace II); $P(A) = 0.25$ (trace III). Assignment of A-centered triads as shown [A = methyl methacrylate (MMA) units, B = methacrylic acid (MAA-d_5) units; i = isotactic placement; s = syndiotactic placement; the superscripted plus sign indicates that the forward and reverse forms, which have the same chemical shift, are involved]. (Reprinted by permission of Ref. 13.)

0.74; trace II, $P(A) = 0.53$; trace III, $P(A) = 0.25$. Figure 10.10 shows the assignment of the A-centered triads, where A is methyl methacrylate. In Figure 10.10 the α-CH$_3$ and β-CH$_2$ resonances of atactic MMA-d_5/MAA copolymers are shown as the mole fraction of MMA-d_5 changes: trace I, $P(A) = 0.75$; trace II, $P(A) = 0.49$; trace III, $P(A) = 0.25$. The assignments of the B-centered triads are shown. The use of MMA-d_5 leads to the elimination of all A-centered triads, and MAA-d_5 has the effect of eliminating all B-centered triads, yielding simpler spectra and a less difficult assignment task. Only six peak positions are observed for the α-CH$_3$ resonance peaks regardless of configurational or compositional factors in the copolymers. Obviously, most of the 20 possible triads overlap. The procedure used for the assignments was to do a partial assignment on the isotactic and syndiotactic copolymers, since for these copolymers only 6 compositional triads of either syndiotactic or isotactic configuration may occur. Additionally, the only assignments considered were those that agreed with the known chemical shifts of the homopolymer triads. When

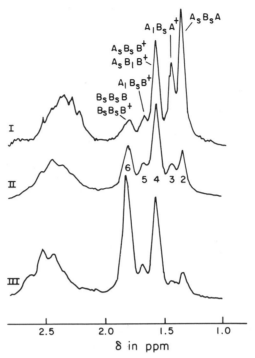

Figure 10.10 ^{1}H NMR spectra of the α-CH$_3$ and β-CH$_2$ resonance regions of atactic MMA-d_5–MAA copolymers in pyridine solutions. Mole fractions of MMA-d_5: $P(A) = 0.75$ (trace I); $P(A) = 0.49$ (trace II); and $P(A) = 0.25$ (trace III). The symbols are the same as for Figure 10.9. (Reprinted by permission of Ref. 13.)

the chemical shifts of the isotactic and syndiotactic triads were known, as well as the chemical shifts from 2 heterotactic triads from PMMA and PMAA, the chemical shifts of the remaining 6 heterotactic triads could be estimated from shift increments. The chemical shift increments (in pyridine as solvent) were determined by the following rules:

1. Replacement of a central A by B leads for a given triad to a shift increment Δ_1 that is about the same regardless of whether the flanking units are A or B or whether the triad is syndiotactic, isotactic, or heterotactic (replacement $X_xA_xX \rightarrow X_xB_xX$). $\Delta_1 = 0.13 \pm 0.01$ ppm.

2. Replacement of a flanking A by B in a syndiotactic placement leads for a given triad to a shift increment Δ_2 that is about the same regardless of whether the two remaining units are A or B or whether the triad is syndiotactic or heterotactic (replacement $A_sX_xX \rightarrow B_sX_xX$ or $X_xX_sA \rightarrow X_xX_sB$). $\Delta_2 = 0.225 \pm 0.015$ ppm.

3. Replacement of a flanking A by B in an isotactic placement leads for a given triad to a shift increment Δ_3 that is about the same

regardless of whether the two remaining monomeric units are A or B or whether the triad is isotactic or heterotactic (replacement $A_iX_xX \rightarrow B_iX_xX$ or $X_xX_iA \rightarrow X_xX_iB$). $\Delta_3 = 0.13 \pm 0.01$ ppm.

4. Replacement of a syndiotactic placement by an isotactic placement between a flanking A and a central unit in a given syndiotactic triad results in a shift increment Δ_4 regardless of whether the monomeric units are A or B. Also, the analogous reverse replacement in a given isotactic triad results in a shift increment Δ_4' (replacement $A_sX_sX \rightarrow A_iX_sX$ or $X_sX_sA \rightarrow X_sX_iA$ for Δ_4, and $A_iX_iX \rightarrow A_sX_iX$ or $X_iX_iA \rightarrow X_iX_sA$ for Δ_4'). $\Delta_4' = -0.135 \pm 0.015$ ppm. $\Delta_4 = 0.09 \pm 0.01$ ppm.

5. Replacement of one syndiotactic placement by an isotactic placement between a flanking B and a central unit in a given syndiotactic triad results in a shift increment Δ_5 regardless of whether the other monomeric units are A or B. Also, the analogous reverse replacement in an isotactic triad results in a shift increment Δ_5' (replacement $B_sX_sX \rightarrow B_iX_sX$ or $X_sX_sB \rightarrow X_sX_iB$ for Δ_5, and replacement $B_iX_iX \rightarrow B_sX_iX$ or $X_iX_iB \rightarrow X_iX_sB$ for Δ_5'). $\Delta_5 = 0.00 \pm 0.01$ ppm. $\Delta_5' = 0.00 \pm 0.01$ ppm.

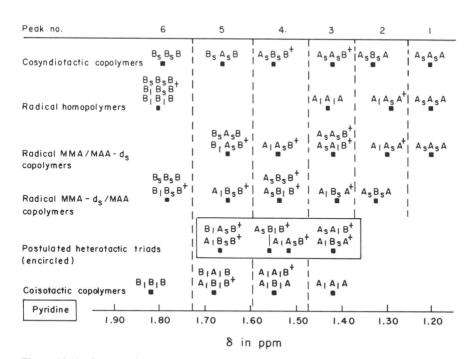

Figure 10.11 Survey of chemical shifts of triads for pyridine solutions at 100°C of the different MMA/MAA polymer types investigated. Postulated chemical shifts of heterocyclic triads in a rectangle. Vertical, dashed lines separate well resolved peaks. (Reprinted by permission of Ref. 13.)

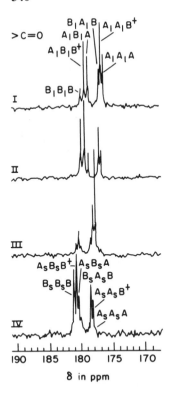

Figure 10.12 ^{13}C NMR spectra of the >C=O resonance of tactic MMA/MAA copolymers in pyridine solution. Isotactic copolymers are shown in trace I [$P(A) = 0.60$] and trace II [$P(A) = 0.33$]. Syndiotactic copolymers are seen in trace III [$P(A) = 0.74$] and trace IV [$P(A) = 0.31$]. (Reprinted by permission of Ref. 13.)

To summarize the chemical shift rules, it is assumed that the change of a structural feature $A \rightleftarrows B$ or $i \leftrightarrows s$ in a given triad does not greatly change the conformational or solvation state of the unchanged part of the triad.

The results of the assignments to the six different peaks are shown graphically in Figure 10.11. The assignments were tested using compositional and configuration statistics, and both were found to be satisfactory. The results were consistent with $\Sigma_s = 0.392$ and $\Sigma_p = 0.037$ for MMA/MAA copolymers. The accuracy of the direct determination of triad probabilities by simulation is sufficient for assignment.

Additionally, ^{13}C NMR spectra of these copolymers were run, and the >C=O resonances are shown in Figure 10.12. Six peaks are clearly visible for the isotactic copolymers, corresponding to the six possible isotactic triads. Six peaks also appear for the syndiotactic copolymers.

NMR techniques still play the prominant role in the measurement of stereoregularity, with ^{13}C NMR further simplifying the task.

10.7.2 IR Methods

The IR spectra of a sterically pure isotactic or syndiotactic polymer with helical structure have unique absorption bands that vary remarkably with variations in the physical state as well as with the introduction of chemical impurities. Bands sensitive to the physical state may be classified into two categories according to their origins (14). *Crystalline* bands arise from intermolecular forces in the crystal lattice where the polymer molecules pack together in a regular three-dimensional arrangement. These crystalline bands are relatively rare, but polyethylene ($730\,cm^{-1}$) and polystyrene ($985\,cm^{-1}$) do exhibit such bands. The other type of band is connected with the intramolecular vibrational coupling within a single chain. These bands correspond to the vibrations of relatively long stereoregular and conformationally regular sequences of isotactic or syndiotactic blocks. The behavior of these regularity bands in the spectra depends on the minimal length n of sequence sufficient for the mode. If n is larger than the average length of a helix sequence in a stereoregular polymer melt or solution, the bands disappear from the spectra of amorphous stereoregular polymers. These latter bands have been classified as *helix* bands.

Finally, we have *regularity* bands, which are bands whose intensity usually does not change significantly in an amorphous polymer but always depends on the type of stereoplacement and the length of the stereosequence (14).

Table 10.9 (15) gives a listing of the sequence bands of polypropylene, and a similar listing for polystyrene is given in Table 10.10, where n is the number of units in sequence for the bands to appear (15). The helix bands show drastic decreases in intensity as the sample goes from a highly crystalline to the amorphous state, indicating that the intensities are closely

Table 10.9   Sequence-Sensitive Bands of Propylene Units

Position of Band (cm^{-1})	Type of Regularity	Classification	n
998	isotactic	helix band	10–12
973	isotactic	regularity band	3–4
841	isotactic	helix band	12
977	syndiotactic	helix band	
962	syndiotactic	regularity band	
867	syndiotactic	helix band	
936	—	band of isolated propylene unit	

Table 10.10   Some Sequence-Sensitive Bands of Styrene Units

Position of the Band (cm^{-1})	Type of Regularity	Classification	n
1085	Isotactic	Helix or regularity band	4
1053	Isotactic	Helix or regularity band	10
985	Isotactic	Crystallinity band	
918	Isotactic	Helix band	8–10
896	Isotactic	Helix band	16
586	Isotactic	Helix or crystallinity band	
565	Isotactic	Crystallinity or regularity band	4–5
558	Isotactic	Helix or regularity band	
1070	Predominantly syndiotactic (in IR spectra of radical PS)		
540	Predominantly syndiotactic (in IR spectra of radical PS)		
1075		Band of isolated styrene units	
555–550		Band of isolated styrene units	

related to the conformational regularity of the polymer chain. However, the helix bands differ with respect to their sensitivity to the regularity or sequence length of the polymer chain. This difference in sensitivity of the helix bands suggests that a certain sequence length of regular structure is necessary to give rise to each helix band and that this critical length varies with the vibrational mode to which the band is assigned. When we can estimate the critical length of each helix band from the correlation between the absorption intensity and the regularity of the polymer chain, it is possible to use this information to investigate variations in tacticity. The problem with this approach is that there are at least three general types of disorder that can disrupt the ideal polymer chain: (1) conformational disorder, (2) configurational disorder, and (3) chemical disorder arising from copolymerization or isomerization of the chain. Unfortunately, the spectral changes that result from all these types of disorder are the same. In every case the appearance of these defects decouples or shortens the sterically regular blocks. Accordingly, we cannot uniquely extract the effect of the regular stereosequence length from the observed data.

One method of overcoming these difficulties is to use copolymers of the normal and deuterated monomers. In this way samples of any chemical composition can be prepared, and the statistical distribution of the sequence length of the normal species can be derived from the knowledge of the kinetics of the copolymerization. Furthermore, the molecular and crystal structures as well as the crystallizability of the normal homopolymer remain unaltered throughout the whole range of compositions. For these deuterated copolymers in the crystalline state, the normal undeuterated monomer sequences are interrupted by the insertion of the deuterated species into the chain. The normal sequences involved in this partially deuterated polymer chain, though they assume the regular conformation, differ in vibrational coupling among the neighboring units from the infinitely long regular sequence of the normal species. Therefore the helix bands are greatly affected by the statistical length of the normal species. On the other hand, when the stereoregular polymer melts or dissolves, shortening of the regular spatial sequence occurs as a result of disordering of the skeletal conformation. In the partially disordered chain the regular sequences are isolated from one another by the monomeric units of the disordered conformation.

Thus the regular sequences in the two cases differ with each other in the structure of the sequence ends. If we neglect the effects of the sequence ends on the vibrations of regular sequences of finite length as an approximation, then the sequences having the same length, both in the deuterated copolymers and in the disordered homopolymers, make the same contribution to the intensities of the helix bands. Accordingly, the investigations of the IR spectra of such deutero copolymers can provide useful information on the sterically irregular chains.

Studies of the copolymers of styrene and $-\alpha-d_1$-styrene and mixtures of IPS and IPS-α-d_1 (19), and their results for the 920 cm^{-1} band, are shown in Figure 10.13. The straight line with the slope 45° for the mixture means that the absorption coefficient per normal styrene unit remains constant throughout the whole range of composition, and therefore the helix bands arise from intramolecular interactions within the chain. On the other hand, the deuterated copolymers give curves that rapidly decrease with mole fraction of styrene. This decrease indicates that the shortening of the normal styrene sequence in the copolymer reduces the averaged absorption coefficient. Different bands give different curves when compared with each other, indicating that the dependence of the absorption intensity on the sequence length differs from band to band.

For long sequences the absorption coefficient per chemical repeating unit remains independent of the sequence length. For short sequences the vibrational modes are different from those of the infinite chain, so the vibrational frequencies of these modes differ as well. According to the theory there exists a dependence between the length of the regular block and the frequency of the bands:

$$\nu_n^2 = \sum_{k=0}^{M} A_k \cos \frac{k\pi}{n+1},$$

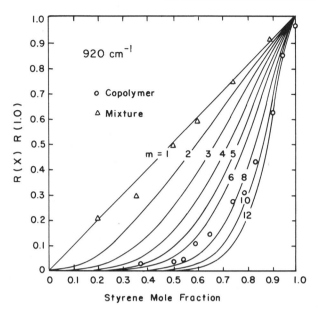

Figure 10.13 Intensity of the 920 cm^{-1} band measured for the highly crystalline samples of styrene and styrene-α-d_1, and mixtures of IPS and IPS-α-d_1. (Reprinted by permission of Ref. 19.)

where ν_n is the frequency of the block of length n, A_k is a constant, and M is usually 1 or 2 (20). One expects a similar smooth dependence of the extinction coefficients on n for helix or regularity bands; that is, the extinction coefficient increases with n and gradually becomes constant (like ν_n) for large N. Normally, one invokes the approximation that the absorption coefficient per normal unit is constant for sequences of length greater than or equal to n and zero for sequences of length less than n. This threshold method is widely used, because the statistical dependence for fractions of units is the sum of sterically regular blocks, beginning from any given value of n, and can be used for comparison of experimental and statistical data. Accordingly, the reduced relative intensity R(X)/R(1.0) equals $F(n) \cdot$ X, where $F(n)$ is the total fraction of normal units that exist in sequences of lengths greater than or equal to n, and X is mole fraction of normal monomer units. Thus for the 998 cm^{-1} band shown in Figure 10.13, $n \cong 10$; that is, a regular sequence of about 10 units is necessary for appearance of this helix band in polystyrene (19). The results for other bands in polystyrene are shown in Table 10.10.

Similar studies have been made for polypropylene (21). The regularity bands are at 1220, 1168, 998, 900, 841, and 809 cm^{-1} in polypropylene, and the band at 1460 cm^{-1} is considered an internal standard. The results for the 998 and 841 cm^{-1} bands are shown in Figure 10.14. The experimental points for the 998 cm^{-1} band correspond to an n value of 10–11. For the 841 cm^{-1}

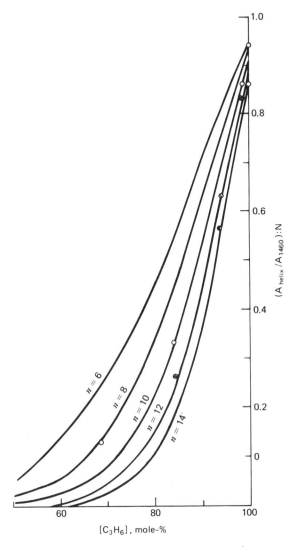

Figure 10.14   Dependance of the relative intensities of the 998 cm^{-1} band (o) and 841 band ($\otimes$) in the spectra of C_3H_6–C_3D_6 copolymers on composition. (Reprinted by permission of Ref. 21.)

band the n value is 12–14. The value of n for the 973 cm^{-1} band is 4, as shown in Figure 10.15 (21).

An index of syndiotactic regularity in polypropylene has been measured by using the absorbance of the 867 cm^{-1} syndiotactic band divided by an internal thickness band (16). An index of isotactic regularity has been used based on the absorbance of the 998 cm^{-1} band divided by the 973 cm^{-1} band (17). For PVC the band at 1428 cm^{-1} corresponds to the deformation vibration of —CH$_2$— groups of syndiotactic TT dyads,

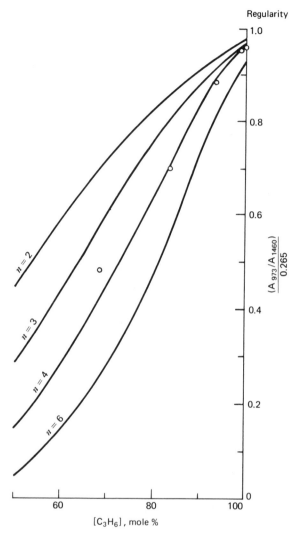

Figure 10.15 Dependance of the relative intensity of the 973 cm^{-1} band in the spectra of C$_3$H$_6$–C$_3$D$_6$ copolymers on composition. (Reprinted by permission of Ref. 21.)

whereas both syndiotactic and isotactic dyads absorb at 1434 cm^{-1}. Therefore the absorbance ratio A_{1428}/A_{1438} is a relative measure of the ratio of syndiotactic to isotactic dyads (18).

All IR methods of measuring stereoregularity are relative and serve to rank the relative tacticity of the samples. Calibration is rarely possible, although model compounds and mixtures of homopolymers have been used. Both methods of calibration are only valid when the chosen analytical bands are highly localized. Calibration with other physical techniques such as NMR has also been useful.

10.8   MOLECULAR MECHANISM OF STEREOSPECIFIC POLYMERIZATION

10.8.1   Free Radical Polymerization

In free radical polymerization the molecular mechanism is based on equilibrium between two different conformationally related, growing chain structures. In Figure 10.16 (23) structure A is such that the conformation of $P_n(C_3—C_4— \cdots)$ trans to $(C_2—C_1)$ is energetically favorable, and the monomer attacks C_1 predominantly from the upper side of the paper. Let us assume that the C_3 configuration is d; accordingly, the configuration around C_1 should be the same as that of C_3 (i.e., d) after the attack of the second monomer.

The growing chain structure B in Figure 10.16 is the inverse. There is

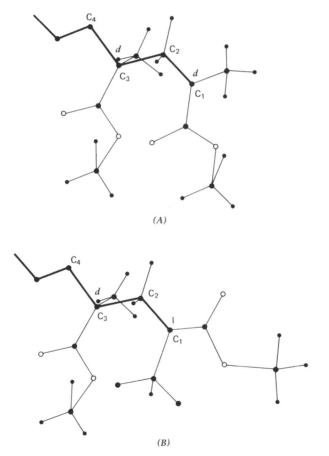

(A)

(B)

Figure 10.16   Model A in propagation stage of radical polymerization of methyl methacrylate and model B in propagation of methyl methacrylate. (Reprinted by permission of Ref. 23.)

some rotation around the C_2—C_1 axis before the growing chain is attacked by the incoming monomer. This rotation results in a change in the configuration around C_1. Since the two structures are in equilibrium, $A \underset{k_2}{\overset{k_1}{\rightleftarrows}} B$, the position of the equilibrium determines the preference for meso or isotactic addition. The factors determining the equilibrium are the interaction between the substituent groups of the terminal unit of C_3 and those of the added monomer of C_1—C_2. Conformation B is considered to be favored relative to A, so syndiotactic addition is favored in free radical polymerization. Syndiotactic addition increases as the polymerization temperature decreases.

10.8.2 Anionic Polymerization

The stereochemical behavior of anionic polymerization is similar to that of free radical polymerization in the sense that the free rotation about the carbon–carbon bonds between the penultimate and terminal units is not restricted by condensation onto a catalyst species. But there is interaction between the substitutent of the penultimate group and the conjugated carbonium of the terminal unit, as indicated by the dependence on penultimate or first order Markov statistics. The most favorable relative conformation of the terminal and penultimate groups is that in which the dipoles are opposite each other.

     In coordinated anionic polymerization the polymers that are produced are nearly sterically pure. Figure 10.17 gives a hypothetical transition state

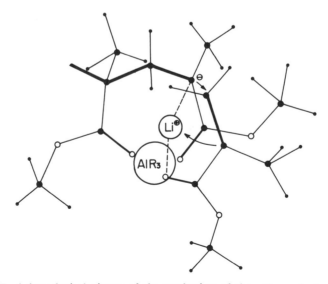

Figure 10.17 A hypothetical picture of the mechanism of the poly merization of methyl methacrylate by coordinated anionic catalyst. (Reprinted by permission of Ref. 23.)

for coordinated anionic polymerization. The stereochemistry is determined by the stability of the chelate ring formed in the vicinity of the growing chain end.

10.8.3 Mechanism of Ziegler–Natta Catalyst

Since the discovery of isotactic polypropylene, the problem of the origin of the stereospecific trend of α-olefin polymerization has received much attention. The problem is to explain the isospecific polymerization that causes an isotactic macromolecule to be formed through a series of successive asymmetric additions of the same sign on the monomer double bond. It is clear that in isospecific polymerization the catalyst surface is chiral and the steric control is caused by the chirality of the catalyst complex. The polymerization mechanism of α-olefins is often defined as anionic-coordinated. The term *anionic* is intended to denote an asymmetric electronic distribution about the metal-polymer radical (P) bond, with a higher electronic density on the polymer radical than on the metal. The term *coordinated* is used to show that the monomer coordinates on the transition metal before insertion into the metal-polymer bond. The first step is the formation of the metal-alkyl bond, followed by coordination and insertion of the monomer. These reactions are written as follows:

$$\alpha\text{-TiCl}_3 + \text{Al}\!-\!\text{R}_3 \rightarrow (+)\text{Ti}\!-\!\text{R} \qquad \text{catalyst formation}$$

$$(+)\text{Ti}\!-\!\text{R} + \text{C}_3\text{H}_6 \rightarrow (+)\text{Ti}\!-\!\text{R} \cdot (\text{C}_3\text{H}_6) \qquad \text{coordination}$$

$$[(+)\text{Ti}\!-\!\text{R} \cdot \text{C}_3\text{H}_6] \rightarrow (+)\text{Ti-}[(d)\text{CH}_2\text{CH}(\text{CH}_3)\!-\!]\text{R} \qquad \text{insertion}$$

$$(+)\text{Ti}[(d)\text{CH}_2\text{CH}(\text{CH}_3)\!-\!]\text{R} + \text{C}_3\text{H}_6 \rightarrow (+)\text{Ti-}[(d)\text{CH}_2\text{CH}(\text{CH}_3)\!-\!]_2\text{R} \qquad \text{propagation}$$

$$\vdots \qquad\qquad \overset{\displaystyle \text{CH}_3}{\underset{\displaystyle |}{}}$$

$$+ (\text{Ti})\!-\![(d)\text{CH}_2\text{CH}(\text{CH}_3)\!-\!]_n\text{R} + \text{transfer} \rightarrow \text{CH}_2\!\!=\!\!\text{C}\!-\!(\text{C}_3\text{H}_6)_n\!-\!\text{R} \qquad \text{termination}$$

When polymerization is carried out on well formed crystals of α-TiCl$_3$, it is observed that polymer formation takes place only on the lateral faces and the edges of crystals, but not on the basal faces. Polymerization is thus occurring on titanium atoms bearing coordination vacancies. The octahedral coordination of the titanium catalyst is achieved with four ligands that are chlorine atoms, one that is the polymeric radical P, and a coordinated monomer molecule. The steric control is achieved through the asymmetry of the titanium atom.

The monomer can open the double bond in two ways: either cis or trans. By studying the polymerization of 1,2-α-disubstituted olefins, it has been established that cis addition is the mode of isotactic propagation. The double bond of a 1,2-disubstituted olefin CHA=CHB gives two diastereoisomer monomeric units (Figure 10.18). The structure of the monomeric units in the polymer (erythro or threo) is found to be consistent

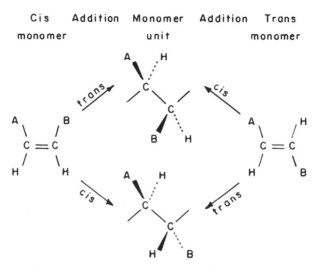

Cis   Addition   Monomer   Addition   Trans
monomer   unit   monomer

Figure 10.18 Addition to the double bond. The attachment of CHA is supposed to occur from the upper side of the paper. Should the attack of CHA occur from the lower side, the antipodic monomer units would be formed.

with cis openings. The cis addition occurs as part of an activated four-center complex (Figure 10.19) in which primary and secondary insertion of the monomer can occur.

Primary insertion is denoted by insertion into a metal primary carbon bond with formation of a new primary carbon-metal bond:

$$M—CH_2—\overset{\overset{\displaystyle CH_3}{|}}{C}H—(C_3H_6)_n—R + C_3H_6 \rightarrow M—CH_2—\overset{\overset{\displaystyle CH_3}{|}}{C}H—(C_3H_6)_{n+1}—R$$

For isotactic polymers the presence of terminal vinylidene unsaturation

$$CH_2{=}\overset{\overset{\displaystyle CH_3}{|}}{C}—(C_3H_6)_n—R$$

is evidence that chain propagation occurs via primary insertion.

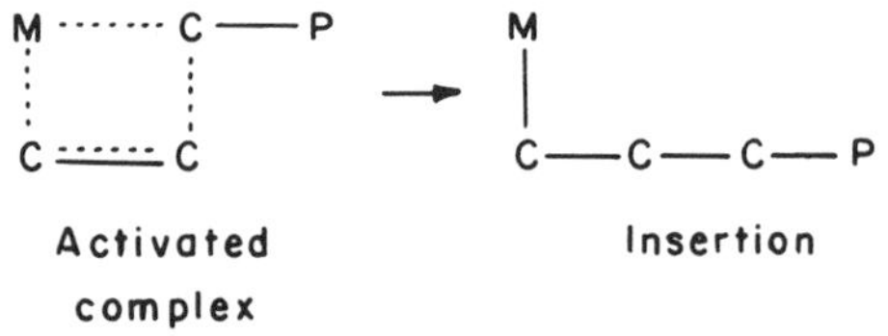

M ······· C —— P   M
                    → |
C ······ C   C — C — C — P

Activated   Insertion
complex

Figure 10.19 The four-center activated complex through which the addition of M–P to the monomer double bond should occur.

Secondary insertion

$$\underset{\text{secondary insertion (in initiation)}}{M-R_5 + C_3H_6 \rightarrow M-\overset{\overset{\displaystyle CH_3}{|}}{C}H-CH_2-R}$$

can also occur. In the presence of syndiospecific catalyst, chain propagation occurs mainly with secondary insertion.

The secondary insertion with formation of an activated four-center
complex explains how syndiotactic propagation occurs through steric interactions of the monomer with the last unit. The activated trans complex
should give rise to weaker nonbonded interactions than the cis complex,
provided that the steric environments above and below M are equal or only
slightly different. The stabler trans complex leads to syndiotactic propagation (Figure 10.20). The formation of activated four-center complexes
via primary insertion leads to a minimization of the nonbonded interactions
between the monomer substituent and the substituent of the last unit in the
two diastereoisomers shown in Figure 10.21. Under these conditions the
steric control depends only on the possible differences of the steric
environment above and below the metal. If the bulkiness above and below
is appreciably different, then the monomer approaches the reactive M—P
bond from a constant direction, and primary insertion leads to primarily
isotactic propagation. These results are summarized in Table 10.11.

For heterogeneous catalysis of the Ziegler–Natta type, the stereospecificity is determined by the active sites on the catalyst surface. For each
compound of a transition metal, the stereospecificity in polymerizing
α-olefins depends on many factors, particularly on the stability of the
chemisorbed catalytic complexes. The stereospecificity is reduced by every
factor that makes such complexes more easily desorbable or dissociable.
The most important factors are (1) an increase in the ionic radius of the
metal linked to the metalloorganic compound; (2) an increase in the length

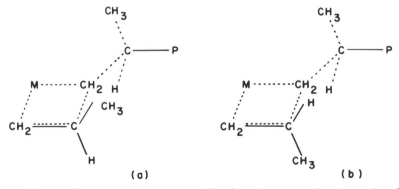

Figure 10.20 Syndiotactic polypropylene resulting from the trans activated complex via cis
ligand migration. (Reprinted by permission of Ref. 24.)

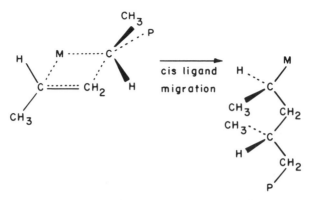

Figure 10.21 Supposed four-center activated complexes for a typical primary insertion of a propylene molecule.

of the alkyls, having at least two carbon atoms, linked to the complex; and (3) irregularities in the crystal lattice. In general, an increase in polymerization temperature reduces stereospecificity, as it increases the dissociation process.

10.8.4 Effect of Polymerization Conditions on Stereoregularity of Polymers

Effect of Temperature on the Stereoregularity of Polymers. For free radical polymerizations of vinyl monomers obeying the terminal model, a decrease in the temperature of the polymerization favors syndiotactic addition, since structure B (Figure 10.16) is preferred. This result can be understood if we write the rate constant for isotactic addition as

$$k_m = \frac{kT}{h} \exp\left(\frac{\Delta S_m^*}{R} - \frac{\Delta H_m^*}{RT}\right),$$

where ΔH_m^* and ΔS_m^* are the activation enthalpies and entropies, respectively. Similarly, the rate constant for syndiotactic addition is

$$k_r = \frac{kT}{h} \exp\left(\frac{\Delta S_r^*}{R} - \frac{\Delta H_r^*}{RT}\right),$$

Table 10.11   Peculiar Features of Isotactic and Syndiotactic Polymerization

	Isotactic Propagation	Syndiotactic Propagation
Addition to the double bond	cis	cis
Monomer insertion	Primary	Secondary
Chiral center of steric control	M	Last unit of the growing chain

so

$$\frac{km}{kr} = \exp\left(\frac{\Delta S_m^* - \Delta S_r^*}{R} - \frac{\Delta H_m^* - \Delta H_r^*}{RT}\right).$$

Therefore the differences in activation enthalpies and entropies are given by

$$\Delta H_m^* - \Delta H_r^* = \Delta(\Delta H_p^*) = \frac{-R\partial \ln[P_m/(1-P_m)]}{\partial(1/T)}$$

and

$$\Delta S_m^* - \Delta S_r^* = \Delta(\Delta S_p^*) = R \ln\left(\frac{P_m}{1-P_m}\right) + \frac{\Delta(\Delta H_p^*)}{T}.$$

Values of P_m range from 0.13 at $-78°$ to 0.36 at $250°$ for free radical poly(methyl methacrylate). An Arrhenius plot of these data yields

$$\Delta(\Delta H_p^*) = 1 \text{ kcal},$$

$$\Delta(\Delta S_p^*) = 1 \text{ eu}.$$

Thus syndiotactic placement appears to be favored by the additional entropy of activation required for isotactic addition, which itself is favored by entropy.

For stereospecific polymerization in homogeneous ionic systems, two types of temperature effects are observed. First, the stereoregularity is greatly affected by the polymerization temperature, the more stereoregular polymer being obtained at low temperatures. Homogeneous cationic polymerization apparently has little or no tendency toward complex formation between the counterion and the monomer substituent, and the stereoregularity is controlled mainly by the steric hindrance of the substituents and the interaction between the growing chain end and a counterion. The second type of temperature effect occurs when the stereoregularity is not affected by the polymerization temperature, with stereoregular polymer being produced at high temperature. For this so-called *coordinate polymerization*, there is a strong interaction between the counterion and the substituent of the monomer. In the first type of polymerization the repulsion energy between the substituent may be about 1–2 kcal/mole, but in the latter polymerizations the energy of complex formation is larger. Therefore the difference in these energies of interactions may cause the difference in temperature dependence.

Effect of Bulkiness of the Substituent. The effect of the bulkiness of the substituent on the stereochemistry of the methacrylates has been examined in both free radical and anionic polymerization. For free radical polymerization the stereoregularity is controlled only by the terminal group, and the preference for syndiotactic addition increases in the order methyl < isopropyl < D-bornyl < tert-butyl ≃ polymethacrylic acid, which approaches an atactic polymer ($P_m = 0.5$). For anionic polymerizations obeying the penultimate model, the effect of the ester group in the penultimate unit is to prevent succession of syndiotactic–syndiotactic dyads. The difference in anionic polymerization may result from the increased interaction between the dipole of the carbonyl group in the penultimate unit and the conjugated carbonion of the terminal unit. The most favorable relative conformation of carbonyl groups of the terminal and penultimate units is presumably that in which the dipoles are opposite each other.

For coordinated anionic polymerization the bulkiness has little effect in determining the chelate coordination. The coordination complex, including lithium and aluminum metals between the carbonyl groups of the terminal and penultimate units and of the attacking monomer, restricts the rotation of the propagating chain and determines the attacking direction of a monomer, resulting in the predominance of isotactic addition.

Effect of Copolymerization on Stereoregularity. The tacticity of poly(vinyl chloride) in copolymers has been studied to determine the role of copolymerization (25). The syndiotacticity of PVC appears to decrease as the comonomer content increases. The comonomer content was kept lower than 10% so that the copolymers could be considered as models of PVC having chain irregularities. As the comonomer value increases, the comonomer effect on the syndiotacticity increases. Thus it appears that steric hindrance plays a role in determining the amount of syndiotacticity that is propagated during copolymerization. The copolymerizations were carried out over a range of temperatures, but no difference in the effect of temperature was observed relative to the homopolymer.

REFERENCES

1. F. R. Bovey, *High Resolution NMR of Macromolecules*, Academic Press, New York, 1972.
2. A. Zambelli, A. L. Segre, M. Farina, and G. Natta, *Makromol. Chem.*, **110**, 1 (1967).
3. H. L. Frisch, C. L. Mallows, and F. A. Bovey, *J. Chem. Phys.*, **45**, 565 (1966).
4. B. D. Coleman and T. G. Fox, *J. Chem. Phys.*, **38**, 1065 (1963).
5. B. D. Coleman and T. G. Fox, *J. Polym. Sci.*, **A1**, 3183 (1963).
6. Yu. V. Kissin, V. N. Tsvetkova, and N. M. Chirkov, *Eur. Polym. J.*, **8**, 529 (1972).
7. F. A. Bovey and G. V. D. Tiers, *Adv. Polym. Sci.*, **3**, 139 (1963).
8. E. Klesper and W. Gronski, *J. Polym. Sci., Polym. Lett.* **7**, 661 (1969).
9. E. Klesper, *J. Polym. Sci., A-1*, **8**, 1191 (1970).
10. W. Ritchey and F. Knoll, *J. Polym. Sci.*, **B4:11**, 853 (1966).
11. J. Schaefer in *Topics in C-13 NMR Spectroscopy*, Vol. I, George Levy, Ed., John Wiley & Sons, New York, 1974, p. 149.
12. J. Schaefer, *Macromolecules*, **4**, 105 (1971).

13. E. Klesper, A. Johnsen, W. Gronski, and F. W. Wehrli, *Makromol. Chem.*, **176**, 1071 (1973).
14. G. Zerbi, F. Campelli, and V. Zamboni, *J. Polym. Sci.*, C-7, 141 (1963).
15. Yu. V. Kissen, *Adv. Polym. Sci.*, **15**, 92 (1974).
16. G. Natta, I. Pasquon, and A. Zambelli, *J. Polym. Sci.*, C-4, 411 (1964).
17. J. Luongo, *J. Appl. Polym. Sci.*, **9**, 502 (1960).
18. H. Germarn, H. Hellwege, and U. Johnson, *Makromol. Chem.*, **60**, 106 (1963).
19. M. Kobayashi, K. Akita, and H. Tadokoro, *Makromol. Chem.*, **118**, 324 (1968).
20. R. Zbinden, *Infrared Spectroscopy of High Polymers*, Academic Press, New York–London, 1969.
21. Yu. V. Kissen and L. A. Rishina, *Eur. Polym. J.*, **12**, 757 (1976).
22. G. Natta, *J. Polym. Sci.*, **34**, 531 (1959).
23. T. Tsuruta, T. Makimoto, and H. Kanai, *J. Macromol. Chem.*, **1**, (1966).
24. A. Zambelli and C. Tosi, *Adv. Polym.*, **15**, 1 (1974).
25. J. Guzman Pecote and J. Millian, *Eur. Polym. J.*, **12**, 295 (1976).

11

Chain Isomerism due to Branching in Polymers

11.1 INTRODUCTION (1)

Although the concept of branching during polymerization developed simultaneously with our knowledge of polymerization reactions, the precise nature and distribution of branches still remain a puzzle. This result arises from the complexity of the branching reactions and the fact that most practical systems have only a few branches.

Nearly all polymers contain chain branches to a greater or lesser degree. There are two types of branches: short chain branches and long chain branches. The short chain branches contain only two or three monomer units, whereas the long chain branches may be nearly as long as the main polymer chain. The distribution of the branches is also important. The short chain branches can be clustered, that is, in groups like knots in a rope, or randomly distributed along the chain. The long chain branches can be in the form of T-shaped molecules, that is, with a single isolated branch as generally occurs with intermolecular chain transfer, or the long chain branches can occur in the form of comblike molecules as a result of grafting of one polymer on the backbone of another. Finally, the long chain branches can occur as starlike molecules; this occurs when the initiator generates several chains simultaneously. The properties of the polymer chain depend on the number of branches, the type of branches, and the distribution of these branches on the polymer chain.

Branches can be introduced on the polymer chain during or subsequent to the polymerization. During free radical polymerization short chain branches are introduced by intramolecular chain transfer (*backbiting*) while long chain branches are introduced by intermolecular chain transfer. For condensation polymerization, branches occur by using a multifunctional (greater than two) monomeric unit. Branches may be

356

introduced chemically by a grafting process, by ionizing radiation, by addition of a divinyl monomer with a vinyl monomer, or by chemical addition.

Branches produce substantial differences in the chemical, physical, mechanical, and rheological properties. The branch points provide weak points that are susceptible to chemical attack such as oxidation and thermal degradation. The mechanical properties of branched molecules of low branch density are only changed slightly. Branched elastomers have lower elasticity and breaking strengths than do linear elastomers. Branched thermoplastics have decreased strength and stiffness relative to linear plastics. Comblike molecules prepared as graft copolymers have good toughness properties. The rheological properties of branched molecules are substantially different due to two effects. The size of branched molecules is smaller in solution than linear molecules of the same molecular weight. As a result the viscosity is lower at low shear. The branches also contribute to a substantial broadening of the molecular weight distribution. This broadening introduces additional deviation from Newtonian flow at high shear.

11.2   SHORT CHAIN BRANCHING IN POLYMERS

11.2.1   Mechanism of Formation

Polyethylene.   Intramolecular chain transfer is generally responsible for the short chain branches in free radical polymerizations. This reaction has been termed backbiting because the radical end abstracts a hydrogen from the backbone of the chain by formation of a transient six-membered ring. This is illustrated as follows:

The six-membered ring is preferred, as the formation of larger transient rings is highly improbable, so backbiting further back on the chain is less likely.

The rate of intramolecular transfer is a unimolecular reaction

$$R_{bb} = k_{bb}[R^*], \tag{11.1}$$

where $[R^*]$ is the concentration of radicals and k_{bb} is the rate constant. The rate of propagation is

$$R_p = k_p[M][R^*], \tag{11.2}$$

so the relative rate of branch formation is

$$\frac{R_{bb}}{R_p} = \frac{k_{bb}}{k_p[M]}. \tag{11.3}$$

Since the concentration of the radical species has canceled, the branch formation should be independent of the degree of polymerization. An increase in the temperature of polymerization favors branch formation over chain propagation, because the extraction of a hydrogen from a molecule requires an energy of activation of the order of 10 kcal/mole, whereas the chain propagation has an energy of activation of 3 kcal/mole. For polyethylene, since the polymerization is carried out under pressure, an increase in pressure increases the monomer concentration, so the number of branches should decrease. These effects of temperature and pressure are illustrated in Table 11.1. The effect of temperature on the branch content of polyethylene has been studied (2), and it was found that the activation energy for chain branching exceeds that of the propagation by 5.0 kcal. The effect of pressure on the branching in free radical polyethylene has been studied (3). The difference in the molar volume between the initial and activated state was found to be -24.4 cc. This result compares favorably with the difference in the molar volume of hexane and cyclohexane, which is -22.7 cc. Finally, the dependence of short chain branching on the degree of polymerization has been studied by a unique fractionation process (4).

The samples were first fractionated by a solvent elution process at a temperature above the melting point of the polymer. The fractions obtained were separated only on the basis of molecular weight. These molecular weight fractions were refractionated by a rising temperature elution

Table 11.1   Effect of Pressure and Temperature on Short Chain Branching in Polyethylene

Pressure (atm)	Temperature (°C)	Number per 1000 C Atoms
800	250	35
3000	250	10
800	130	15
3000	130	5

process that fractionates on the basis of branch content. The branching was independent of the degree of polymerization. The probability of k number of branches occurring on a chain n units long can be written

$$P_{nk} = {}_nC_k p^k q^{n-k}, \tag{11.4}$$

where p is the probability of branch formation at each carbon, q is the probability that propagation occurs rather than branch formation, and ${}_nC_k$ is the number of ways k branches can be distributed over n units. For a polyethylene chain of 2500 units and a k of 2.24 $CH_3/100$ carbon atoms, the probability of branch formation at each carbon is 0.0224 (4).

The distribution of the branches along the chain indicates that the side chains are predominantly C_2 and C_4, although there are significant concentrations of C_5, C_6, C_7, and C_8. The concentration of C_1 and C_3 side chains is very small. These data suggest a multiple intramolecular chain transfer mechanism for the formation of short branches. The initial transfer occurs through a six-membered ring in the transition state to give a secondary radical with a butyl group on the end of the chain. Since this secondary radical is less reactive than a primary one, for energetic as well as steric reasons, the next stage is the addition of a monomer unit to give a new primary radical. It is suggested that further intramolecular transfer by this radical through a six-membered-ring transition state is favored because the main chain hinders the approach of the monomer and provides a high local polymer concentration.

Suppose that P_1 is the probability of the initial transfer relative to propagation, and P_2 is the probability of a secondary transfer along the chain in each direction, with only six-membered-ring transition states being considered and all C–H bonds being taken as equivalent. Then, if only one secondary transfer occurs and the ethyl–butyl ratio is 2:1 for a polymer with 25 methyl groups/1000 carbons, the following values for P_1 and P_2 are found: $P_1 = 0.028$, $P_2 = 0.4$. This can be represented schematically as

$$\sim CH_2 - CH_2 - CH_2 - CH - CH_2 - CH_2 - CH_2 - CH_3$$

$$\underset{\overset{|}{CH_2}}{\overset{|}{CH_2}}$$

$$\overset{P_1 P_2}{\diagup} \quad \underset{\overset{|}{CH_2}}{CH_2} \quad \overset{P_1 P_2}{\diagdown}$$

$$\sim CH_2 - CH - CH_2 - CH - Bu \qquad \sim CH_2 - CH_2 - CH_2 - CH - CH_2 - CHEt$$

$$\qquad\qquad |\qquad\qquad\qquad\qquad\qquad\qquad\qquad\qquad |$$

$$\qquad\qquad Et \qquad\qquad\qquad\qquad\qquad\qquad\qquad Et$$

$$P_1(1 - 2P_2)$$

Propagation $\tag{11.5}$

This second transfer through a six-membered ring in the transitional state gives a secondary radical with either two ethyl side chains or one ethyl and one butyl side chain as part of a 2-ethyl hexyl group. The ethyl and butyl groups must occur in "clusters" along the main polymer chain either as such or in the form of complex isoalkyl side chains. Of course, it is also possible for additional multiple intramolecular transfer reactions to occur. A total of 53 different branched structures can be generated with multiple intramolecular transfer reactions, but experimental evidence is lacking to verify the existence of all these structures.

Poly(vinyl chloride).   The short chain branches in PVC are primarily CH_2Cl groups with a few chlorobutyl groups. The butyl groups can be formed by a back-biting mechanism similar to that for polyethylene:

$$
\begin{array}{l}
\text{—CH}_2 \quad \text{CH}_2 \ \text{Cl} \ \text{CH}_2 \longrightarrow \text{—CH}_2 \quad \text{CH}_2 \ \text{Cl} \ \text{CH}_2 \\
\qquad\ \ \searrow \diagup \quad \diagdown |\diagup \quad \diagdown \qquad\qquad \searrow \diagup \quad \diagdown |\diagup \quad \diagdown \\
\qquad\quad \text{CH} \qquad \text{C} \qquad \text{CHCl} \qquad \text{CH} \qquad \text{C} \qquad \overset{\bullet}{\text{CH}} \\
\qquad\quad | \qquad\quad | \qquad\quad | \qquad\qquad | \qquad\quad | \qquad\quad | \\
\qquad\quad \text{Cl} \qquad \text{H} \qquad \text{CH}_2 \qquad \text{Cl} \qquad \text{CH}_2 \qquad \text{Cl} \\
\qquad\qquad\qquad\ \searrow \overset{\bullet}{\text{CH}} \diagup \qquad\qquad\qquad\ \text{CHCl} \\
\qquad\qquad\qquad\qquad | \qquad\qquad\qquad\qquad\ | \\
\qquad\qquad\qquad\qquad \text{Cl} \qquad\qquad\qquad\qquad \text{CH}_2 \\
\qquad\qquad\qquad\qquad\qquad\qquad\qquad\qquad\ \text{CH}_2\text{Cl} \quad (11.6)
\end{array}
$$

Methyl branches are probably generated by an intramolecular process, and two possibilities have been suggested. An occasional head-to-head monomer addition

$$
\begin{array}{l}
\text{—CH}_2\text{—CH—CH—}\overset{\bullet}{\text{CH}}_2 \Longrightarrow \text{—CH}_2\text{—CH—}\overset{\bullet}{\text{CH}}\text{—CH}_2\text{Cl} \\
\qquad\qquad |\quad\ \ | \qquad\qquad\qquad\qquad\quad | \\
\qquad\quad \text{Cl}\quad \text{Cl} \qquad\qquad\qquad\qquad\ \text{Cl} \\
\qquad\qquad\qquad\qquad\qquad\qquad\qquad\ |\quad H_2C\!=\!CHCl \\
\qquad\qquad\qquad\qquad\qquad\qquad\qquad\ \downarrow \\
\qquad\qquad\qquad\qquad \text{—CH}_2\text{—CH—CH—CH}_2\text{—CH}\bullet \\
\qquad\qquad\qquad\qquad\qquad\quad |\qquad\quad |\qquad\qquad\quad | \\
\qquad\qquad\qquad\qquad\qquad\ \text{Cl}\quad \text{CH}_2\text{Cl}\qquad \text{Cl} \quad (11.7)
\end{array}
$$

can be followed either by an ordinary monomer addition or a radical rearrangement. However, this process is limited by the presence of very few head-to-head/tail-to-tail units. Alternatively, it has been suggested that a rearrangement of a normal growing chain radical by a 1,2-hydrogen shift can also form a pendant chloromethyl group (5):

$$
\begin{array}{l}
\text{—CH}_2\text{—}\overset{\bullet}{\text{CH}} \longrightarrow \ \overset{\bullet}{\text{CH}} \xrightarrow{\ H_2C=CHCl\ } \text{—CH—CH}_2\text{—}\overset{\bullet}{\text{CH}} \\
\qquad\quad | \qquad\qquad\ | \qquad\qquad\qquad\qquad | \qquad\qquad\quad | \\
\qquad\quad \text{Cl} \qquad\quad \text{CH}_2\text{Cl} \qquad\qquad\quad \text{CH}_2\text{Cl} \quad \text{Cl} \quad (11.8)
\end{array}
$$

An alternative source of chloromethyl branches has been proposed (6) involving a mutual interaction of two growing polymer chains:

$$
\text{(11.9)}
$$

The basic mechanism of chloromethyl branch formation requires further study. In PVC the number of branches does not change significantly with polymerization temperature (in the range 40–75°C).

11.2.2 Characterization of Short Chain Branching in Polymers

To measure the number of branches, three requirements must be met. First, there must be a sufficient number of branches. Second, the branches must be uniquely identifiable either with respect to the end group or the method of bonding to the main chain. Finally, this unique character must be detectable by a sensitive analytical method.

IR Spectroscopy of Short Chain Branching. IR spectroscopic techniques have been used to measure the number of branches in polyethylene samples. The spectra of a low density (branched) polyethylene and a high density (unbranched) polyethylene in the region of interest are shown in Figure 11.1. The band at 1378 cm^{-1} arises from the methyl groups. In addition to the 1378 cm^{-1} band, three interfering bands attributable to methylenes are present: one at 1368 cm^{-1}, another at 1352 cm^{-1}, and one at 1304 cm^{-1}, the latter standing somewhat apart from the rest of the group. The methyl band lies on the side of the methylene 1368 cm^{-1} band. Indeed, in the high density polyethylene the methyl band is lost in the side of the 1368 cm^{-1} band and is hard to detect by simple inspection. All methods for measuring the absorbance of the methyl band must provide some way for the elimination of this interference. This is particularly critical in the case of the high density polyethylene.

A difference spectrum is the most accurate way to eliminate an interfering absorbance; the IR instrument itself performs the subtraction of the sample and reference absorbance at equal frequencies. Willbourn (7)

SPECTRA OF TYPICAL POLYETHYLENES

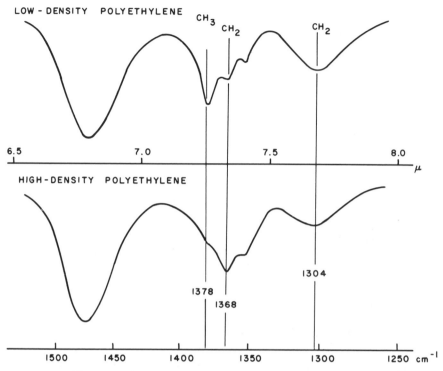

Figure 11.1 IR spectra of low and high density polyethylene in the 1500–1250 cm^{-1} region.

used as a compensating material a polymer (linear polyethylene) completely lacking in the structural features being examined, namely branches. The sample being examined and the compensating polymethylene sample are both solid, the latter being wedge shaped. The aim is to compensate completely for the 1365/1350 cm^{-1} methylene doublet, as shown in Figure 11.2. By varying the amount of reference material in the reference beam, the methylene absorption can be completely eliminated, so the methyl absorption becomes a strong symmetric absorption, as shown in Figure 11.2. The intensity of the 1378 cm^{-1} band is directly proportional to the methyl group concentration. A complication enters, however. The absorbance at 1378 cm^{-1} depends on whether the methyl is attached to an ethyl or butyl side group. The absorbance of a simple pendent methyl group is greater than that of a methyl group that is part of a pendent ethyl group, which in turn is greater than the optical densities of methyl groups attached to the still longer pendent alkyl groups. The absorptivities of the methyl groups at the end of the pendent propyl and butyl groups are the same within experimental error and, as would be expected, are the same as the optical densities of the methyl groups at the ends of the higher

WILLBOURN METHOD

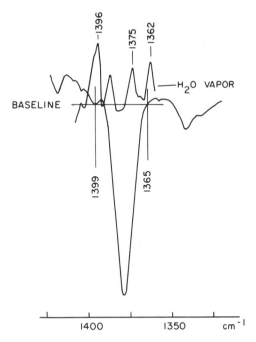

Figure 11.2 IR difference spectrum of low density polyethylene obtained by the Willbourn wedge method in the 1500–1250 cm^{-1} region.

n-alkanes. The differences in the optical densities of the methyl groups in going from methyl to ethyl to longer branch lengths are quite substantial: 1.55:1.25:1, respectively, and it is, of course, necessary to recognize these differences to quantitatively determine the degree of branching.

For low levels of branching even the methyl end groups of the pure polymethylene interfere. To eliminate the reference standard, a "self-compensation" technique has been developed. The 1368 and 1350 cm^{-1} methylene bands are amorphous bands; that is, they arise from the amorphous portions of the polymer. The self-compensation technique is based on the fact that the amorphous content, and with it the relative intensity of the methylene amorphous bands, can be varied by heat treatment. The methyl band, permanently associated with the amorphous phase, by contrast remains constant. A thick film is pressed and carefully annealed so that it has only a small amount of amorphous material and the intensities at 1368 and at 1350 cm^{-1} are small. Another thinner film is pressed and shock-cooled to give it considerable amorphous material. With sufficient care the two films (one quenched and one annealed) of different thickness can be prepared so that they will have the same absorbance at 1368 cm^{-1}. When the annealed film is placed in the sample beam and the quenched film

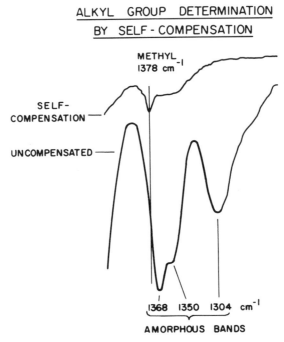

Figure 11.3 IR spectra of high density polyethylene obtained by the self-compensation method.

in the reference beam of the spectrophotometer, the resulting spectrum looks like the one shown in Figure 11.3. The methyl absorbance measured at 1378 cm^{-1} is that of a hypothetical film having a weight per unit area that is the difference between those of the annealed and quenched films. The methyl content is characteristic of the sample.

Some information about the length of the branches can also be obtained from the IR spectra by examining the frequency range arising from the methyl rocking mode in the 750 cm^{-1} region. It is necessary to compensate to eliminate the strong 720/730 cm^{-1} doublet. A band at 745 cm^{-1} can be assigned to butyl side chains and, with the proper calibration and careful compensation procedures, the number of butyl groups can be measured (7). The weak absorption in the spectrum of polyethylene at 770 cm^{-1} has been ascribed to ethyl branches (7). Confirmation of this assignment is possible from a study of hydrogenated polybutadiene, in which the band is intense.

IR spectroscopy can also be used for the determination of the short chain branching in poly(vinyl chloride). It is necessary to hydrogenate the polymer and reduce the chloromethyl groups to methyl before the IR measurements are performed as with polyethylene.

NMR Analysis of Short Chain Branching in Polymers. The use of ^{13}C NMR to study branching in polyethylene and poly(vinyl chloride) has yielded important results. The chemical shift behavior of linear and branched alkanes has been studied intensively. In order to designate backbone and branch carbon resonances for a branched molecule, we use the following system (8):

$$-CH_2-\underset{\epsilon}{CH_2}-\underset{\delta}{CH_2}-\underset{\gamma}{CH_2}-\underset{\beta}{CH_2}-\underset{\alpha}{CH}-\underset{\alpha}{CH_2}-\underset{\beta}{CH_2}-\underset{\gamma}{CH_2}-\underset{\delta}{CH_2}-\underset{\epsilon}{CH_2}-$$

$$CH_2 \quad n$$
$$|$$
$$CH_2 \quad n-1$$
$$\vdots \quad \vdots \quad n = \text{branched length}$$
$$CH_2 \quad 2$$
$$|$$
$$CH_3 \quad 1$$

The branch ^{13}C chemical shifts, listed for each carbon as a function of branch length (8), are tabulated in Table 11.2.

Different resonances are expected for each carbon in a branch up to a branch length of 10 carbons, but branches of 1–5 carbon atoms can be mutually distinguished experimentally. Later (8), the ^{13}C NMR spectra of model branched polymers were compared as shown in Figure 11.4, and the agreement is excellent. The ^{13}C spectra of low density polyethylene with the appropriate assignments are given in Figure 11.5 (9). Ethyl (1.2/1000

Table 11.2 Calculated ^{13}C Chemical Shifts for Branches 1–10 Carbon Atoms in Length with Parameters of Grant and Paul[a]

Branch Length	Chemical Shift (ppm)[b] (Grant and Paul parameters carbon number in branch)									
	10	9	8	7	6	5	4	3	2	Methyl
10	34.70	27.80	30.29	29.98	29.87	29.76	29.45	31.94	22.54	13.45
9	—	34.70	27.80	30.29	29.98	29.76	29.45	31.94	22.54	13.45
8	—	—	34.70	27.80	30.29	29.87	29.45	31.94	22.54	13.45
7	—	—	—	34.70	27.80	30.18	29.56	31.94	22.54	13.45
6	—	—	—	—	34.70	27.69	29.87	32.05	22.54	13.45
5	—	—	—	—	—	34.59	27.38	32.36	22.65	13.45
4	—	—	—	—	—	—	34.28	29.87	22.96	13.56
3	—	—	—	—	—	—	—	36.77	20.47	13.87
2	—	—	—	—	—	—	—	—	27.37	11.38
1	—	—	—	—	—	—	—	—	—	19.66

[a] Shift in ppm relative to tetramethylsilane.
[b] The chemical shifts enclosed by dotted lines will probably be obscured by overlap with the 30 ppm major methylene resonance.

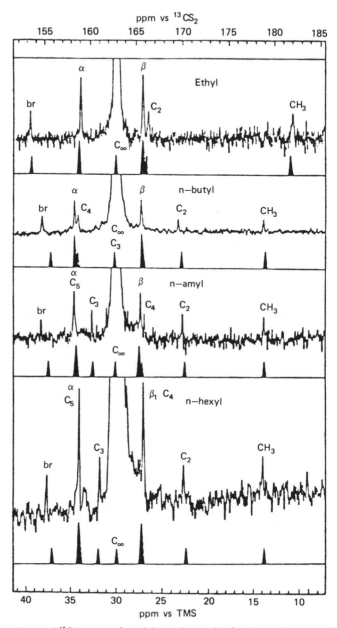

Figure 11.4 Observed ¹³C spectra of model copolymers having (top to bottom) ethyl, *n*-butyl, *n*-amyl, and *n*-hexyl branches. The predicted spectra for each of the model copolymers are shown below their observed spectra. (Reprinted by permission of Ref. 8.)

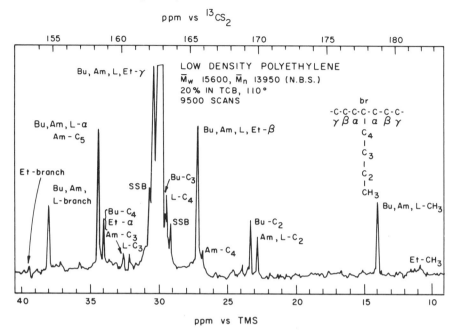

Figure 11.5   The 25 MHz ^{13}C spectrum of low density polyethylene, 20% in 1,2,4-trich-lorobenzene at 110. The diagram at the upper right shows the nomenclature employed for the carbons associated with a branch. The end carbon (i.e., C_1) is designated as CH_3; Et = ethyl, Bu = *n*-butyl, Am = *n*-amyl, and L = "long" in the sense described here. "SSB" designates spinning side bands to the principal methylene resonance. (Reprinted by permission of Ref. 9.)

CH_2), *n*-butyl (5.2/1000 CH_2), *n*-amyl (1.5/1000 CH_2), and "long" (0.9/1000 CH_2) are observed for the NBS standard (SRM 1476) that was fractionated. The presence of ethyl groups supports the multiple backbiting mechanism. In Figure 11.6 the predicted spectra for the tetrafunctional *n*-butyl branch, the 1,3-paired ethyl branches, and the 5-ethylhexyl branch are compared with the experimental spectrum (10). It is apparent that there is an insufficient number of multiple branched structures to be detected under these conditions. By extending the pulse time, higher sensitivity is observed and the branch types arising from multiple intramolecular chain transfer are observed as shown in Figure 11.7 (11).

One might suspect that ^{13}C NMR could directly determine the short chain branches in poly(vinyl chloride), but the spectrum is so complex that the relatively few branches do not generate sufficient resonance intensity to be clearly separable from configurational and other structural irregularities. But, in the same manner as previously discussed with the IR technique, the reduction of PVC to polyethylene with LiAlH$_4$ removes the configurational complexity. The spectrum of a reduced commercial poly(vinyl chloride) is shown in Figure 11.8 (10). The pattern of minor peaks corresponds closely to that of methyl branches rather than *n*-butyl. The methyl frequency is

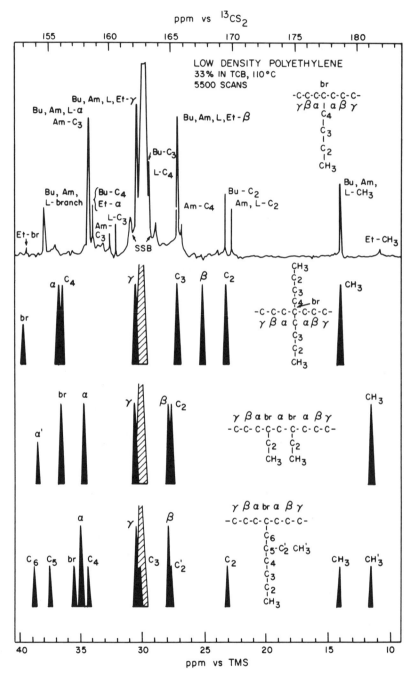

Figure 11.6 Experimental ^{13}C spectrum for an unfractionated low density polyethylene (33% in 1,2,4-trichlorobenzene) compared to predicted spectra for (top to bottom) a tetrafunctional *n*-butyl branch, 1,3-paired ethyl branches, and a 5-ethylhexyl branch. (Reprinted by permission of Ref. 10.)

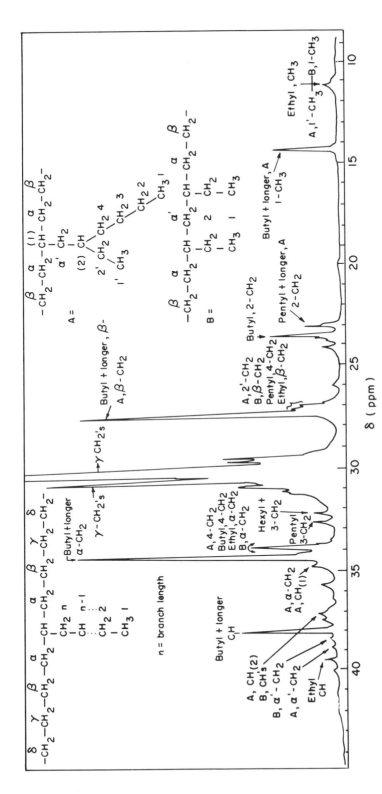

Figure 11.7 ^{13}C NMR spectrum of a 20% solution of low density polyethylene in trichlorobenzene. (Reprinted by permission of Ref. 11.)

369

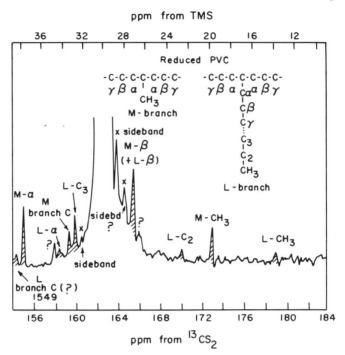

Figure 11.8   The 25 MHz ^{13}C NMR spectrum of a reduced commercial poly(vinyl chloride) (observed using a 20% solution of 1,2,4-trichlorobenzene at 135°C). (Reprinted by permission of Ref. 10.)

3.2 CH_3/1000 CH_2. Resonances appear at all expected positions for long branches, that is, n-amyl, n-hexyl, or longer with a frequency of about 1 CH_3/1000 CH_2. However, these may arise from the reduced paraffinic chain end.

^{13}C NMR spectroscopy is an absolute method and should provide a more reliable total branch frequency and also a more reliable branch distribution than the IR method. However, the peak intensities in ^{13}C NMR spectra are dependent on the relaxation times of the carbon atoms. In the pulsed Fourier transform technique, variations in intensity occur if the pulse cycle time is shorter than the relaxation times of the carbon atoms. The alkyl branch carbons have different relaxation times than the main chain carbons; in particular, the terminal methyl groups have the longest relaxation time. A substantial increase in intensity occurs upon increasing the pulse interval from 1 to 5 sec, indicating relaxation effects. These uncertainties in the increased intensities of the resonances of carbon atoms in different environments result in the calculated branch frequency being less reliable than the branch type.

In the most recent NMR work on low density polyethylene, the status of the field can be summarized by two quotes from a recent paper (45):

It becomes quite obvious that when the branching is examined in terms of specific types, each polymer is essentially different.

and

These results have pointed out quite forcibly that there must be other, more complex types of short chain branching whose identification should be actively sought.

Detection of Branches by Analysis of Radiolysis Products.   An approach to the problem of determining the structure of branches is to analyze the hydrocarbon product fragments resulting from chain scission induced by

Table 11.3   Types of Branch Clusters Obtained by Multiple Intramolecular Transfer (12)

Branch Structure	Groups Indicated by Irradiation
First transfer $-C-C_4$	C_4
Second transfer $-C-C-C-C_2$ $\quad$ C_2	C_2
$-C-C-C-C_4$ $\quad\quad C_2$	$C_2, C_4, n\text{-}C_7, iso\text{-}C_8$
Third transfer $-C-C-C-$ $\quad C_2 \quad C_2$	C_2
$-C-C-C-C_2$ $\quad C_2 \quad C_2$	$C_2, n\text{-}C_5, iso\text{-}C_6$
$-C-C-C-C-C-C_4$ $\quad\quad C_2 \quad C_2$	$C_2, C_4, n\text{-}C_7, iso\text{-}C_8$ $iso\text{-}C_{11}, iso\text{-}C_{12}$
$-C-C-C-C_4$ $\quad C_2 \quad C_2$	C_2, C_4

Table 11.4 G Values for Hydrocarbon Radiolysis Products[a] from LDPE-1 (C'_n indicates alkene)

Irradiation Temperature (°C)	Hydrocarbon															
	C_1	C_2	C'_2	C_3	C'_3	C_4	C'_4	C_5	C'_5	C_6	C'_6	C_7	C'_7	C_8	C_9	C_{10}
25	1.03	3.4	0.015	0.38	b	2.9	b	0.37	0.01	0.35	0.005	0.2	b	0.25	0.05	b
150	1.39	6.8	0.5	0.84	0.08	5.9	0.05	0.85	0.05	0.72	b	0.45	b	0.29	0.2	0.05

[a] G values $\times 10^2$.
[b] Yields <0.005.

Table 11.5 Branch Detection Efficiencies (12)

Branch Species	Branch Detection Efficiency
n-Ethyl	2.4
n-Butyl	1.0
n-Pentyl	0.78
n-Hexyl	0.67
n-Heptyl	0.56

the high energy (12). The branches are effectively removed as complete units by the high energy irradiation and appear in the gaseous products as the corresponding paraffins. The polymers are irradiated with x-rays, and the volatile products trapped and analyzed. If the scission occurs exclusively at the tertiary carbons, the expected products are as given in Table 11.3 (12). The volatile products for a branched polyethylene polymerized by the high pressure method are shown in Table 11.4 (11).

The G values are the number of moles per 100 EV of energy absorbed for temperatures of 25–30° and 150°C. It appears that, unless the irradiated polymer is heated above its melting point, the yields of the less volatile hydrocarbon products are reduced. Another factor for consideration is the assumption that the efficiency of the elimination of alkyl branches is independent of the length of the branch. Actually, the yields of branches varies considerably, as shown in Table 11.5 (12).

Additionally, the radiolytic fragmentation of the main chain produces a series of products, and a correction must be applied. Thus

$$G(hc) = K[(\text{No. of branches}/1000 \text{ cations}) + \chi \, (\text{No. of chain ends}/1000 \text{ C})],$$

where K is a proportionality constant, and χ is the relative probability of products of the hydrocarbon from the main chain.

The factor χ can be determined from a straight chain hydrocarbon like n-docasane. With these corrections the radiolysis results for branched polyethylene agree with the ^{13}C NMR results (11). However, further work on the nature of the fragmentation process is required before these results can be accepted unequivocally.

11.3 LONG CHAIN BRANCHING IN POLYMERS

Long chain branches are defined as branches whose length is nearly the same as the length of the backbone. To further classify systems, we must distinguish between cases where all branches possess equal length and cases with a random distribution of lengths of branches. Branched molecules can be divided on the basis of the character of the joint point of branches into (1) comb-shaped, in which the branches are attached to the main spine, (2) statistical molecules, in which the polymer does not possess a main spine, and (3) star-shaped branched molecules, in which all the branches are attached to a single point.

The number of branches N_b is correlated with the number of branch points b and with the functionality of branching f according to the equation

$$N_b = (f-1)b + 1, \tag{11.10}$$

which applies to any arbitrary character of branching (14).

The measure used for quantitative characterization of branching is the density of branching, λ, which is equal to the fraction of monomers possessing branch points compared to the total number of units:

$$\lambda = \alpha(b/n), \tag{11.11}$$

where n is the degree of polymerization and α is a constant that depends on the functionality of branching (14).

Branching is a precursor to cross-linking. When the branch units end in other branches, it is possible for the density of branching to reach a critical value so that a gel is formed. A gel is defined as a spatial network and is insoluble. The number of cross-links required for formation of a gel is very small. For a monodispersed polymer only one cross-link for every two molecules results in a gel. Surprisingly, less cross-linking is required for polymers with a random molecular weight distribution; only one cross-link for every four molecules gives gelation. The cross-link density at the gel point is

$$\rho_c = \frac{1}{(\overline{DP}_w)}, \tag{11.12}$$

where ρ_c is the fraction of units in the polymer which are cross-linked (13).

11.3.1 Mechanism of Long Branch Formation

Chain Transfer in Free Radical Polymerization. Branches are formed by intermolecular hydrogen transfer during radical polymerization. The reactions are

$$M_n^* + \text{wwCH}_2\text{—CHR—CH}_2\text{ww} \rightarrow M_n H + \text{wwCH}_2^*\text{CR—CH}_2\text{ww}$$

or

$$M_n^* + P \rightarrow M_n H + M^*$$

The relative rate of branch formation can be found by noting that the rate of branch formation is

$$R_{lb} = k_{lb}[M_n^*][P], \tag{11.12}$$

where [P] is the concentration of polymer.

The rate of propagation is given by

$$R_p = k_p[M][M_n^*], \tag{11.13}$$

and the relative rate becomes

$$\frac{R_{lb}}{R_p} = \frac{k_{lb}}{k_p} \frac{[P]}{[M]}. \tag{11.14}$$

The extent of branching is favored as the concentration of polymer formed is increased, so branching increases with degree of conversion. Branching is favored when a high concentration of initiator is used, since the steady state concentration of radicals is higher and hydrogen abstraction from the dead polymer occurs more often. The branches formed are shorter as is the main chain. An increase in the temperature of the polymerization favors branching, since the energy of activation for the branching reaction is higher than for propagation. For example, no long chain branching is found for poly(vinyl acetate) when it is polymerized at $-30°C$, but at $70°C$ 3 per 1000 carbons are found.

When long chain branching arises in free radical polymerization, this reaction does not affect the number-average DP if the radicals produced propagate normally. The actual effect of the long chain branching is to divide a single molecule into two portions and attach one portion to a dead polymer and the other portion remains unattached. Thus chain transfer increases the yield of large polymers and increases the weight fraction of smaller polymers. This necessarily broadens the molecular weight distribution and increases the weight-average DP and the polydispersity index $\bar{M}_w/\bar{M}_n$.

For a continuous reactor an expression for the molecular weight distribution for chain transfer with termination by deproportionation (15) is given by

$$W_n = \frac{(1-\beta)a^2 n}{[1+\beta an]^{(1+1/\beta)}}, \tag{11.15}$$

where β is the fraction of the polymer contained in the branches and $1/a$ is the number-average number of monomer units added by a radical before it is destroyed by termination or transfer. The mean number of branches per monomer unit is

$$\rho = a\beta \tag{11.16}$$

The weight and number average DP are

$$\overline{DP}_n = \frac{1}{a(1-\beta)}, \tag{11.17}$$

$$\overline{DP}_w = \frac{2}{a(1-2\beta)}.$$

The branching reaction can be reduced by the addition of chain transfer agents or by polymerizing in a solvent with chain transfer tendency. These agents compete for the radical with the dead polymer chain, and consequently less branching occurs.

Nearly all polymers prepared by free radical polymerizations have some degree of long chain branching. Polyethylene, poly(vinyl acetate), and polystyrene all exhibit long chain branching.

Polyfunctional Condensation Homopolymerization. In multifunctional condensation primary chains are formed which consist of repeat units of three types: (1) a finite or dangling unit bearing unreacted functionalities, (2) loops or units formed by intramolecular cross-linking, and (3) units involved in cross-link formation between primary chains. The fraction of each type depends on the conversion and the degree of polymerization. Schematically, we can represent the stepwise homopolymerization of A_f by

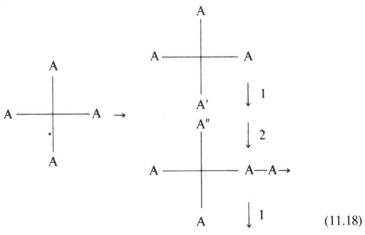

$$(11.18)$$

As the initial polymerization proceeds, finite branched chains are formed with residual unreacted functionalities. The fraction of unreacted functionalities depends on the stage of the reaction. The molecular weight increases through branching. With increasing degree of reaction or branching the molecular weight distribution gets broader.

A particularly simple approach to the complex problem of calculating the average molecular weights of nonlinear polymers involves the use of the law of conditional expectation (16). Let Y be a random variable, A be an event, and B its complement. Then the expectation (or average value) $E(Y)$ is given by the law of total probability for expectations:

$$E(Y) = E(Y/A)P(A) + E(Y/B)P(B), \qquad (11.19)$$

where $E(Y/A)$ is the conditional expectation given the event A has occurred.

Consider the condensation of A_f moles of monomer bearing f groups above. Pick any A, for example A', labeled above (eq. 11.18). The question involves the determination of the weight $W_{A'}^{out}$, attached to A' looking in the 1 direction (out from the parent). Obviously, $W_A^{out} = 0$ if A' has not reacted. If A' has reacted, then $W_{A'}^{out}$ equals $W_{A''}^{in}$, the weight attached to A'' looking along direction 2 (into the parent molecule of A''). Therefore

$$W_{A'}^{out} = \begin{cases} 0 \text{ if A' does not react,} \\ W_{A''}^{in} \text{ if A' does react (with A'').} \end{cases} \tag{11.20}$$

According to eq. 11.19,

$$E(W_{A'}^{out}) = E(W_{A'}^{in}/A \text{ reacts})P(A \text{ reacts}) +$$
$$E(W_{A'}^{out}/A \text{ does not react})P(A \text{ does not react})$$
$$= E(W_{A''}^{in})p + 0(1-p) = pE(W_{A''}^{in}), \tag{11.21}$$

where p is the extent of reaction, given by

$$p = \frac{A - A_t}{A}, \tag{11.22}$$

where A is the number of initial moles of A type and A_t is the number of moles after reaction time t. The weight $E(W_{A''}^{in})$ is the molecular weight of A_f plus the sum of the expected weights on each of the remaining $f - 1$ arms. Thus

$$E(W_A^{in}) = M_{A_f} + (f - 1)E(W_A^{out}), \tag{11.23}$$

and the repetitive nature of the molecule leads us back to the starting point. The total molecular weight W_{A_f} is the weight attached to one of A_f's arms counting in both directions (1 and 2):

$$W_{A_f} = W_A^{in} + W_A^{out}.$$

The average molecular weight attached to a random A_f is

$$\bar{M}_w = E(W_{A_f}) = E(W_A^{in}) + E(W_A^{out}),$$

so substitution yields

$$\bar{M}_w = M_{A_f}\left(\frac{1+p}{1-p(f-1)}\right),$$

or

$$\bar{X}_w = \frac{1+p}{1-p(f-1)}. \tag{11.24}$$

This is a weight average, because picking an A_f group at random corresponds to picking a unit mass and then finding the expected weight of the molecule of which it is a part. The average molecular weight for a tetrafunctional network as a function of the extent of reaction is given in Figure 11.9 (17). When the weight-average molecular weight diverges, the system forms a gel or infinite network. Since this condition corresponds to

$$1 \le p(f-1),$$

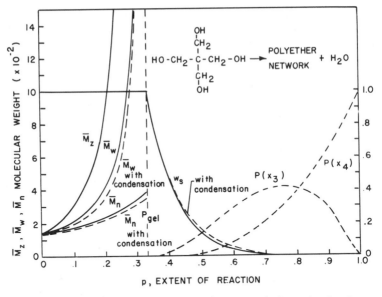

Figure 11.9 Calculated properties for the polyether network formation by the stepwise polymerization of pentaerythritol. (Reprinted by permission of Ref. 17.)

the critical degree of conversion is

$$p_c = \frac{1}{f-1}. \tag{11.25}$$

The weight fraction of soluble material, W_s, characterizes the state of the network. Up to the gel point all of the molecules are finite, so the weight fraction of solubles, W_s, is unity. Beyond the gel point molecules are rapidly incorporated into the network and W_s decreases quickly. For a multifunctional homopolymerization a randomly chosen A_f is part of the sol if all f of its arms lead to finite chains. The calculation is similar to the preceding one for weight-average molecular weight. We need to know the probability that following direction 1 (looking out from the molecule) leads to a finite or dangling chain. Let F_A^{out} be the event that 1 is the start of a finite chain. Then using eq. 11.19 yields

$$P(F_A^{\text{out}}) = P(F_A^{\text{out}}/A \text{ reacts})P(A \text{ reacts}) +$$

$$P(F_A^{\text{out}}/A \text{ does not react})P(A \text{ does not react})$$

$$= P(F_A^{\text{in}})p + 1(1-p) = pP(F_A^{\text{in}}) + 1 - p, \tag{11.26}$$

where F_A^{in} is the event in direction 2 that is the start of a finite chain. Of course, $F_A^{\text{out}} = 1$ for $P(A \text{ does not react})$ because the chain terminates, that is, is finite. Since all the F branches must be considered,

$$P(F_A^{\text{in}}) = P(F_A^{\text{out}})^{f-1}, \tag{11.27}$$

and the repetitive nature of this simple branched molecule leads us back to the starting situation. Combining eqs. 11.26 and 11.27, we can solve for $P(F_A^{out})$:

$$pP(F_A^{out})^{f-1} - P(F_A^{out}) - p + 1 = 0 \qquad (11.28)$$

or, in terms of $P(F_A^{in})$,

$$[pP(F_A^{in}) + 1 - p]^{f-1} = P(F_A^{in}). \qquad (11.29)$$

Equation 11.29 is in the form of a probability generating function of a binomial random variable with parameters $f - 1$ and p. Our event of finite chain corresponds to extinction, and the probability of extinction is the unique solution of eqs. 11.28 and 11.29 in the interval (0, 1) if it exists and 1 otherwise. Physically, when $P(F_A^{out}) = 1$, the system has not gelled. For postgel relations the root is between 0 and 1. The solutions are

$$\text{for } f = 3, \qquad P(F_A^{out}) = \frac{1-p}{p};$$

$$\text{for } f = 4, \qquad P(F_A^{out}) = \left(\frac{1}{p} - \frac{3}{4}\right)^{1/2} - \frac{1}{2}. \qquad (11.30)$$

The weight fraction of solubles W_s can now be calculated as a function of conversion. A randomly chosen A_f molecule is a part of the sol if all f of its arms lead to finite chains. Thus

$$W_s = P(F_A^{out})^f, \qquad (11.31)$$

which from eq. 11.30 yields

$$\text{for } f = 3, \qquad W_s = \frac{(1-p)^3}{p^3}.$$

Figure 11.9 shows W_s as a function of extent of reaction for $f = 4$. Note how rapidly the sol fraction disappears after gel formation. The number-

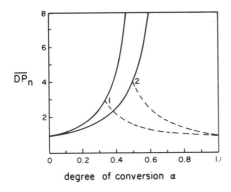

Figure 11.10 Number-average degree of polymerization plotted against the conversion degree for a random homopolymerization. Plot 1: trifunctional system; plot 2: tetrafunctional system. The dashed lines denote values related to the sol fraction after gelation.

average degree of polymerization for the sol fraction has also been calculated and is shown in Figure 11.10 (18).

With increasing conversion the molecular weight gets broader, and at the gel point the number-average degree of condensation is still finite, whereas the weight-average degree of conversion is infinite. After the gel point the gel fraction increases rapidly and the molecular weight in the sol gets lower because large species are attached to the gel preferentially. At high conversions only very small species (oligomers and monomers) are present in the sol fraction.

The cross-link density or concentration of effective junction points in the infinite network can also be calculated (17). An A_f chosen at random acts as an effective junction point if three or more arms lead to the infinite network. If only one arm is infinite, three A_f molecules are just dangling on the network; if two are infinite, A_{f_i} forms part of a chain connecting two effective junction points but is not an effective junction point. In general the probability that an A_{f_i} monomer will be an effective cross-link of degree M is

$$P(X_{M_i f_i}) = \binom{f_i}{M} P(F_A^{out})^{f_i - M} [1 - P(F_A^{out})]^M.$$

Figure 11.9 shows the calculated value of A_f as a trifunctional cross-link $[P(X_3)]$ and as a tetrafunctional cross-link $[P(X_4)]$. Observe that trifunctional cross-links form initially, then give way to tetrafunctionals that become unity at complete reaction (assuming no side reactions or small rings are formed). One can now calculate the cross-link density, which is the internal concentration of $[A_{f_i}]_0$ times the probability $P(X_{M_i f_i})$ summed over $f_i = M$ to the highest functionality:

$$[X_M] = \sum_{f_i = M}^{f_k} [A_{f_i}]_0 P(X_{M_i f_i}). \tag{11.32}$$

The total cross-link density $[X]$ is the sum of the individual $[X_M]$'s from $M = 3$ to f_k. Since p goes to 1 in the limit of complete reaction, $[X_M] = [A_{f_m}]_0$.

The above treatment is based on the assumption of an ideal network formed such that (1) all functional groups of the same type are equally reactive, (2) all groups react independently of one another, and (3) no intramolecular reactions occur in finite species.

It is possible to treat the problem for the substitution effect, that is, the change in functionality as a result of the state of other groups (19), and to calculate the critical conversion at the gel point.

The effect of intramolecular cross-linking on network formation and the gel point was not taken into account. In reality, intermolecular cross-linking is always accompanied by closed loop formation. As seen above, in a network a cross-link is intermolecular if at least three independent paths issue from it to the surface of the sample. Some cross-links must be intramolecular, because formation of new paths during cross-linking is limited by the volume of the links. For A_f the number of links formed as n steps is $f(f-1)^n$ if all functionalities have reacted intermolecularly, whereas the available volume is proportional to $(n+1)^3$, so that for $n \to \infty$ the ratio $[\sum_{i=0}^{n} f(f+1)^i]/(n+1)^3$ diverges if $f > 2$. A prediction of the number and form of cyclic structures in a network of arbitrary cross-linking density is at the moment impossible. An increase in their number is expected with increased conversion and with increasing dilution of the system.

Polyfunctional Condensation Copolymerization. The general multifunctional copolymerization can be represented schematically as

$$A_{f1} + A_{f2} + \cdots + A_{fk} + B_{g1} + B_{g2} + B_{gl} \to \text{Network}.$$

or in the form

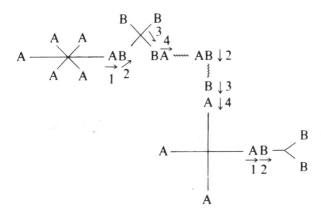

We need the expected weight along $\underline{1}$ with consideration of reaction with all possible B_{gj}'s. By generalization of eq. 11.21,

$$E(W_A^{out}) = E(W_A^{out}|A \text{ does not react})P(A \text{ does not react}) +$$

$$\sum_{j=1}^{l} E(W_A^{out}|A \text{ reacts with } B_{gj})P(A \text{ reacts with } B_{gj}) =$$

$$0(1 - p_A) + \sum_{j=1}^{l} E(W_{B_{gj}}^{in})p_A b_{gj} = p_A \sum_{j=1}^{l} b_{gj} E(W_{B_{gj}}^{in}). \qquad (11.33)$$

In this case we need to describe the mole fraction of B's on B_{gj}:

$$b_{gj} = \frac{g_j B_{gj}}{\sum_{j}^{l} g_j B_{gj}}.$$

In direction 2 there is a relation for each B_{gj}:

$$E(W_{B_{gj}}^{in}) = M_{B_{gj}} + (g_j - 1)E(W_B^{out}). \qquad (11.34)$$

The expected weights along directions 3 and 4 are derived similarly:

$$E(W_B^{\text{out}}) = p_B \sum_i a_{f_i} E(W_{A_{f_i}}^{\text{in}}), \tag{11.35}$$

$$E(W_{A_{f_i}}^{\text{in}}) = M_{A_{f_i}} + (f_i - 1)E(W_A^{\text{out}}), \tag{11.36}$$

where a_{f_i} is the mole fraction of all A's on A_{f_i},

$$a_{f_i} = \frac{f_i A_{f_i}}{\sum_i f_i A_{f_i}},$$

and p_B is related to p_A as follows:

$$p_B = \frac{\sum_i f_i A_{f_i}}{\sum_j g_j B_{gj}} p_A = rp. \tag{11.37}$$

Solving this system of equations we obtain

$$E(W_A^{\text{out}}) = \frac{p_A M_b + p_A p_B (g_e - 1)M_a}{1 - p_A p_B (g_e - 1)(f_e - 1)}, \tag{11.38}$$

$$E(W_B^{\text{out}}) = \frac{p_B M_a + p_A p_B (f_e - 1)M_b}{1 - p_A p_B (g_e - 1)(f_e - 1)},$$

where

$$f_e = \sum_i f_i a_{f_i},$$

$$g_e = \sum_j g_j b_{gj},$$

$$M_a = \sum_i M_{A_{f_i}} a_{f_i},$$

$$M_b = \sum_j M_{B_{gj}} b_{gj}.$$

If $W_{A_{f_i}}$ is the weight of the molecule to which a random A_{f_i} belongs, and $W_{B_{gj}}$ is defined similarly, then it follows that

$$E(W_{A_{fi}}) = M_{A_{fi}} + f_i E(W_A^{\text{out}}),$$

$$E(W_{B_{gi}}) = M_{B_{gi}} + g_j E(W_B^{\text{out}}).$$

As before, to find $\bar{M}_w$, we take a unit of mass at random and compute the expected weight of the molecule to which it belongs:

$$\bar{M}_w = \sum_i \omega_{A_{fi}} E(W_{A_{fi}}) + \sum_j \omega_{B_{gi}} E(W_{B_{gi}}),$$

where

$$\omega_{A_{fi}} = \frac{M_{A_{fi}} A_{fi}}{\sum_i M_{A_{fi}} A_{fi} + \sum_j M_{B_{gj}} B_{gj}}$$

and $\omega_{B_{gj}}$ is defined similarly. Substituting and rearranging, we obtain

$$\bar{M}_w = \frac{p_B m_a' + p_A m_b'}{p_B m_a + p_A m_b} + \frac{p_A p_B [p_A(f_e - 1)M_b^2 + p_B(g_e - 1)M_a^2 + 2M_a M_b]}{(p_B m_a + p_A m_b)[1 - p_A p_B(f_e - 1)(g_e - 1)]}.$$

$$(11.39)$$

The M_w is demonstrated as a function of extent of reaction for the $A_3 + B_2$ copolymer in Figure 11.11 and the $A_4 + B_2$ system in Figure 11.12, where

$$m_a = \sum_i M_{A_{fi}} \frac{a_{fi}}{f_i},$$

$$m_a' = \sum M_{A_{fi}}^2 \frac{a_{fi}}{f_i},$$

with similar expressions for m_b and m_b' involving the Bg_j's. This equation is a general equation and holds for nearly all nonlinear stepwise copolymerizations (16). For the general case $\bar{M}_w$ becomes infinite when

$$(p_A p_B)_{gel} = r p_{gel}^2 = \frac{1}{(f_e - 1)(g_e - 1)}.$$

$$(11.40)$$

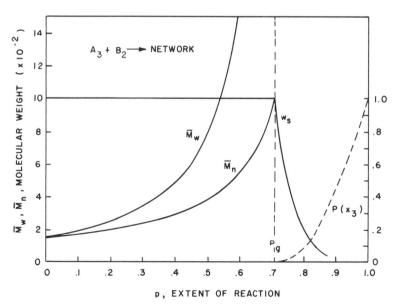

Figure 11.11 Calculated properties for the urethane network formation by the stepwise polymerization of 2-hydroxymethyl-2-ethyl-1,3-propanediol and 1,6-hexamethylene diisocyanate, with $r = 1$. (Reprinted by permission of Ref. 17.)

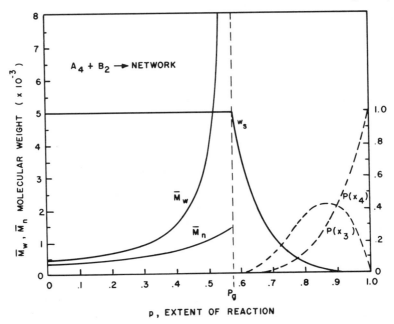

Figure 11.12 Calculated properties for silicone rubber formation by the hydrosilation reaction of $(HCH_3PhSiO)_4Si$ and vinyl terminated poly(dimethylsiloxane), $M_{B_2} = 5000$ (monodisperse), with $r = 1$. (Reprinted by permission of Ref. 17.)

Figure 11.11 shows the calculated properties for an $A_3 + B_2$ system, and Figure 11.12 the calculated properties for a $A_4 + B_2$ system (17).

The weight fraction of sol can be calculated in the same fashion as for polyfunctional homopolymerization except that each species must be weighted by its mole fraction of functional groups, a_{f_i} and b_{g_j}:

$$P(F_A^{out}) = p_A P(F_B^{in}) + 1 - p_A,$$

$$P(F_B^{in}) = \sum_j b_{g_j} P(F_B^{out})^{g_j - 1},$$

$$P(F_B^{out}) = p_B P(F_A^{in}) + 1 - p_B,$$

$$P(F_A^{in}) = \sum_i a_{f_i} P(F_A^{out})^{f_i - 1}.$$

Combining equations yields

$$p_A \sum_j b_{g_j} [1 - p_B + p_B \sum_i a_{f_i} P(F_A^{out})^{f_i - 1}]^{g_j - 1} - P(F_A^{out}) - p_A + 1 = 0.$$

$$(11.41)$$

Useful special cases include mixtures with only B_2 present:

$$rp^2 \sum_i a_{f_i} P(F_A^{out})^{f_i-1} - P(F_A^{out}) - rp^2 + 1 = 0.$$

For the system $A_3 + A_2 + A_1 + B_2$ the desired root is

$$P(F_A^{out}) = \frac{(1 - (\alpha_3 + \alpha_2))rp^2}{\alpha_3 rp^2},$$

and for $A_4 + A_3 + A_2 + A_1 + B_2$,

$$P(F_A^{out}) = \frac{[\alpha_3^2 - \alpha_4(3\alpha_4 + 2\alpha_3 + 4\alpha_2 - 4/rp^2)]^{1/2}}{2\alpha_4} - \frac{1}{2}$$

provided they are between 0 and 1.

We can readily generalize eq. 11.31 to a mixture of A_{f_i} and B_{g_i} by weighting each species by its mass fraction in the mixture, $\omega_{A_{f_i}}$:

$$\omega_s = \sum_{i=1}^{h} \omega_{A_{f_i}} P(F_A^{out})^{f_i} + \sum_{f=1}^{l} \omega_{B_{g_j}} P(F_B^{out})^{g_j} \tag{11.42}$$

Figure 11.11 illustrates the ω_s for an $A_3 + B_2$ copolymerization, while Figure 11.12 illustrates an $A_4 + B_2$ copolymerization.

Cross-linking   Copolymerization   Reactions   with   Divinyl   Monomers (20).   Among the cross-linking chain reactions the vinyl-divinyl copolymerization is the most common way of preparing networks. Branch points can be introduced by copolymerization of a divinyl monomer, leaving independent vinyls of the same reactivity with a monovinyl monomer with a different reactivity. The vinyls are called *independent* if the reactivity of the remaining vinyl is not changed after the first one has reacted. In general, the vinyls in the divinyl monomer can have either the same or different reactivities. The vinyl-divinyl copolymerization can be described by the normal copolymerization equation

$$\frac{-dM_1}{dM_2} = \left(\frac{M_1}{M_2}\right)\frac{r_1 M_1 + M_2}{r_2 M_2 + M_1},$$

where M_1 and M_2 are molar concentrations of unreacted vinyls belonging to the mono- and divinyl monomer, respectively, and r_1 and r_2 are the reactivity ratios.

The cross-link density is expressed as

$$\rho = 2\frac{X_2}{m_1 + m_2 + 2X_2}, \tag{11.43}$$

where X_2 is number of moles of doubly reacted divinyl molecules, m_1 is the number of reacted monovinyl molecules, and m_2 is the number of divinyl molecules with one vinyl reacted.

Let P_ν be the probability that a double bond of the divinyl monomer has reacted; then

$$M_2 = (1 - P_\nu)^2 M_2^0,$$

$$m_2 = (1 - P_\nu)P_\nu M_2^0,$$

$$X_2 = \frac{P_\nu^2 M_2^0}{2},$$

where M_2 and M_2^0 are the actual and initial numbers of moles of vinyl in the unreacted divinyl monomer. The degree of conversion of the divinyl monomer, α_2, is

$$\alpha_2 = \frac{M_2^0 - M_2}{M_2^0}.$$

Let α be the degree of conversion with respect to all vinyls in the system, which gives upon substitution for the crosslink density

$$\rho = \frac{2f_2^0[1 - \alpha_2/2 - (1 - \alpha_2)^{1/2}]}{\alpha}, \qquad (11.44)$$

where f_2^0 is the initial fraction of vinyls belonging to the divinyl monomer. The relationship between α_2 and α is

$$\alpha_2 = \frac{2\alpha \bar{f}_2}{f_2^0} - \alpha^2 \left(\frac{\bar{f}_2}{f_2^0}\right)^2, \qquad (11.45)$$

where $\bar{f}_2$ is the average composition of the copolymer expressed as the mole fraction of reacted vinyls of the divinyl monomer; $\bar{f}_2$ is obtained by integration of the copolymerization equation. In this way one can obtain all the information necessary for calculating the cross-link density. The composition of the monomers, the composition of the polymer, and the cross-linking density as a function of conversion in vinyl-divinyl copolymerization are shown in Figure 11.13 (20). The degrees of polymerization,

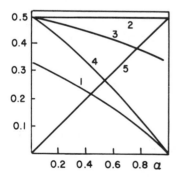

Figure 11.13 Composition of the monomers, composition of the polymer, and the cross-linking density as a function of conversion in vinyl-divinyl copolymerization: $r_1 = r_2 = 1$, $f_1^0 = 0.5$. (1) Mole fraction of the divinyl monomer in the monomer mixture; (2) mole fraction of reacted monovinyl units from all vinyl units reacted; (3) mole fraction of all divinyl molecules in the polymer; (4) mole fraction of divinyl molecules that reacted with only one vinyl bond in the polymer; (5) the cross-linking density. (Reprinted by permission of Ref. 20.)

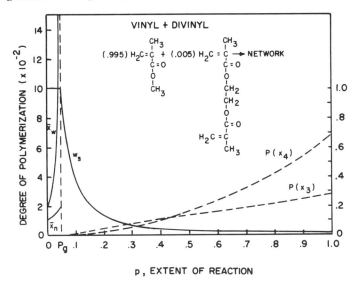

Figure 11.14 Calculated properties versus extent of reaction of C=C groups for chain addition polymerization of methyl methacrylate with 0.5 mole-% ethylene dimethacrylate, $q = 0.999$. (Reprinted by permission of Ref. 17.)

the sol fraction, and the number of tri- and tetrafunctional links as a function of cure are shown in Figure 11.14 (17).

The complexity of the copolymerization involving a divinyl monomer with unequal or dependent groups is at least that of a three component system. Alkyl methacrylate and dienes are examples of monomers with groups that are not equally reactive and dependent. The copolymerization of methyl methacrylate with glycol dimethacrylate is ideal. Polymerization of styrene with glycol dimethacrylate deviates slightly from ideality. When the reactivity of the divinyl group exceeds that of the vinyl group very much, as, for example, with vinyl chloride-glycol dimethacrylate, in the last stages of copolymerization the monovinyl groups react almost exclusively, the molecular weight distribution is very broad, and a part of the polymer remains unattached to the network even at $\alpha = 1$. In the opposite case, for example, acrylonitrile-divinyl adipate, pendent vinyls are involved in the reaction almost exclusively, so toward the end of the reaction only very short chains are formed.

Cross-linking of Polymer Chains. An important class of networks is formed by cross-linking long polymer chains after polymerization. Such cross-linking is often termed *vulcanization*, and most often these reactions occur through unsaturation. If all the reactive groups are of the same type, then the polymerization can be modeled as a mixture of species A_{f_i}, where f_i is quite high.

If the reactive groups are uniformly distributed along the chains, then

$$M_{A_{f_i}} = f_i M_c, \qquad (11.46)$$

where M_c is the weight between cross-linkable sites. For the cross-linking of a mixture of A_{f_i}, eq. 11.39 becomes

$$\bar{M}_w = \frac{m'_a}{m_a} + \frac{\rho M_a^2}{m_a[1 - \rho(f_e - 1)]}, \qquad (11.47)$$

where ρ is the extent of cross-linking or the fraction of repeat units that are crosslinked. Using eq. 11.44 to substitute for m_a, m'_a, and M_a gives

$$\bar{DP}_w = \frac{\bar{M}_w}{M_c} = \frac{f_e(1 - \rho)}{1 - \rho(f_e - 1)}. \qquad (11.48)$$

The definition of f_e is

$$f_e = \frac{\sum_i f_i^2 A_{f_i}}{\sum_i f_i A_{f_i}} = \frac{\sum M_{A_{f_i}}^2 A_{f_i}}{M_c \sum M_{A_{f_i}} A_{f_i}} = \overline{DP}_{w_0},$$

where $\overline{DP}_{w_0}$ is the weight-average degree of polymerization of the original polymer. Substituting gives

$$\overline{DP}_w = \frac{\overline{DP}_{w_0}(1 - \rho)}{1 - \rho(\overline{DP}_{w_0} - 1)}. \qquad (11.49)$$

The results as a function of average number of cross-links formed per chain are shown in Figure 11.15 (17).

To derive the condition for the gel point in random cross-linking of existing chains of arbitrary molecular weight distribution, one selects at random a chain having n monomer units. If one unit of the selected chain happens to be cross-linked, the probability that this unit is a part of the selected chain is equal to the weight fraction W_n of the n-mer. The expected number of additional cross-linked units in the n-mer is then $\rho(n - 1)$. The mean expected number of additional cross-linked units in a chain is

$$\epsilon = \sum \rho W_n(n - 1) = \rho(\overline{DP}_w - 1). \qquad (11.50)$$

For an infinite network it is necessary that $\epsilon > 1$. At the gel point ϵ equals unity, which leads to

$$\rho_c = \frac{1}{\overline{DP}_w - 1} \simeq \frac{1}{\overline{DP}_w}. \qquad (11.51)$$

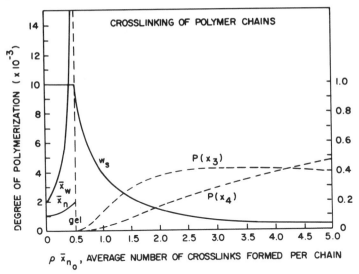

Figure 11.15 Calculated properties versus degree of cross-linking for starting chains of $\overline{DP}_n = 1000$ and most probable distribution. (Reprinted by permission of Ref. 17.)

If not all units of the primary chain are capable of cross-linking, the fraction of them capable, s, gives for the gel point

$$\rho \simeq \frac{1}{s\overline{DP}_w}.$$

The number of cross-linked units per primary molecule is equal to $\gamma = \rho \overline{DP}_n$. At gelation $\gamma = 1$ for monodispersed and $\frac{1}{2}$ for most probable. To calculate the W_s it is convenient to consider the cross-linking of polymers with a most probable distribution of cross-links as a stepwise reaction of A_2 with a small amount of A_4 to a high degree of conversion, $\rho = P_0$, so the equations derived (eq. 11.41) can be used with $\rho = a_4$ as the variable. Thus we use eq. 11.41 for the case $A_4 + A_2$ with $b_2 = p_B = 1$ to indicate high conversion. The result is

$$P(F_A^{\text{out}}) = \left[\frac{1}{\rho(\bar{x}_{n_0} - 1)} + \frac{1}{4}\right]^{1/2} - \frac{1}{2}$$

provided $\rho(\bar{x}_{n-1}) > \frac{1}{2}$. Substituting this result in eq. 11.42 gives

$$W_s = P(F_A^{\text{out}})^2\{1 - \rho[P(F_A^{\text{out}})^2 - 1]\},$$

or

$$W_s \approx \left\{\left[\frac{1}{\rho(\bar{x}_n - 1)} + \frac{1}{4}\right]^{1/2} - \frac{1}{2}\right\}^2. \tag{11.52}$$

This result is illustrated in Figure 11.15.

Graft Polymerization (21). When active sites are formed on the backbone of a polymer chain, another monomer can be polymerized on the backbone, leading to a comblike graft polymer. Radical sites can be generated with free radicals, light, electrons, or mechanical energy. Ionic sites can also be generated, and anionic and cationic polymerizable monomers can be used to form graft polymers. Additionally, one can use functional groups on the polymer as handles for branching or grafting. For example, an alcohol group can be used to react with ethylene oxide to give a polyethylene oxide graft polymer.

11.3.2 Branching Factors for Model Branched Polymers

It has been known that a branched molecule extends over a smaller average volume in solution than a linear molecule of the same molecular weight. The unperturbed dimensions are fundamental molecular properties for investigating the form of the polymer chain in dilute solution. The rotational isomeric model (22) has been used to calculate the dimensions of polymer chains as a function of the degree of polymerization for linear polymer chains. The vector $\bar{r}$ that connects the two ends of the chain is given by

$$\bar{r} = \sum_{i=1}^{n} \bar{l}_i,$$

where $\bar{l}_i$ is the skeletal bond vector. The scalar length r_{ij} connecting atoms i and j is defined as

$$r_{ij}^2 = \sum_{i'=i+1}^{j} l_{i'}^2 + 2 \sum_{i < i' < j \leq j'} \bar{l}_{i'} \bar{l}_{j'}.$$

The radius of gyration, S, is defined as the root-mean-square distance of the collection of atoms from their common center of gravity; so

$$S^2 = (n+1)^{-1} \sum_{0}^{n} S_i^2, \tag{11.53}$$

where S_i is the distance of atom i from the center of gravity of the chain in a specified configuration. The relationship between the radius of gyration and the interatomic distance r_{ij} is (22)

$$S^2 = (n+1)^{-2} \sum_{0 \leq l < j \leq n} r_{ij}^2.$$

For an ideal, freely jointed chain of n bonds

$$\langle r^2 \rangle_0 = nl^2,$$

where l is the bond length. The radius of gyration is given by

$$\langle S^2 \rangle = \frac{nl^2}{6}.$$

The rotational isomeric model has been used to calculate the mean square radius of gyration for branched polypeptides (23). The calculations for star-branched models are based on

$$S^2 = (n + 1)^{-1}\left(\sum_{k=1}^{b} \sum_{i=1}^{n_k} S_i^2 + S_0^2\right),$$ (11.54)

where S_0 is the kth subchain coming from the branch point. The branch point has the symbol 0, and each structural unit in the kth subchain is counted as $1, 2, \ldots$, and n_k for the b branches. The vector from the center of mass of the molecule to the ith atom of the subchain is S_i.

The radius of gyration ratio $\langle S^2 \rangle_0 / n l^2$ from starlike molecules is shown in Figure 11.16 for the case where the branch lengths are uniform, and in Figure 11.17 for branch lengths of random length (23). The radius of gyration increases asymptotically with the degree of polymerization n like the linear chain. Also, the radius of gyration is always smaller than that of a linear chain of identical molecular weight and is extremely sensitive to the number of branches, as shown in Figure 11.18. For the same number of branches the radius of gyration is always larger for the random chain length branches than the uniform distribution.

For comblike branched polymers the radius of gyration ratio has been calculated for branch lengths of uniform length (Figure 11.19) and random length (Figure 11.20) as a function of the degree of polymerization (24). The radius of gyration increases asymptotically with degree of polymerization and branch number. The radius of gyration for the same number

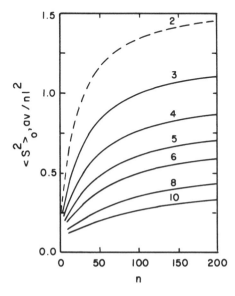

Figure 11.16 The radius of gyration ratio plotted against the degree of polymerization for starlike polypeptides having uniform distribution of subchain lengths. The number on each curve indicates the number of branches. (Reprinted by permission of Ref. 23.)

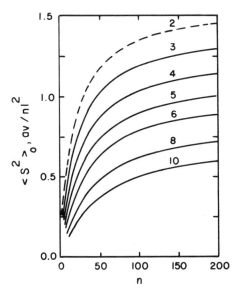

Figure 11.17 The radius of gyration ratio plotted against the degree of polymerization for starlike polypeptides having random distribution of subchain lengths. The number on each curve indicates the number of branches.

of branches is larger for comb molecules with uniform branch lengths than for comb molecules whose branch lengths are randomly distributed in length. The radius of gyration ratio decreases asymptotically with number of branch points.

A branching parameter, g, is defined by

$$g = \frac{\langle S^2 \rangle_{0B}}{\langle S^2 \rangle_{0L}}, \tag{11.55}$$

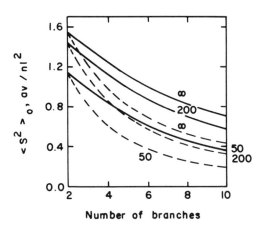

Figure 11.18 The radius of gyration ratio plotted against the number of branches for starlike polypeptides having random distribution (solid lines), and having uniform distribution of subchain length (broken lines). The number on each curve indicates the degree of polymerization. (Reprinted by permission of Ref. 23.)

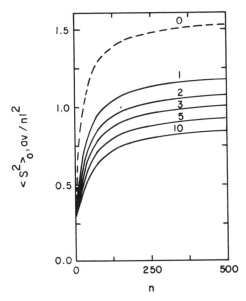

Figure 11.19 The radius of gyration ratio plotted against the degree of polymerization for comblike polypeptides of $f = 3$ having uniform distribution of subchain lengths. The number on each curve indicates the number of branching points. (Reprinted by permission of Ref. 24.)

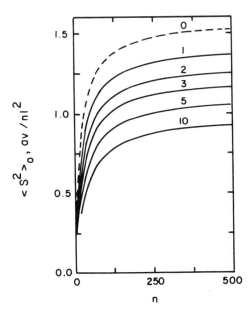

Figure 11.20 The radius of gyration ratio plotted against the degree of polymerization for comblike polypeptides of $f = 3$ having random distribution of subchain lengths. The number on each curve indicates the number of branching points. (Reprinted by permission of Ref. 24.)

where $\langle S^2 \rangle_{0B}$ is the radius of gyration for a branches molecule and $\langle S^2 \rangle_{0L}$ is the radius of gyration for the linear molecule of equivalent degree of polymerization.

For a molecule with λ random branches g can be determined geometrically and is given by

$$g = \sum_{\lambda} \left(\frac{3n_\lambda^2}{n^2} - \frac{2n_\lambda^3}{n^3} \right)^2, \tag{11.56}$$

where n_λ is the number of segments in branch λ.

If the branch points possess functionality f and branches of fixed length,

$$g = \frac{3}{f} - \frac{2}{f^2} \quad \text{(star-shaped)},$$

so for $f = 3$, $g = \frac{7}{9}$, and when $f = 4$, $g = \frac{5}{8}$.

The g factor depends on several factors including the character of attachment. The g factor for statistically branched macromolecules is less than for comb-shaped molecules. The branch length distribution affects g,

Table 11.6 Numerical Values of the g Factor^a (14)

f	b	Statistically Branched Polymers		Comb-Shaped Branched Polymers		f	b	Statistically Branched Polymers		Comb-Shaped Branched Polymers	
		g_1	g_2	g_1	g_2			g_1	g_2	g_1	g_2
3	1	0.900	0.788	0.900	0.778	4	1	0.800	0.625	0.800	0.625
	2	0.829	0.712	0.829	0.712		2	0.691	0.545	0.691	0.545
	3	0.771	0.668	0.774	0.668		3	0.618	0.496	0.618	0.496
	4	0.730	0.633	0.733	0.638		4	0.566	0.460	0.569	0.465
	5	0.691	0.605	0.703	0.617		5	0.525	0.432	0.534	0.443
	6	0.663	0.581	0.679	0.601		6	0.493	0.410	0.508	0.428
	7	0.636	0.560	0.660	0.589		7	6.466	0.391	0.488	0.416
	8	0.613	0.512	0.645	0.580		8	0.443	0.374	0.472	0.407
	9	0.592	0.525	0.632	0.572		9	0.423	0.360	0.459	0.400
	10	0.573	0.510	0.622	0.566		10	0.406	0.317	0.448	0.394
	12	0.541	0.485	0.605	0.556		12	0.378	0.326	0.431	0.385
	15	0.505	0.451	0.587	0.546		15	0.344	0.301	0.414	0.375
	20	0.454	0.414	0.567	0.535		20	0.305	0.270	0.395	0.365
	25	0.418	0.381	0.555	0.529		25	0.277	0.248	0.381	0.359
						8	1	0.53	0.35	—	—
							2	0.41	0.28	—	—
							3	0.34	0.24	—	—

^{a} Here g_1 represents random distribution of lengths of branches; g_2, fixed lengths of branches.

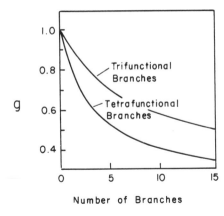

Figure 11.21 Change in the value of g with number of branches for a monodispersed polymer in a θ solvent. Branching is random and branches are randomly spaced.

with the g factor for macromolecules of variable branch length being greater than those of fixed length. The theoretical calculations for different models of branching have been made, and the results are shown in Table 11.6 (25). Figures 11.21 and 11.22 show graphical representations of the dependence of the g factor on functionality and types of distribution for trifunctional branch points (25).

A test of the theoretical results can be accomplished by measuring $\langle S^2 \rangle_b$ and $\langle S^2 \rangle_l$ at the θ temperature for branched and linear molecules of the same molecular weight and by comparing the ratio with the theoretical values. Here the numbers and length of the branches need to be known. Unfortunately, when this is done, the experimental value of the ratio

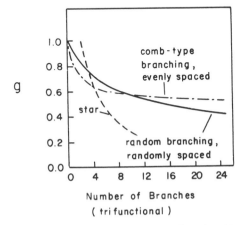

Figure 11.22 Change in the value of g with number of trifunctional branches for three types of branched polymers.

$\langle S^2 \rangle_b / \langle S^2 \rangle_l$ is higher than the theoretical value of g. No complete resolution of this problem exists. A discussion of the various factors that could explain this discrepancy has been given by Small (26).

It should be mentioned that only long chain branching leads to a noticable decrease of the dimensions of macromolecules. Short chain branches have little effect ($\sim 1\%$) except at very high branch density.

11.3.3   Characterization of Long Chain Branching (27, 28)

For the complete characterization of a branched polymer one requires the location of the branch points, the number of branch points, the branch lengths, and the primary chain lengths. It is only rarely that such complete characterization of long chain branching is possible, and this is usually for "ideal" model polymers prepared under special conditions. Even for these special model polymers, side reactions and exchange reactions can lead to difficulties that require efficient separation procedures. Consequently, only qualitative or semiquantitative measurements of long chain branching are possible.

If one assumes a completely random branching, only two parameters are required. One is the probability α that a chain starting from a branch point ends in another branch point. Thus if a branch terminates in an end group, the probability is $\alpha(1-\alpha)$. Also required is the probability q of continuation of a chain equal, in kinetic terms, to the average propagation rate divided by the average termination rate. The probability that a chain branch has n units is $(1-q)q^{n-1}$. The branching index λ is given by

$$\lambda = \frac{2^{1/2}(1-\alpha)^{1/2}}{M_0} \frac{(1-q)}{q}. \tag{11.57}$$

In general, the density of branching is so low and the differences between branch points and the normal repeat units so small, that chemical and spectroscopic methods are not helpful. Sensitive end group analysis techniques can sometimes yield number-average estimates of the number of branches if they are coupled with measurements of the number-average degree of polymerization. In a few cases the long chain branches can be clipped from the primary chains, and the $\bar{M}_n$ of the branched and the severed primary chains M'_n can be measured. The average number of branches N_b is then calculated as

$$\bar{N}_b = \frac{\bar{M}'_n - \bar{M}_n}{\bar{M}'_n} \tag{11.58}$$

This technique has been used for the determination of branched poly(vinyl acetate). The acetyl groups are removed from poly(vinyl acetate) by hydrolysis with methanolic potassium hydroxide. The resulting linear poly(vinyl alcohol) can be reacetylated to measure $\bar{M}'_n$. Ozonolysis followed by hydrolysis can be used to sever the branches or cross-links in polybutadiene. Unfortunately, the number of systems susceptible to measurements of these kinds is low. However, as the sensitivity of the spectroscopic techniques improves, perhaps more systems can be studied in this direct manner. Until then, it will be necessary to rely on the techniques that utilize the differences in the solution properties to estimate long chain branching. As will be seen, the techniques are based on differences in the hydrodynamic volumes between branched and linear polymers. However, at least two measurements are required, as well as some knowledge of the topology of the branch structure. Additionally, from the study of model branched systems, the expected agreement between theory and experiment is not realized. All of the measurements of long chain branching give relative values and should only be used for comparative purposes.

Viscosity Measurements of Chain Branching. The intrinsic viscosity of a branched polymer is, in general, lower than that of the corresponding linear chain of high molecular weight. In terms of the radius of gyration, the viscosity ratio at the θ temperature can be written

$$\frac{[\eta]_{oB}}{[\eta]_{oL}} = \frac{\phi_{oB}}{\phi_{oL}}\left[\frac{\langle S^2\rangle_{oB}}{\langle S^2\rangle_{oL}}\right]^{3/2}\left[\frac{\alpha_B}{\alpha_L}\right], \tag{11.59}$$

where ϕ_{oB} and ϕ_{oL} are the hydrodynamic interaction constants for branched and linear chains, and α_B and α_L are the expansion factors for the branched and linear polymers, respectively.

If we assume that branched and linear molecules have the same limiting value of ϕ and that the θ temperature is unaffected by branching, then ϕ_{oB}/ϕ_{oL} and (α_B/α_L) equal unity. Thus

$$\frac{[\eta]_{oB}}{[\eta]_{oL}} = \left[\frac{\langle S^2\rangle_{oB}}{\langle S^2\rangle_{oL}}\right]^{3/2}. \tag{11.60}$$

Under these circumstances the effect of branching on the intrinsic viscosity is determined completely by its effect on the radius of gyration. Using the usual definition of the branching factor g given in eq. 11.55,

$$g = \frac{\langle S^2\rangle_B}{\langle S^2\rangle_L},$$

we have

$$\frac{\eta_{oB}}{\eta_{oL}} = g^{3/2}. \tag{11.61}$$

In addition to the intrinsic viscosity of the sample, its viscosity-average molecular weight must be known. Recently, GPC measurements have been used for this purpose. The theoretical expressions for g for the various models have been given previously.

However, experimentally it has been found that the viscosity of the branched molecules is generally lower relative to the linear equivalent than predicted by the above equation. At least part of the ambiguity arises from the occurrence of a broad molecular weight distribution for branched polymers. Also, the hydrodynamic properties of chain molecules are apparently sensitive not only to the radius of gyration but, in addition, depend on some of the structural details of the branched molecule. Theoreticians differ, and values of the exponents for $\frac{1}{2}$ to $\frac{3}{2}$ have been suggested (28). The experimental studies have also tended to be ambiguous, since some work supports the $g^{3/2}$ dependence and others the $g^{1/2}$. Apparently, no single functional relationship between $[\eta]_{oB}/[\eta]_{oL}$ and g satisfactorily accounts for all branched structures.

For the class of polymers consisting of combs and stars, a relationship of the following form (28) has been suggested:

$$g' = g^{m(\lambda)}, \tag{11.62}$$

where λ is the fraction of material in the backbone of the molecule (zero for stars). In this relation $m(\lambda) = \frac{1}{2}$ for $\lambda = 0$ (stars) and $m(\lambda) = \frac{3}{2}$ in the limit as $\lambda \to 1$; the relation

$$m(\lambda) = \frac{1}{2} + \lambda \tag{11.63}$$

can be used.

Light-Scattering Measurements of Long Chain Branching. The light-scattering technique is based on the measurement of the angular dependence of the intensity of scattered radiation, near zero scattering angle, extrapolated to zero concentration. From such measurements $\langle S^2 \rangle$ can be determined and related to the degree of branching through the g factors for the type of branched system. Unfortunately, a number of problems arise to complicate these measurements. First, the $\langle S^2 \rangle$ measurement is a z-average measurement for polydispersed systems and cannot be related to the linear molecule without a knowledge of the molecular weight distribution. Secondly the light scattering measurements need to be very precise or the calculated results are meaningless.

Gel Permeation Chromatography Measurements of Long Chain Branching. Gel permeation chromatography (GPC) is a technique of molecular size separation accomplished on a gel column, using a liquid chromatography apparatus. The term *gel permeation* is derived from the method of separation on a column consisting of highly cross-linked polystyrene gel with a liquid structure. Separation occurs on the basis of the permeability of the gel. Molecules larger than maximum pore size pass through the column in the interstitial volume. Molecules smaller than the maximum enter the gel and are size separated. The smaller molecules require more solvent to be eluted. The GPC curve for a linear homopolymer reflects its molecular weight distribution. The conversion of the elution volume V to a molecular weight requires a calibration curve. An unequivocal $V = f(M)$ relationship can be obtained if monodispersed fractions of the polymer are available for a range of molecular weights. However, for most polymers such fractions are not available although it has been established that the elution volume V_i in GPC is proportional to the hydrodynamic volume of the polymer molecule (29):

$$V_i = [\eta]_i M_i. \qquad (11.64)$$

Thus "universal" hydrodynamic volume calibration curves can be constructed by plotting the product of the intrinsic viscosity $[\eta]$ and the weight-average molecular weight for a narrow fraction versus elution volume for a given column set in a specific solvent at a given temperature. A linear relationship is usually obtained if

$$\log[\eta]_i[M]_i = f_i(V_i). \qquad (11.65)$$

If the separation is by the effective hydrodynamic volume, it is reasonable to assume that it applies to the case of branched polymers. A single curve of $[\eta]M$ versus V is sufficient to describe the behavior of linear, star, and comb model polystyrenes (29). Thus

$$M_l[\eta]_l = [M]_b[\eta]_b. \qquad (11.66)$$

However, for branched systems the GPC curve is not only a function of molecular weight but also of the degree and type of branching.

Since the $[\eta]$ of the branched polymer is less than for a linear polymer of equivalent molecular weight, one can write

$$[\eta]_{br} = g^m[\eta]_l \qquad (11.67)$$

For a linear molecule

$$[\eta]_l = KM^a, \qquad (11.68)$$

where K and a are the Mark-Houwink coefficients. Then the elution volume for a branched molecule is

$$V_{br} = g^m KM^{a+1}. \qquad (11.69)$$

When the polymer is a randomly branched monodispersed polymer with tetrafunctional branching units,

$$g^m = \left[\left(1 + \frac{\lambda M_i}{6}\right)^{1/2} + \frac{4\lambda M_i}{3\pi}\right]^{-1/2}. \qquad (11.70)$$

The method requires the calculation of an intrinsic viscosity from the GPC curve, the universal calibration curve, and a value of λ from

$$[\eta]_{cal} = \sum_i \frac{f_i V_i}{M_i}. \qquad (11.71)$$

This calculated intrinsic viscosity is compared to that of the experimentally determined intrinsic viscosity $[\eta]_{obs}$. An iterative process is employed which seeks a value of λ such that $[\eta]_{cal} = [\eta]_{obs}$ (30, 31). Finally, one can also calculate the various branched average molecular weights.

The experimental method of determining the degree of long chain branching involves the following:

1. Obtain a linear sample of polymer.
2. Fractionate the standard and determine the $\bar{M}_w$, the intrinsic viscosity in the solvents, and the GPC elution volume for each fraction.
3. Determine the Mark-Houwink parameters K and a for the linear polymer:

$$\eta = KM^a.$$

4. Establish that the polymer fits the universal calibration

$$\log[\eta]_i M_i = f(V_i).$$

5. For the branched polymer calculate the $[\eta]$ from the GPC data using the parameters and the theoretical model applicable to the branching topology.
6. Iterate the value of λ until the calculated $[\eta]$ equals the experimentally observed $[\eta]_{obs}$. Calculate the desired molecular weigh averages using the determined value of λ.

The sources of error in the GPC-$[\eta]$ method are associated with (1) band spreading in the column, (2) a nonlinear plot of $\log[\eta]M$ over a broad molecular weight range, (3) assumption of the type of branched molecule used for the theoretical value of g, and (4) the question of whether the

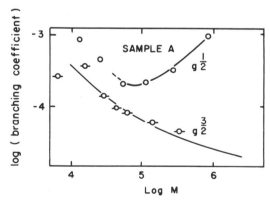

Figure 11.23 Molecular weight distribution of branching coefficient for polyethylene: (O) $g^{1/2}$ method; (-O-) $g^{3/2}$ method.

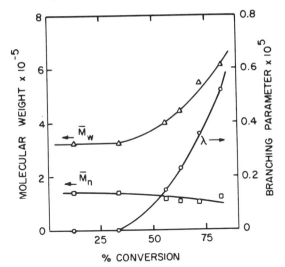

Figure 11.24 Branching parameter as a function of conversion for modified polychloroprene. (Reprinted by permission of Ref. 35.)

expansion of hydrodynamic volume of branched molecules in a good solvent may not be the same as for linear molecules.

When the concurrent GPC and $[\eta]$ method was applied to polyethylene (30–34), it was revealed that the long branching is a function of the molecular weight.

The results for low density polyethylene are given in Figure 11.23 for different values of g as well as for the fractionated samples. The distribution of branches as a function of molecular weight is probably associated with the distribution of residence times of the molecules in the reactor such that different molecules have different probabilities of growing branches. The increase of λ at lower molecular weight has also been associated with short chain branches (33).

The GPC-$[\eta]$ technique has also been applied to polychloroprene (35), and the results are shown in Figure 11.24. The branching index is an extreme function of the conversion and increases as the polymerization progresses.

Sedimentation Measurements of Long Chain Branching (36, 37). The sedimentation constants for a polymer are given by

$$S_d = \frac{K_S M}{R}, \tag{11.72}$$

where M is the molecular weight, R is the hydrodynamic radius at the θ temperature, and K_S contains the partial specific volume of the polymer and the density and volume of the solvent. For linear and branched polymers of the same molecular weight

$$\frac{(S_d)_l}{(S_d)_b} = \frac{R_{br}}{R_l} = h. \tag{11.73}$$

Calculations of h have been made for the various types of branched structures (38). The principal advantage of the sedimentation measurements would appear to be its independence from light scattering or viscometric measurements.

Sedimentation measurements are also made in conjunction with viscosity, light scattering, or GPC measurements (36, 37).

Another approach is to make concurrent sedimentation and diffusion constant measurements (39). This method was applied to the copolymerization of styrene-divinyl benzene and a monotonic variation of branching with molecular weight was not observed. The method is based on the following relationships. The sedimentation constant is approximated as

$$S_d = K_s h^{-1} M^{0.5}, \tag{11.74}$$

and the diffusion constant as

$$D = K_D h^{-1} M^{-0.5}. \tag{11.75}$$

For a randomly branched polymer with tetrafunctional branches

$$h^{-1} = 0.605(\lambda M)^{1/4} + 0.474(\lambda M)^{-1/4},$$

so

$$S_d M^{-1/4} = 0.605 K_s \lambda^{1/4} M^{1/2} + 0.474 K_s \lambda^{-1/4},$$

$$DM^{3/4} = 0.605 K_D \lambda^{1/4} M^{1/2} + 0.474 K_D \lambda^{-1/4}. \tag{11.76}$$

If $S_d M^{-1/4}$ and $DM^{3/4}$ are plotted against $M^{1/2}$, the slope and intercept give the branch point density λ. Experimentally (39), the plots deviate from straight lines at the lower and higher molecular weight regions. These deviations are associated with experimental error. This method is restricted to polymers having only one peak in the molecular weight distribution curve, since the diffusion method cannot be applied to a sample with two peaks in the differential distribution curve of molecular weight.

11.3.4 Characterization of Cross-linked Structures

The characterization of network polymers is particularly difficult for several reasons. First, the number of topologically possible structures is very large. For example, a tetrafunctional network has 34 different ways in which the chains can be connected to various junctions. Second, the chemical nature of the junction point is very similar to the backbone structures, making it difficult to analyze separately. Third, low concentrations of cross-links are usually sufficient to make useful networks sufficiently accurate, and sensitive analyses are required. Finally, analytical methods requiring solution of the polymer are not useful due to network insolubility. The structural elements making up a network are shown in Figure 11.25 (40). Uncross-linked chains do not form part of the cross-linked system, but are present in all systems. The single element junctions can be connected to two different chains (s) or with one end free (f), and void functionalities (v). Two element junctions occur when two chains connect at the same pair of junctions. It is possible for the two element functionalities to exist as a loop (l), that is, a chain connected with both ends to the same cross-link, or a doublet (d), where two chains connected to the same pair of junctions. Triplets (t), quadruplets (q), and higher multiplet pair groups of three, four, or more chains connecting the same pair of cross-links can occur with higher functionalities.

The network is characterized by the types of junctions (J) with a given configuration comprised of these structural elements. The individual junction types are given in Figure 11.26 for trifunctional systems. The junction types are labeled with the symbols of the structural elements (S, F, v, etc.) and in powers indicating the number of identical elements in the

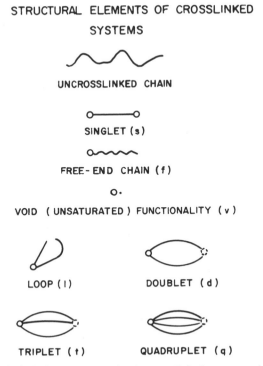

STRUCTURAL ELEMENTS OF CROSSLINKED

SYSTEMS

UNCROSSLINKED CHAIN

SINGLET (s)

FREE-END CHAIN (f)

VOID (UNSATURATED) FUNCTIONALITY (v)

LOOP (l)   DOUBLET (d)

TRIPLET (t)   QUADRUPLET (q)

Figure 11.25 Topological elements appearing in cross-linked systems with maximum functionality $f = 4$. (Reprinted by permission of Ref. 40.)

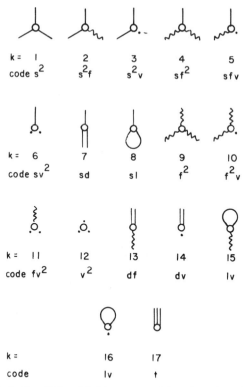

Figure 11.26 Topologically distinguishable types of junctions in trifunctional systems. (Reprinted by permission of Ref. 40.)

same junction. Thus S^3 denotes the junction with three singlet chains. The J types of cross-links can also be classified into the number of independent paths (m) connecting each junction to other junctions. The classification in terms of the independent path S is shown in Table 11.7 for a cross-linked system of functionality 3. The number of junction types is the number of parameters required for the description of the structure of the network.

Table 11.7   The Groups of Junctions with 0, 1, 2, and 3 Paths in a Cross-Linked System with Maximum Functionality $s = 3$

No. of Paths, m	Junction Types (code)	Fraction of Junctions in the Group, $\bar{n}^{(m)}$
0	$f^3, f^2v, fv^2, v^3, lf, lv, t$	$\bar{n}^{(0)} = n_9 + n_{10} + n_{11} + n_{12} + n_{15} + n_{16} +$ $+ n_{17}$
1	$sf^2, sfv, sv^2, sl, df, dv$	$\bar{n}^{(1)} = n_4 + n_5 + n_6 + n_8 + n_{13} + n_{14}$
2	s^2f, s^2v, sd	$\bar{n}^{(2)} = n_2 + n_3 + n_7$
3	s^3	$\bar{n}^{(3)} = n_1$

Table 11.8 Structural Characteristics of Cross-Linked Systems with Maximum Functionality, s

Maximum Functionality, s	Structural Elements[a]	No. of Structural Elements, $E(s)$	No. of Junction Types, $J(s)$
1	s, f, v	3	3
2	s, f, v, d, l	5	8
3	s, f, v, d, l, t	6	17
4	s, f, v, d, l, t, q	7	34
5	$s, f, v, d, l, t, q, m5$	8	62
6	$s, f, v, d, l, t, q, m5, m6$	9	109
7	$s, f, v, d, l, t, q, m5, m6, m7$	10	182
8	$s, f, v, d, l, t, q, m5, m6, m7, m8$	11	296
9	$s, f, v, d, l, t, q, m5, m6, m7, m8, m9$	12	466

[a] s = singlet, f = free-end chain, v = void functionality, l = loop, d = doublet, t = triplet, q = quadruplet, mk = higher multiplets.

The number of structural elements possible for each functionality is easily deduced and is given in Table 11.8.

Characterization of Cross-linked Polymers by Gel Content Analysis. When the number of cross-links exceeds a certain critical value, a fraction of the polymer molecules are involved in the three-dimensional network. Since this gel portion is insoluble, it can be isolated from the soluble portion.

The gel content method requires a determination of the weight fraction of nonextractable polymer. Experimentally, problems exist at low and high gel levels. At low gel content loss of some of the gel into the extracting solvent is a problem, and at high gel content the attainment of equilibrium is difficult. Finally, the drying procedure needs to be carefully controlled to assure complete drying. However, the gel content analysis requires no special use of the theory of rubber elasticity and does not suffer from the problem of entanglements as pseudo-cross-links.

Characterization of Networks by Elasticity Analysis (41, 42). It is possible to utilize results from rubber elasticity theory to determine the cross-link density. The equilibrium stress in extension for a swollen unfilled elastomer is given by

$$\tau = nRT\phi^{-1/3}(\alpha - \alpha^{-2}),$$

where τ is the stress, n is the number of network chains per cubic centimeter, R is the gas constant, T is the absolute temperature, ϕ is the volume fraction of elastomer in swollen sample, and α is the extension ratio. For filled elastomer

$$\tau = n\nu_r RT\phi^{-1/3}(\alpha - \alpha^{-2}),$$

where ν_r is the volume fraction of polymer. This expression is simplified if measurements are made at very small deformations under compression:

$$(\alpha - \alpha^{-2}) = \frac{3\Delta h}{h_s} = \frac{3\Delta h\phi^{1/3}}{h_0},$$

where h_0 is the height of undeformed, unswollen sample, h_s is the height of undeformed, swollen sample, and Δh is the deformation for a given stress. Then

$$\tau = \frac{F}{A_0} = n\nu_r RT\phi^{-1/3}\frac{3\Delta h\phi^{1/3}}{h_0},$$

where A_0 is the cross-sectional area of unswollen sample. Differentiating with respect to changes in height dh gives

$$\frac{1}{A_0}\frac{dF}{dh} = \frac{3n\nu_r RT}{h_0}.$$

If a measured sample is swollen to equilibrium and known weights are applied to give measured deformations, a force-deflection plot may be obtained. This is a straight line with slope $dF/dh = S_\epsilon$:

$$n = \frac{h_0 S_\epsilon}{3\nu_r RTA_0}.$$

If $\langle DP_c \rangle$ is the average number of repeating chain units between cross-links when M_0 is the average molecular weight between cross-links,

$$n = \frac{\rho}{\bar{M}_c} = \frac{\rho}{\langle DP_c \rangle M_C},$$

$$\langle DP_c \rangle = \frac{\rho 3\nu_r RTA_0}{M_c h_0 S_\epsilon},$$

where M_c is the average molecular weight of repeating unit.

Problems arise in using this technique from the cross-links arising from trapped entanglements. Thus a large overestimation of cross-link concentration occurs. Attempts to correct for these entanglements do lower the measured values but they are not necessarily correct.

Characterization of Networks by Swelling Analysis.   When a cross-linked polymer is placed in a solvent, the swelling is a function of the cross-link density, that is, the greater the cross-link density the less the swelling. The volume ν_2 of a swollen network is given by (44)

$$-\ln(1 - \nu_2) - \nu_2 - \chi\nu_2^2 = (V_1\phi M_c)\nu_2^{1/3},$$

where M_c is the number-average molecular weight of the chains between crosslinks, V_1 is the molar volume of the swelling liquid, and χ is the interaction parameter.

Hence one can measure M_c by measuring swelling volumes provided that the Flory interaction parameter is known. The interaction parameter is especially independent of polymer fraction ν_2 with good solvents and it increases linearly with increasing ν_2 for poorer solvents. However, theoretical studies of swelling of cross-linked polymers suggest that χ is a complex function of network parameters and the free energy of interaction (44). In addition, real polymer networks are not ideal and have a variety of imperfections, as shown earlier. Although a variety of studies have attempted to account for these imperfections, work remains in this area.

Measurements of the volume of swelling should be considered as relative measures of cross-linking.

REFERENCES

1. M. Morton, *Chemical Reactions of Polymers*, E. M. Fettes, Ed., Interscience Publishers, New York, 1964, p. 811.

2. A. G. Morrel, *Disc. Faraday Soc.*, **22**, 153 (1956).

3. J. Mochi, T. Tamura, M. Hagiwara, M. Gotodu, and T. Kagiya, *J. Polym. Sci.*, **4**, 283 (1966).

4. K. Shiroyama, T. Okuda, and S. Kita, *J. Polym. Sci.*, **3**, 907 (1965).

5. K. B. Abbas, F. A. Bovey, and F. C. Shilling, *Makromol. Chem.* (Suppl. 1), **1**, 227 (1975).

6. G. Park, *J. Macromol. Sci.—Phys.*, **B14**, 151 (1977).

7. A. H. Willbourn, *J. Polym. Sci.*, **34**, 569 (1959).

8. J. C. Randall, *J. Polym. Sci., Polym. Phys. Ed.*, **11**, 275 (1973).

9. F. A. Bovey, F. C. Schilling, F. L. McCrackin, and H. L. Wagner, *Macromolecules*, **9**, 76 (1976).

10. F. A. Bovey, *Structural Studies of Macromolecules by Spectroscopic Methods*, K. J. Ivin, Ed., John Wiley & Sons, London, 1976, p. 181.

11. T. N. Bowner and J. H. O'Donnell, *Polymer*, **18**, 1033 (1977).

12. D. A. Boyle, W. Simpson, and J. D. Waldron, *Polymer*, **2**, 323, 335 (1961).

13. P. J. Flory, *Principles of Polymer Chemistry*, Cornell University Press, Ithaca, New York, 1953.

14. V. A. Grechanovskii, *Rubber Chem. Technol.*, **45**, 519 (1972).

15. J. K. Beasley, *J. Am. Chem. Soc.*, **75**, 6123 (1953).

16. C. W. Macosko and D. R. Miller, *Macromolecules*, **9**, 199 (1976).

17. D. R. Miller and C. W. Macosko, *Macromolecules*, **9**, 206 (1976).

18. P. Luby, *J. Polym. Sci.*, **53C**, 23 (1975).

19. M. Gordon and G. R. Scantlebury, *Trans. Faraday Soc.*, **60**, 604 (1964).

20. K. Dusek and W. Prins, *Adv. Polym. Sci.*, **6**, 1 (1969).

21. J. A. Manson and L. H. Sperling, *Polymer Blends and Composites*, Plenum Press, New York, 1976.
22. P. J. Flory, *Statistical Mechanics of Chain Molecules*, Interscience Publishers, New York, 1969.
23. M. Oka and A. Nakajima, *Polym. J.*, **9**, 573 (1977).
24. M. Oka and A. Nakajima, *Polym. J.*, **9**, 583 (1977).
25. M. L. Miller, *The Structure of Polymers*, Reinhold, New York, 1964.
26. P. A. Small, *Adv. Polym. Sci.*, **18**, 1 (1975).
27. W. W. Grassley, *Characterization of Macromolecular Structures*, D. McIntyre, Ed., National Academy of Sciences, Washington, D.C., 1968.
28. G. C. Berry and E. F. Casassa, *Macromol. Rev.*, **4**, 1 (1970).
29. Z. Grubisic, P. Rempp, and M. Benoit, *J. Polym. Sci.*, **B5**, 753 (1967).
30. E. E. Drott and R. A. Mendelson, *J. Polym. Sci.*, **A-2, 8**, 1361 (1970).
31. E. E. Drott and R. A. Mendelson, *J. Polym. Sci.*, **A-2, 8**, 1375 (1970).
32. M. Kurata, H. Okamoto, M. Iwama, M. Abe, and T. Homma, *Polym. J.*, **3**, 739 (1972).
33. S. Nakano and Y. Goto, *J. Appl. Polym. Sci.*, **20**, 3313 (1976).
34. G. R. Williamson and A. Cervenka, *Eur. Polym. J.*, **10**, 295 (1974).
35. M. M. Coleman and R. E. Fuller, *J. Macromol. Sci.—Phys.*, **B11**, 419 (1975).
36. L. H. Tung, *J. Polym. Sci.*, **A-2, 7**, 47 (1969).
37. L. H. Tung, *J. Polym. Sci.*, **A-2, 9**, 759 (1971).
38. M. Kurata and M. Fukatsu, *J. Chem. Phys.*, **41**, 2934 (1964).
39. H. Matsuda, I. Yamada, M. Okabe, and S. Kuroiwa, *Polym. J.*, **9**, 527 (1977).
40. A. Ziabicki and J. Walasak, *Macromolecules*, **11**, 471 (1978).
41. E. F. Cluff, E. K. Gladding, and R. Pariser, *J. Polym. Sci.*, **45**, 344 (1960).
42. K. E. Polmanteer and J. D. Helmer, *Rubber Chem. Technol.*, **38**, 123 (1965).
43. P. F. Flory and J. Rehner, Jr., *J. Chem. Phys.*, **11**, 512 (1943).
44. K. Dusek, *J. Polym. Sci.*, **39C**, 83 (1973).
45. D. E. Axelson, G. C. Levy, and L. Mandelkern, *Macromolecules*, **12**, 41 (1979).

Index

409